Stochastic Reliability Modeling, Optimization and Applications

Stochastic Reliability Modeling, Optimization and Applications

Editors

Syouji Nakamura
Kinjo Gakuin University, Japan

Toshio Nakagawa
Aichi Institute of Technology, Japan

World Scientific

NEW JERSEY · LONDON · SINGAPORE · BEIJING · SHANGHAI · HONG KONG · TAIPEI · CHENNAI

Published by

World Scientific Publishing Co. Pte. Ltd.
5 Toh Tuck Link, Singapore 596224
USA office: 27 Warren Street, Suite 401-402, Hackensack, NJ 07601
UK office: 57 Shelton Street, Covent Garden, London WC2H 9HE

British Library Cataloguing-in-Publication Data
A catalogue record for this book is available from the British Library.

STOCHASTIC RELIABILITY MODELING, OPTIMIZATION AND APPLICATIONS
Copyright © 2010 by World Scientific Publishing Co. Pte. Ltd.
All rights reserved. This book, or parts thereof, may not be reproduced in any form or by any means, electronic or mechanical, including photocopying, recording or any information storage and retrieval system now known or to be invented, without written permission from the Publisher.

For photocopying of material in this volume, please pay a copying fee through the Copyright Clearance Center, Inc., 222 Rosewood Drive, Danvers, MA 01923, USA. In this case permission to photocopy is not required from the publisher.

ISBN-13 978-981-4277-43-3
ISBN-10 981-4277-43-6

Printed in Singapore.

Preface

One small research group with an organizer Toshio Nakagawa of Aichi Institute of Technology and fifteen members was started in Nagoya, Japan on February, 1989. This group was named *Nagoya Computer and Reliability Research* (NCRR) with the objective of presenting and writing research papers studied by each member. The NCRR has no rule and no duty, however, has a strong desire to study computer and reliability problems in one's daily life and place of work. The NCRR has continued for twenty years unexpectedly, and during this interval, each member has presented actively many papers in the following international conferences:

- The First Australia-Japan Workshop on Stochastic Models in Engineering, Technologies and Management, July 14–16, 1993, Gold Coast, Australia.
- UK-Japanese Workshop on Stochastic Modeling in Innovative Manufacturing, July 20–21, 1995, Cambridge, UK.
- The Second Australia-Japan Workshop on Stochastic Models in Engineering, Technologies and Management, July 17–19, 1996, Gold Coast, Australia.
- The First Euro-Japanese Workshop on Stochastic Risk Modelling for Finance, Insurance, Production and Reliability, September 7–9, 1998, Brussels, Belgium.
- The First Western Pacific/Third Australia-Japan Workshop on Stochastic Models in Engineering, Technology and Management, September 23–25, 1999, Christchurch, New Zealand.
- International Conference on Applied Stochastic System Modeling, March 29–30, 2000, Kyoto, Japan.
- The Second Euro-Japanese Workshop on Stochastic Risk Modelling

for Finance, Insurance, Production and Reliability, September 18–20, 2002, Chamonix, France.
- 9th ISSAT International Conference Reliability and Quality in Design. August 6-8, 2003, Waikiki, Hawaii, USA.
- 10th ISSAT International Conference Reliability and Quality in Design, August 5-7, 2004, Las Vegas, Nevada, USA.
- 2004 Asian International Workshop on Advanced Reliability Modeling, August 26–27, 2004, Hiroshima, Japan.
- International Workshop on Recent Advances in Stochastic Operations Research, August 25–26, 2005, Canmore, Canada.
- 2006 Asian International Workshop on Advanced Reliability Modeling, August 24–26, 2006, Busan, Korea.
- 13th ISSAT International Conference Reliability and Quality in Design, August 2-4, 2007, Seattle, Washington, USA.
- 2008 Asian International Workshop on Advanced Reliability Modeling, October 23–25, 2008, Taichung, Taiwan.

Several good papers selected from the above conferences have published in some research journals and on book forms.

In memory of twenty th anniversary, we make a plan of publishing the book written by each member titled on *Stochastic Reliability Modeling, Optimization and Applications* from World Scientific Publishing. The book is composed of three parts: Reliability Theory, Computer System Reliability, and Reliability Applications. Throughout this book, we formulate stochastic models by applying mainly renewal, Markov renewal and cumulative processes in stochastic processes, and analyze them by using the techniques of reliability theory. Furthermore, we obtain the availability, the expected cost and the overhead as an objective function, and discuss analytically optimal policies which minimize them. Such methods and results would be more useful for studying and applying other reliability models in practical fields.

Part I consists of three chapters, focusing on discussing optimal policies for standard reliability models: Chapter 1 presents a generalization of basic concepts of binary state to multistate coherent systems, and gives the existence theorems of series and parallel systems, IFRA and NBU theorems and their properties of coherent systems. The treatment of k-out-of-n systems is newly presented in this chapter. Chapter 2 considers three replacement policies for cumulative damage models in which an item is replaced at time, shock number, damage level, and at failure, whichever occurs first. Another three replacement policies for the failure interaction and the opportunistic models are taken up, and optimal policies are discussed analytically, using the techniques of cumulative processes. Chapter 3 extends the standard inspection model to four inspection policies for a two-unit system, a system with two types of inspection, a system with self-testing, and a system with a finite operation time. The expected cost rates for each model are obtained, and optimal policies which minimize them are derived by using the methods of inspection policies.

Part II consists of five chapters, focusing on analyzing a various of computer systems by using reliability theory: Chapter 4 discusses hybrid state saving schemes with realistic restrictions. Optimal checkpoint intervals are derived by using approximations for overhead evaluation. The implemented logic circuit simulator and numerical examples are also presented. Chapter 5 formulates three stochastic models of a system with networks by using Markov renewal processes. Optimal policies which minimize the expected cost rates are derived analytically, and numerical examples are presented. Chapter 6 studies three stochastic models of a communication system with the recovery techniques of checkpoint and rollback, a mobile communication system with recovery scheme and a communication system with window flow control scheme. Some reliability measures of each model are obtained by using Markov renewal processes, and optimal policies which minimize them are discussed. Chapter 7 considers two backup models of a database system. The expected cost rates of each model are obtained, and optimal backup policies are discussed by using theory of cumulative processes. Chapter 8 considers two-level recovery schemes of soft and hard checkpoints, and multiple and modified modular systems with sequential checkpoint times, increasing error rates and random processing times. Optimal checkpoint intervals of each model which minimize the total overhead are derived by using reliability theory.

Part III consists of two chapters, focusing on maintenances of miscellaneous systems and optimization problems in management science: Chapter 9 takes up optimal maintenance models of five different systems such as missile, phased array radar, FADEC, co-generation system, and power plant. Some reliability measures of each model are obtained by using reliability theory, and optimal maintenance policies which minimize them are discussed. Chapter 10 formulates three stochastic models associated with the monetary facility, and considers optimization problems which maximize an expected liquidation of holdings, minimize the prepayment risk for a monetary risk, and determine a loan interest rate of banks. Optimal policies for each model are derived analytically, using reliability theory.

We are strongly convinced that the NCRR has run over twenty years by great help and hearty assistance of many peoples: We wish to thank to Professor Shunji Osaki, Nanzan University, Professor Prabhakar Murthy, The University of Queensland, Professor Hoang Pham, Rutgers University, Professor Shigeru Yamada, Tottori University, and Professor Tadashi Dohi, Hiroshima University for having presented the papers at some national conferences. We wish to express our special thanks to Professor Kazumi Yasui, Aichi Institute of Technology, Professor Hiroaki Sandoh, Osaka University, and the other members of NCRR; Mr. Takehiko Nishimaki, Professor Shinichi Koike, Professor Yoshihisa Harada, and Mr. Esturo Shima for their cooperation and valuable discussions. Furthermore, we would like to thank Kinjo Gakuin University Research Grant for the support and Editor Chelsea Chin, World Scientific Publishing Co. Pte. Ltd for providing the opportunity to write this book.

Syouji Nakamura
Toshio Nakagawa
Nagoya, Toyota

List of Contributors

F. Ohi	Nagoya Institute of Technology, Japan
T. Satow	Tottori University, Japan
S. Mizutani	Aichi University of Technology, Japan
M. Ohara	Tokyo Metropolitan University, Japan
M. Arai	Tokyo Metropolitan University, Japan
S. Fukumoto	Tokyo Metropolitan University, Japan
M. Imaizumi	Aichi Gakusen University, Japan
M. Kimura	Gifu City Women's College, Japan
C. H. Qian	Nanjing University of Technology, China
K. Naruse	Nagoya Sangyo University, Japan
S. Maeji	Kinjo Gakuin University, Japan
K. Ito	Kinjo Gakuin University, Japan
S. Nakamura	Kinjo Gakuin University, Japan

Contents

Preface   v

List of Contributors   ix

RELIABILITY MODELS   1

1. Multistate Coherent Systems   3
   FUMIO OHI

   1   Introduction   3
   2   Coherent Systems   6
   3   k-out-of-n systems   13
   4   Modules of Coherent Systems   17
   5   Probabilistic Aspect of Coherent Systems   21
   6   Hazard Transform of Multistate Coherent Systems . . .   25
   7   Concluding Remarks   32
   References   33

2. Cumulative Damage Models   35
   TAKASHI SATOW

   1   Introduction   35
   2   Standard Cumulative Damage Model   38
       2.1   Shock Arrival   38
       2.2   Cumulative Damage   39
       2.3   Cumulative Damage Model   40
   3   Failure Interaction Models   41

		3.1	Age and Damage Limit Model	42
		3.2	Numerical Examples	44
		3.3	Shock Number and Damage Limit (N, k) Model	45
		3.4	Numerical Examples	47
		3.5	Age and Shock Number Model	49
		3.6	Numerical examples	50
		3.7	Conclusions	51
	4	Oppotunistic Replacement Model		52
		4.1	Age Model	53
		4.2	Numerical examples	55
		4.3	Damage Limit Model	57
		4.4	Numerical Examples	59
		4.5	Conclusions	60
	References			61

3. Extended Inspection Models   63

SATOSHI MIZUTANI

	1	Introduction		63
	2	Periodic Policy for a Two-Unit System		65
		2.1	Model and Assumptions	65
		2.2	Optimal Policy	68
		2.3	Numerical Examples	68
	3	Periodic Policy for a System with Two Types of Inspection		69
		3.1	Model and Assumptions	70
		3.2	Optimal Policy	72
		3.3	Numerical Example	73
	4	Periodic Policy for a System with Self-Testing		73
		4.1	Model and Assumptions	74
		4.2	Optimal Policy	76
		4.3	Numerical Examples	78
	5	Optimal Policies for a Finite Interval		79
		5.1	Periodic Inspection Policy	80
		5.2	Sequential Inspection Policy	81
		5.3	Numerical Examples	82
	6	Conclusions		83
	References			84

COMPUTER SYSTEMS 87

4. Stochastic Analyses for Hybrid State Saving and Its
   Experimental Validation   89
   *MAMORU OHARA, MASAYUKI ARAI and
   SATOSHI FUKUMOTO*

   1 Introduction 89
   2 Time Warp Simulation and Hybrid State Saving 91
   3 Analytical Model 97
   4 Implementation of A Concrete Application: A Parallel
      Distributed Logic Circuit Simulator 106
   5 Numerical Examples 109
   6 Concluding Remarks 115
   References 116

5. Reliability Analysis of a System Connected with Networks   117
   MITSUHIRO IMAIZUMI

   1 Introduction 117
   2 Optimal Reset Number of a Microprocessor System with
      Network Processing 119
      2.1 Model and Analysis 120
      2.2 Optimal Policies 124
      2.3 Numerical Example 126
   3 Reliability Analysis for an Applet Execution Process .. 127
      3.1 Model and Analysis 128
      3.2 Optimal Policy 133
      3.3 Numerical Example 134
   4 Reliability Analysis of a Network Server System with DoS
      Attacks 135
      4.1 Model and Analysis 135
      4.2 Optimal Policy 141
      4.3 Numerical Example 142
   5 Conclusions 143
   References 144

6. Reliability Analysis of Communication Systems ... 147

   MITSUTAKA KIMURA

   1. Introduction ... 147
   2. Communication System with Rollback Recovery ... 149
      - 2.1 Reliability Quantities ... 150
      - 2.2 Optimal Policy ... 153
   3. Mobile Communication System with Error Recovery Schemes ... 157
      - 3.1 Reliability Quantities ... 158
      - 3.2 Optimal Policy ... 162
   4. Communication System with Window Flow Control Scheme ... 163
      - 4.1 Reliability Quantities ... 164
      - 4.2 Optimal Policy ... 170
   5. Conclusions ... 172

   References ... 173

7. Backup Policies for a Database System ... 177

   CUN-HUA QIAN

   1. Introduction ... 177
   2. Cumulative Damage Model ... 180
      - 2.1 Standard Cumulative Damage Model ... 181
      - 2.2 Modified Cumulative Damage Model ... 183
   3. Comparison of Backup Schemes and Policies ... 184
      - 3.1 Incremental and Cumulative Backups ... 184
      - 3.2 Incremental Backup Policy ... 186
      - 3.3 Total and Cumulative Backups ... 187
      - 3.4 Cumulative Backup Policy ... 190
   4. Periodic Backup Policies ... 195
      - 4.1 Periodic Incremental Backup Policy ... 196
      - 4.2 Periodic Cumulative Backup Policies ... 198
   5. Conclusions ... 202

   References ... 203

8. Optimal Checkpoint Intervals for Computer Systems ... 205

   KENICHIRO NARUSE and SAYORI MAEJI

   1. Introduction ... 205

	2	Two-level Recovery Schemes	207
		2.1 Performance Analysis	208
		2.2 Expected Overhead	212
	3	Error Detection by Multiple Modular Redundancies	213
		3.1 Multiple Modular System	214
		3.2 Performance Analysis	214
	4	Sequential Checkpoint Intervals for Error Detection	217
		4.1 Performance Analysis	218
		4.2 Modified Model	222
	5	Random Checkpoint Models	226
		5.1 Performance Analysis	227
		5.2 Majority Decision System	235
	6	Conclusion	237
	References		237

RELIABILITY APPLICATIONS   241

9. **Maintenance Models of Miscellaneous Systems**   243

   KODO ITO

	1	Introduction	243
	2	Missile Maintenance	245
		2.1 Expected Cost	247
		2.2 Optimal Policies	248
		2.3 Concluding Remarks	250
	3	Phased Array Radar Maintenance	251
		3.1 Cyclic Maintenance	252
		3.2 Delayed Maintenance	255
		3.3 Concluding Remarks	256
	4	Self-diagnosis for FADEC	257
		4.1 Double Module System	259
		4.2 Triple Module System	261
		4.3 N Module System	263
		4.4 Concluding Remarks	264
	5	Co-generation System Maintenance	264
		5.1 Model and Assumptions	265
		5.2 Expected Cost	266
		5.3 Optimal Policy	267

		5.4	Concluding Remarks	269
	6	Aged Fossil-fired Power Plant Maintenance		269
		6.1	Model 1	271
		6.2	Model 2	274
		6.3	Concluding Remarks	276
	References			276

10. Management Policies for Stochastic Models with Monetary Facilities — 279

SYOUJI NAKAMURA

	1	Introduction		279
	2	Liquidation Profit Policy		280
		2.1	Model 1	282
		2.2	Model 2	283
	3	Prepayment Risk		286
		3.1	Model 1: Interval Estimation	288
		3.2	Model 2: Linear Estimation	289
		3.3	Optimal Policies	289
		3.4	Numerical Example	290
	4	Loan Interest Rate		292
		4.1	Expected Earning without Bankruptcy	293
		4.2	Expected Earning with Bankruptcy	294
		4.3	Numerical Examples	297
	References			299

PART 1
RELIABILITY MODELS

Chapter 1

Multistate Coherent Systems

FUMIO OHI

Department of Mechanical Engineering
Nagoya Institute of Technology,
Gokiso-cho, Showa-ku, Nagoya, 466-8555, Japan
E-mail: ohi.fumio@nitech.ac.jp

1 Introduction

A basic problem in the study of reliability systems is to explain relationships among the operating performances of systems and the components consisting the systems. Using Boolean functions, Mine [14] introduced the concept of monotone systems, in which all the state spaces of components and the systems were assumed to be $\{0,1\}$, so were also called binary state systems, where 0 and 1 denote the failure and the functioning states, respectively. Monotone system means that the more the number of functioning components is, the higher the performance level of the system consisted of the components is.

Mathematical aspects of these binary state monotone systems were explained [3, 4, 8]. Barlow and Proschan [1] have summarized the reliability studies of the binary state monotone systems. Pham [22] has edited the recent work about reliability engineering, and in this handbook, we can find out formulae useful for solving practical reliability problems.

In many practical situations, however, systems and their components could take many other performance levels, from the perfectly functioning state to the complete failure state. Thus, reliability models of multistate systems and components are required for more practical treatment of real reliability systems.

Such multistate systems were introduced in the context of cannibalization [10,11], but these works were not concerned with mathematical aspects of the systems. More mathematical studies of multistate systems were carried out by [2,7]: Barlow and Wu [2] defined multistate coherent systems based on the minimal path and cut sets of binary state systems, and discussed some properties of the multistate systems. El-Neweihi, Proschan and Sethuraman [7] defined the multistate systems assuming that all the state spaces of the systems and their components could be expressed as $\{0, 1, \cdots, M\}$. Their results were very analogous to those of binary systems. Huang, Zuo and Fang [12] introduced the multistate consecutive k-out-of-n systems and provided algorithms to evaluate the performance probabilities of the systems. Zuo, Hang and Kuo [28] defined a multistate coherent systems assuming all the state spaces of the systems and components were the same finite totally ordered sets as [7], and also they presented a definition of multistate k-out-of-n systems in the context. The definitions were technical and then applicable to the real situations.

This chapter is concerned with a mathematical generalization of the concepts of binary state monotone systems mainly based on the work of [17]. Section 2 presents a definition of multistate systems, assuming that state spaces of systems and their components need not to be the same, and are mathematically finite totally ordered sets. We discuss series and parallel coherent systems and obtain an existence theorem which justifies the usual formulae of the series and parallel systems, *i.e.*, $\min_{1 \le i \le n} x_i$ and $\max_{1 \le i \le n} x_i$ in the theory of binary state systems. A definition of dual systems is also presented in this section, using the concept of dual ordered sets.

In Section 3, we present newly a definition of multistate k-out-of-n:G systems and show some properties of them. In the theory of binary state systems, the dual systems of k-out-of-n:G systems are well known to be $n - k + 1$-out-of-n:G systems. But, in our context, the similar proposition generally no longer holds, since the state spaces of components and the system are arbitrarily finite ordered sets and have less restriction than those of binary state systems. But, the duality holds generally for the maximum and minimum k-out-of-n:G systems.

In Section 4, we treat modules of multistate systems. The modules are practically familiar concept for us to follow when constructing a large system. A system is generally composed of many systems of smaller size each of which is called a module and is also composed of many systems. In other words, practical systems have a hierarchic structure and each layer of the hierarchy consists of modules, and each module also consists of modules of smaller size. From the reliability point of view, we are interested in algebraic and probabilistic relations between systems and modules. In other words, how the reliability of the system is determined by the reliabilities of the modules.

The examination of the concept of modules occupies mathematically and practically important part in the theory of binary state systems, where an elegant theorem holds, called three modules theorem [6], but the similar proposition can no longer hold in our multistate cases.

In Section 5, we examine stochastic aspects of multistate systems. The IFR(increasing failure rate), IFRA(increasing failure rate average) and NBU(new better than used) stochastic processes are defined, and IFRA and NBU closure theorems of [23] are then proved in a slightly different situation.

In Section 6, a generalization of the concept of hazard transforms is presented and, using it, we prove that the preservation of IFR property determines the structure of a multistate system as a series system.

Throughout this work, we may recognize that a basic theory of reliability systems treats the algebraic and stochastic relationship between a product partially ordered set and a partially ordered set through an increasing mapping from the former to the latter. There practically exist some examples in which a component has two deteriorating states for which we cannot say the one state is better/worse than the other [27]. So, a model of reliability systems including partially ordered sets as the state spaces is useful, and Yu, Koren and Guo [27] presented more general multistate systems and some properties of them, but which were not a thorough treatment.

This article considers only totally ordered finite sets except for the probabilistic examination, but the basic concepts of reliability systems defined for totally ordered case, as series and parallel systems, k-out-of-n:G systems, modules and stochastic properties, are thought to be easily extended to the partially ordered case. The thorough generalization to the case of arbitral ordered sets, though, is remained to be an open problem.

Notations

We use the following notations: A finite set $C = \{1, 2, \cdots, n\}$ is the set of the components consisting a system and Ω_i ($i \in C$) is the state space of the ith component defined to be a finite totally ordered set in Definition 2.1 and is not necessarily the binary set $\{0, 1\}$. The state space of a system composed of the components is also defined to be a finite totally ordered set.

1) For $A \subset C$, the product set of Ω_i $(i \in A)$ is denoted by $\Omega_A = \prod_{i \in A} \Omega_i$. When $A = \{i\}$, $\Omega_A = \Omega_i$.
2) An element of Ω_A $(A \subset C)$ is denoted by $\boldsymbol{x}^A$ and also simply $\boldsymbol{x}^A = \boldsymbol{x}$ if there is no confusion. When $A = C$, $\boldsymbol{x} \in \Omega_C$ is precisely written as $\boldsymbol{x} = (x_1, \cdots, x_n)$, $x_i \in \Omega_i$ $(i = 1, \cdots n)$.
3) For a subset $A \subset C$, $A' = C \backslash A = \{\, i \mid i \in C,\ i \notin A \,\}$.
4) For $B \subset A \subset C$, P_B is the projection mapping from Ω_A to Ω_B. For $\boldsymbol{x} \in \Omega_C$, $x_i = P_{\{i\}}(\boldsymbol{x})$.
5) For $B \subset A \subset C$ and $V \subset \Omega_A$, $P_{\Omega_B} V = \{\, P_{\Omega_B} \boldsymbol{x} \mid \boldsymbol{x} \in V \,\}$.
6) Let $\{\, B_j \mid 1 \leq j \leq m \,\}$ be a partition of $A \subset C$. Then, for $\boldsymbol{x}_j \in \prod_{i \in B_j} \Omega_i$ $(1 \leq j \leq m)$, $\boldsymbol{x} = (\boldsymbol{x}_1, \cdots, \boldsymbol{x}_m)$ is an element of Ω_A such that $P_{\Omega_{B_j}} \boldsymbol{x} = \boldsymbol{x}_j$. Then for every $\boldsymbol{x} \in \Omega_A$ $(A \subset C)$, $\boldsymbol{x} = (\boldsymbol{x}^{B_1}, \cdots, \boldsymbol{x}^{B_m})$, where $\boldsymbol{x}^{B_i} = P_{B_i}(\boldsymbol{x})$ $(i = 1, \cdots, m)$.
7) $(k_i, \boldsymbol{x}) \in \Omega_A$ $(A \subset C,\ i \in A)$ is an element of Ω_A such that $k \in \Omega_i$ and $\boldsymbol{x} \in \prod_{j \in A \backslash \{i\}} \Omega_j$.
8) $|A|$ is the cardinal number of a set A.

2 Coherent Systems

Definition 2.1. A system composed of n components (a system of order n) is a triplet (Ω_C, S, φ) satisfying the following conditions:

1) $C = \{1, \cdots, n\}$ is the set of the components.
2) Ω_i $(i \in C)$ and S are finite totally ordered sets.
3) φ is a surjection from $\Omega_C = \prod_{i=1}^n \Omega_i$ to S.

Ω_i is the state space of the i-th component and S is the one of the system composed of the n components. The orders on Ω_i $(i \in C)$ and S are denoted by a common symbol $\leq$. m_i and M_i denote the minimum and maximum elements of Ω_i, respectively, and we also use m and M to show the minimum and maximum elements of S, respectively. The order $\leq$ on the product sets $\Omega_C = \prod_{i=1}^n \Omega_i$ is defined as the followings ; for $\boldsymbol{x} = (x_1, \cdots, x_n)$ and $\boldsymbol{y} = (y_1, \cdots, y_n)$ of Ω_C, $\boldsymbol{x} \leq \boldsymbol{y}$ means $x_i \leq y_i$ ($\forall i \in C$), and $\boldsymbol{x} < \boldsymbol{y}$ means $x_i < y_i$ ($\forall i \in C$).

Throughout this chapter, we assume $\Omega_i = \{0, 1, \cdots, N_i\}$ $(i \in C)$ and $S = \{0, 1, \cdots, N\}$, since any finite totally ordered set is isomorphic to some finite set $\{0, 1, \cdots, L\}$ of nonnegative integers $0, 1, \cdots, L$.

We use the notation for a system (Ω_C, S, φ):

$$V_s(\varphi) = \{\, x \mid \varphi(x) = s,\ x \in \Omega_C\},\ s \in S,$$

which is the inverse image of $s \in S$ respect to φ, then for $s \neq t$, $V_s(\varphi) \cap V_t(\varphi) = \phi$ holds. From the surjective property of φ, we have $V_s(\varphi) \neq \phi$ for every $s \in S$. The symbol $MIV_s(\varphi)$ means the set of the minimal elements of $V_s(\varphi)$ ($s \in S$).

When there is no confusion, we write V_s in place of $V_s(\varphi)$ and a system (Ω_C, S, φ) is simply called a system φ.

Definition 2.2. A system φ is called increasing iff for x and y of Ω_C, $x \leq y$ implies $\varphi(x) \leq \varphi(y)$.

For $x_i \in \Omega_C$ ($1 \leq i \leq m$), $\vee_{i=1}^k x_i = x_1 \vee \cdots \vee x_k$ and $\wedge_{i=1}^k x_i = x_1 \wedge \cdots \wedge x_k$ mean the supremum and the infimum of $\{x_1, \cdots, x_k\} \subseteq \Omega_C$, respectively. The same symbols $\vee$ and $\wedge$ are also used for every subset of S with the same meanings. For an increasing system φ,

$$\varphi(\wedge_{i=1}^k x_i) \leq \wedge_{i=1}^k \varphi(x_i), \quad \vee_{i=1}^k \varphi(x_i) \leq \varphi(\vee_{i=1}^k x_i),$$

and furthermore,

$$\varphi(m_1, \cdots, m_n) = m, \quad \varphi(M_1, \cdots, M_n) = M,$$

because of the surjective and increasing properties of φ.

Definition 2.3.

1) Ω_i (or the component i) is said to be relevant to the system φ iff for every r and s of S such that $r \neq s$, there exist k and l of Ω_i and $x \in \prod_{j=1, j \neq i}^n \Omega_j$ such that $(k_i, x) \in V_r$ and $(l_i, x) \in V_s$. In this case, Ω_i is simply called relevant
2) A system φ is said to be relevant iff every Ω_i ($i \in C$) is relevant to the system.

Definition 2.4. A system φ is said to be a coherent system iff the system φ is increasing and relevant.

In the sequel of this section, we examine series and parallel systems and show the existence theorem of series and parallel systems. We start with a definition of series and parallel systems.

Definition 2.5.

1) A system φ is called a series system iff $\inf V_s \in V_s$ holds for every $s \in S$. In other words, every V_s has the minimum element.
2) A system φ is called a parallel system iff $\sup V_s \in V_s$ holds for every $s \in S$. In other words, every V_s has the maximum element.

Lemma 2.1.

(i) For a coherent series system φ, $\inf V_s < \inf V_t$ for every s and t of S such that $s < t$.

(ii) For a coherent parallel system φ, $\sup V_s < \sup V_t$ for every s and t of S such that $s < t$

Proof. We prove only (i), since the proof of (ii) is similar to (i). Letting s and t of S be $s < t$, from the relevant property of φ, there exist $(k_i, \boldsymbol{x})$ and $(l_i, \boldsymbol{x})$ such that $(k_i, \boldsymbol{x}) \in V_s$ and $(l_i, \boldsymbol{x}) \in V_t$.

If $(\inf V_t)_i \leq k$ holds, $\inf V_t \leq (k_i, \boldsymbol{x})$ since $\inf V_t \leq (l_i, \boldsymbol{x})$ and $(\inf V_t)^{C \setminus \{i\}} \leq \boldsymbol{x}$, and hence, $\varphi(\inf V_t) = t \leq \varphi(k_i, \boldsymbol{x}) = s$, which contradicts to the assumption $s < t$. Then, $k < (\inf V_t)_i$.

Thus $(\inf V_s)_i < (\inf V_t)_i$ holds for every $i \in C$. □

The following proposition gives us characterizations of series and parallel coherent systems:

Proposition 2.1. For a coherent system φ, we have the following equivalent relations:

(i) φ is a series system iff $\varphi(\boldsymbol{x} \wedge \boldsymbol{y}) = \varphi(\boldsymbol{x}) \wedge \varphi(\boldsymbol{y})$ holds for every $\boldsymbol{x}$ and $\boldsymbol{y}$ of Ω_C.

(ii) φ is a parallel system iff $\varphi(\boldsymbol{x} \vee \boldsymbol{y}) = \varphi(\boldsymbol{x}) \vee \varphi(\boldsymbol{y})$ holds for every $\boldsymbol{x}$ and $\boldsymbol{y}$ of Ω_C.

Proof. We prove only (ii), since (i) is similarly proved.
Sufficiency: $\sup V_s \in V_s$ holds for every $s \in S$, because

$$\varphi(\sup V_s) = \sup\{\varphi(\boldsymbol{x}) \mid \boldsymbol{x} \in V_s\} = s,$$

from the sufficiency of the equality and that V_s is a finite set.
Necessity: Without loss of generality, we assume $\boldsymbol{x} \in V_s$, $\boldsymbol{y} \in V_t$ and $s \leq t$. $\boldsymbol{y} \leq \boldsymbol{x} \vee \boldsymbol{y} \leq (\sup V_s) \vee (\sup V_t)$ follows from $\boldsymbol{x} \leq \sup V_s$ and $\boldsymbol{y} \leq \sup V_t$. On the other hand, $\sup V_s \vee \sup V_t = \sup V_t$ from Lemma 2.1. Then, $\boldsymbol{y} \leq \boldsymbol{x} \vee \boldsymbol{y} \leq \sup V_t$ holds. Now the parallel system φ implies $\sup V_t \in V_t$ and

$y \in V_t$ from the assumption. Then, the increasing property of φ leads to $\varphi(x \vee y) = t$. Noticing $\varphi(x) \vee \varphi(y) = t$, we conclude the proof. □

Table 1 Coherent series

		Ω_2		
		0	1	2
	0	0	0	0
Ω_1	1	0	1	1
	2	0	1	2

Table 2 Increasing series

		Ω_2		
		0	1	2
	0	0	0	0
Ω_1	1	0	1	1
	2	2	2	2

Example 2.1. Tables 1 and 2 are examples of series coherent and series increasing systems, respectively. Table 1 gives us an intuitive explanation of the equality in (i) of Proposition 2.1. Table 2 shows that the equalities in Proposition 2.1 do not necessarily hold when the coherent property is not assumed. □

Theorem 2.1. (Existence theorem of series and parallel coherent systems) Let Ω_i $(i \in C)$ and S be totally ordered finite sets.

(i) A series coherent system (Ω_C, S, φ) exists iff $|S| \leq \min_{i \in C} |\Omega_i|$.

(ii) A parallel coherent system (Ω_C, S, φ) exists iff $|S| \leq \min_{i \in C} |\Omega_i|$.

Proof. The necessity of the conditions (i) and (ii) are evident from Lemma 2.1, then we prove the sufficiency part of each case.

Sufficiency in (i): Let $\inf V_s$ and V_s $(s \in S)$ be

$$\inf V_s = (s, \cdots, s),$$
$$V_s = \{x \mid \inf V_s \leq x, \inf V_t \not\leq x \ (t > s), \ x \in \Omega_C\},$$

and we construct $\varphi : \Omega_C \to S$ as $\varphi(x) = s$ for $x \in V_s$. Then, this system φ is easily shown to be a series increasing system. Noticing that $(t, \cdots, t) \in V_t$ and $(t, \cdots, t, s, t, \cdots, t) \in V_s$ holds for every s and t of S such that $s < t$, the relevant property of the system φ is evident,

Sufficiency in (ii): Letting $\sup V_s$ and V_s $(s \in S)$ be

$$\sup V_s = (s, \cdots, s), \text{ for } s < N,$$
$$\sup V_N = (N_1, \cdots, N_n),$$
$$V_s = \{x \mid x \leq \sup V_s, \ x \not\leq \sup V_r \ (r < s), \ x \in \Omega_C\},$$

we assume φ as $\varphi(x) = s$ for $x \in V_s$. Then, this system φ is a coherent parallel system. □

Theorem 2.1 tells us a necessary and sufficient condition for us to construct series and parallel coherent systems but not their uniqueness. In fact, we can easily construct several series coherent systems for given Ω_i ($i \in C$) and S satisfying the condition of Theorem 2.1. The following proposition, however, shows us the existence of the maximum and minimum series and parallel coherent systems.

Proposition 2.2. Suppose that $|S| \leq \min_{i \in C} |\Omega_i|$ holds.

(i) There exist the minimum series coherent system $(\Omega_C, S, \varphi_{smin})$ and the maximum series coherent system $(\Omega_C, S, \varphi_{smax})$ satisfying

$$\forall \boldsymbol{x} \in \Omega_C, \quad \varphi_{smin}(\boldsymbol{x}) \leq \psi(\boldsymbol{x}) \leq \varphi_{smax}(\boldsymbol{x}).$$

for every series coherent system (Ω_C, S, ψ).

(ii) There exist the minimum parallel coherent system $(\Omega_C, S, \varphi_{pmin})$ and the maximum parallel coherent system $(\Omega_C, S, \varphi_{pmax})$ satisfying

$$\forall \boldsymbol{x} \in \Omega_C, \quad \varphi_{pmin}(\boldsymbol{x}) \leq \psi(\boldsymbol{x}) \leq \varphi_{pmax}(\boldsymbol{x}).$$

for every parallel coherent system (Ω_C, S, ψ).

Proof.

(i) *Existence of φ_{smax}*: We prove that φ_{smax} is the series coherent system constructed in the proof of Theorem 2.1. Let ψ be a series coherent system satisfying $\varphi_{smax}(\boldsymbol{x}) < \psi(\boldsymbol{x})$ for some $\boldsymbol{x} \in \Omega_C$, and without loss of generality, we assume $\varphi_{smax}(\boldsymbol{x}) = s$ and $\psi(\boldsymbol{x}) = t$, where $s < t$. For some $i \in C$, $(\boldsymbol{x})_i = s$ must hold from the construction of φ_{smax}. Then, $(\inf V_t(\psi))_i \leq s < t$ holds. On the other hand, from Lemma 2.1, we have $(\inf V_p(\psi))_i < p$ for any $p \leq t$, then $(\inf V_0(\psi))_i < 0$, which contradicts to $(\inf V_0(\psi))_i = 0$. Hence, $\varphi_{smax}(\boldsymbol{x}) \geq \psi(\boldsymbol{x})$ holds for every $\boldsymbol{x} \in \Omega_C$.

Existence of φ_{smin}: An argument similar to the proof of the existence of φ_{smax} easily takes us to that the coherent series system constructed by the following is the φ_{smin}:

$$\inf V_s = (N_1 - (N - s), \cdots, N_n - (N - s)) \quad s \in S,$$
$$\inf V_0 = (0, \cdots, 0),$$
$$V_s = \{\, \boldsymbol{x} \mid \boldsymbol{x} \leq \sup V_s,\ \boldsymbol{x} \not\leq \sup V_r\ (r < s),\ \boldsymbol{x} \in \Omega_C \,\},$$
$$\varphi_{smin}(\boldsymbol{x}) = s \quad \text{for } \boldsymbol{x} \in V_s.$$

(ii) It is easily shown by the argument similar to (i) that φ_{pmax} is as constructed in the proof of Theorem 2.1, and φ_{pmin} is as the following:

$$\sup V_s = (N_1 - (N-s), \cdots, N_n - (N-s)) \quad s \in S,$$
$$V_s = \{\, \boldsymbol{x} \mid \boldsymbol{x} \leq V_s,\ \boldsymbol{x} \not\leq \sup V_r\ (r < s),\ \boldsymbol{x} \in \Omega_C \,\},$$
$$\varphi_{pmin}(\boldsymbol{x}) = s, \quad \text{for } \boldsymbol{x} \in V_s. \quad \Box$$

From Proposition 2.2, when $|S| = |\Omega_i|$ ($i \in C$), series and parallel coherent systems are uniquely determined, and we may express each of them as

$$\varphi_{smin}(\boldsymbol{x}) = \varphi_{smax}(\boldsymbol{x}) = \min\{x_1, \cdots, x_n\} = \wedge_{i=1}^n x_i,$$
$$\varphi_{pmin}(\boldsymbol{x}) = \varphi_{smax}(\boldsymbol{x}) = \max\{x_1, \cdots, x_n\} = \vee_{i=1}^n x_i.$$

These are the usual formulae of series and parallel systems [1], [7].

Example 2.2. The following Tables 2.3–2.8 are examples of φ_{smax}, φ_{smin}, φ_{pmax}, φ_{pmin}, and coherent systems φ_1 and φ_2:

Table 2.3. φ_{smin}

	Ω_2			
	0	1	2	3
Ω_1 0	0	0	0	0
1	0	0	1	1
2	0	0	1	2

Table 2.4. φ_{smax}

	Ω_2			
	0	1	2	3
Ω_1 0	0	0	0	0
1	0	1	1	1
2	0	1	2	2

Table 2.5. φ_{pmin}

	Ω_2			
	0	1	2	3
Ω_1 0	0	0	1	2
1	1	1	1	2
2	2	2	2	2

Table 2.6. φ_{pmax}

	Ω_2			
	0	1	2	3
Ω_1 0	0	1	2	2
1	1	1	2	2
2	2	2	2	2

Table 2.7. φ_1

	Ω_2			
	0	1	2	3
Ω_1 0	0	0	0	0
1	0	0	0	1
2	0	0	1	2

Table 2.8. φ_2

	Ω_2			
	0	1	2	3
Ω_1 0	0	1	2	2
1	1	2	2	2
2	2	2	2	2

Relating to these examples, we note that $\varphi_1(\boldsymbol{x}) \leq \varphi_{smin}(\boldsymbol{x})$ and $\varphi_{pmax}(\boldsymbol{x}) \leq \varphi_2(\boldsymbol{x})$ for every $\boldsymbol{x} \in \Omega_C$.

In the theory of binary state systems, any coherent system is bounded from the below by series system and from the above by parallel system.

These examples, however, show us that the similar results no longer hold in our theory of multistate systems.

Table 2.9. φ_1

	Ω_2		
Ω_1	0	1	2
0	0	0	0
1	0	0	1
2	0	1	2

Table 2.10. φ_2

	Ω_2		
Ω_1	0	1	2
0	0	1	2
1	1	2	2
2	2	2	2

Furthermore, noticing that $\varphi_1(x_1, 0) = \varphi_1(x_1, 1)$ for every $x_1 \in \Omega_1$ and $\varphi_2(x_1, 2) = \varphi_2(x_1, 3)$ for every $x_1 \in \Omega_1$, we may merge some states of a component to one state. For the case of the system φ_1, the state space of the second component is essentially $\{1, 2, 3\}$, and the state space of the second component is essentially $\{0, 1, 2\}$ for the system φ_2. Then, we have transformed coherent systems which are equivalent to the original coherent systems. □

In this chapter we present the following propositions.

Proposition 2.3. Let φ be an increasing system satisfying $(s, \cdots, s) \in V_s(\varphi)$ for every $s \in S$. Then, using the φ_{smax} and φ_{pmax} of Proposition 2.2,

$$\forall \boldsymbol{x} \in \Omega_C, \quad \varphi_{smax}(\boldsymbol{x}) \leq \varphi(\boldsymbol{x}) \leq \varphi_{pmax}(\boldsymbol{x}).$$

Proof. Let $\varphi(\boldsymbol{x}) < \varphi_{smax}(\boldsymbol{x})$ hold for some $\boldsymbol{x} \in \Omega_C$, and assume $\varphi_{smax}(\boldsymbol{x}) = t$. From the construction of φ_{smax}, $\boldsymbol{x} \geq \boldsymbol{t} = (t, \cdots, t)$. Then, $\varphi(\boldsymbol{t}) \leq \varphi(\boldsymbol{x}) < \varphi_{smax}(\boldsymbol{x}) = t$ and $\varphi(\boldsymbol{t}) \neq t$, which contradicts to $\boldsymbol{t} \in V_t(\varphi)$. Hence, $\varphi_{smax}(\boldsymbol{x}) \leq \varphi(\boldsymbol{x})$ for every $\boldsymbol{x} \in \Omega_C$.

A similar argument proves $\varphi(\boldsymbol{x}) \leq \varphi_{pmax}(\boldsymbol{x})$ for every $\boldsymbol{x} \in \Omega_C$. □

We close this section with an alternative definition of systems and a definition of dual systems, which will be used in Sections 4 and 3.

Definition 2.6. (An alternative definition of systems) A system composed of n components is a triplet $(\Omega_C, S, \mathcal{V})$ satisfying the following conditions:

1) $C = \{1, \cdots, n\}$.
2) Ω_i $(i \in C)$ and S are totally ordered finite sets.
3) $\mathcal{V} = \{V_s; s \in S\}$ is a partition of Ω_C.

The equivalence of Definitions 2.1 and 2.6 is evident. Using Definition 2.6, we may express increasing and relevant properties of systems.

Proposition 2.4.

(i) A system $(\Omega_C, S, \mathcal{V})$ is increasing iff $y \not\leq x$ holds for every $x \in V_s$ and $y \in V_t$, whenever $s < t$.
(ii) A system $(\Omega_C, S, \mathcal{V})$ is relevant iff $(P_{\Omega_C \setminus \{i\}} V_s) \cap (P_{\Omega_C \setminus \{i\}} V_t) \neq \phi$ for every $i \in C$ and $s, t \in S$ such that $s \neq t$.

Next, we present a definition of dual systems and some remarks.

Definition 2.7. The dual system of a system $(\prod_{i=1}^{n} \Omega_i, S, \varphi)$ is the system $(\prod_{i=1}^{n} \Omega_i^D, S^D, \varphi^D)$ defined as the following:

1) Ω_i^D $(i = 1, \cdots, n)$ and S^D are the dual ordered sets of Ω_i $(i = 1, \cdots, n)$ and S, respectively.
2) $\varphi^D : \prod_{i=1}^{n} \Omega_i^D \to S^D$ is defined as

$$x \in \prod_{i=1}^{n} \Omega_i^D, \ \varphi^D(x) = \varphi(x).$$

Denoting the dual orders commonly by $\leq_D$,

$$k, l \in \Omega_i^D, \ k \leq_D l \iff k \geq l, \ (i = 1, \cdots, n),$$

$$x, y \in \prod_{i=1}^{n} \Omega_i^D, \ x \leq_D y \iff x \geq y,$$

$$s, t \in S, \ s \leq_D t \iff t \geq s.$$

The state spaces, say, S and S^D differ only by the order and the elements consisting the sets are the same. For example, the maximum element of S is the minimum element of S^D. It is easily verified that if a system is increasing, then the dual system is also increasing. The similar proposition holds for the coherent properties of the dual systems.

In the theory of binary state systems, it is well known that the dual systems of series(parallel) systems are parallel(series) systems. The same proposition holds in our coherent context, but, we treat these properties in a wider context in the next section of k-out-of-n systems.

3 k-out-of-n systems

The treatment of k-out-of-n systems in this section is newly presented here. For a system (Ω_C, S, φ) and $A \subset C$, we define $\varphi_A : \prod_{i \in A} \Omega_i \to S$ as the

following:
$$x \in \prod_{i \in A} \Omega_i, \quad \varphi_A(\boldsymbol{x}) = \varphi(\boldsymbol{x}, \boldsymbol{0}^{A'}).$$
Then, we have a system (Ω_A, S, φ_A).

Definition 3.1. A coherent system (Ω_C, S, φ) of order n, i.e., $C = \{1, \cdots, n\}$, is called:

1) A coherent k-out-of-n:G system, when the following conditions hold:
   (1-i) $\forall A \subset C$ such that $|A| = k$, (Ω_A, S, φ_A) is coherent series.
   (1-ii) $\forall \boldsymbol{x} \in \Omega_C$, $\varphi(\boldsymbol{x}) = \max_{A:A \subset C, |A|=k} \varphi_A(\boldsymbol{x}^A)$.
2) A coherent k-out-of-n:F system, when the following conditions hold:
   (2-i) $\forall A \subset C$ such that $|A| = k$, (Ω_A, S, φ_A) is coherent parallel.
   (2-ii) $\forall \boldsymbol{x} \in \Omega_C$, $\varphi(\boldsymbol{x}) = \min_{A:A \subset C, |A|=k} \varphi_A(\boldsymbol{x}^A)$.

Reminding Proposition 2.2, we notice that the condition about the state spaces, $|S| \leq \min_{1 \leq i \leq n} |\Omega_i|$, should be hold, when we consider the coherent k-out-of-n:G(F) systems.

Proposition 3.1.

(i) φ is a coherent series system of order n iff φ is a coherent n-out-of-n:G system.
(ii) φ is a coherent series system of order n iff φ is a coherent 1-out-of-n:F system.
(iii) φ is a coherent parallel system of order n iff φ is a coherent 1-out-of-n:G system.
(iv) φ is a coherent parallel system of order n iff φ is a coherent n-out-of-n:F system.

Proof. The equivalent relationship of (i) and (ii) is clear from Definition 3.1 1) and 2), respectively. We prove only (iii), since (ii) is similarly proved.
For a 1-out-of-n:G system φ, we have from Definition 3.1 1),
$$\varphi(\boldsymbol{x} \vee \boldsymbol{y}) = \max_{1 \leq i \leq n} \varphi(x_i \vee y_i, \boldsymbol{0}^{\{i\}'})$$
$$= \max_{1 \leq i \leq n} \left(\varphi(x_i, \boldsymbol{0}^{\{i\}'}) \vee \varphi(y_i, \boldsymbol{0}^{\{i\}'}) \right)$$
$$= \max_{1 \leq i \leq n} \varphi(x_i, \boldsymbol{0}^{\{i\}'}) \vee \max_{1 \leq i \leq n} \varphi(y_i, \boldsymbol{0}^{\{i\}'})$$
$$= \varphi(\boldsymbol{x}) \vee \varphi(\boldsymbol{y}),$$
because Ω_i and S are totally ordered sets. Then, φ is a parallel system.

Let φ be a coherent parallel system. From Lemma 2.1, for $s < t$, $\sup V_s < \sup V_t$, so to say,

$$\forall i \in C, \ (\sup V_0)_i < (\sup V_1)_i < \cdots < (\sup V_N)_i,$$

reminding that we assumed $S = \{0, 1, \cdots, N\}$ and $\Omega_i = \{0, 1, \cdots, N_i\}$ ($i \in C$). Then,

the minimum element of $V_0 = (0, \cdots, 0)$,

the set of the minimal elements of V_t

$$= \left\{ \left(\sup V_{t-1} \right)_i + 1, \mathbf{0}^{\{i\}'} \right) \Big| i = 1, 2, \cdots n \right\}, \ 1 \leq t \leq N,$$

and $\varphi(\boldsymbol{x}) = \max_{1 \leq i \leq n} \varphi_{\{i\}}(x_i)$. Hence, the coherent parallel system φ is 1-out-of-n:G system. □

Theorem 3.1. (Minimal elements of k-put-of-n:G systems) Let (Ω_C, S, φ) be a coherent k-out-of-n:G system of order n, and

$$\forall \boldsymbol{x} \in \Omega_C, \ \varphi(\boldsymbol{x}) = \max_{A: A \subset C, \ |A|=k} \varphi_A(\boldsymbol{x}^A),$$

where (Ω_A, S, φ_A) is a coherent series system. Then, for the minimal elements of $V_s(\varphi)$,

$$MIV_s(\varphi) = \bigcup_{A: A \subset C, \ |A|=k} \{ \ (\boldsymbol{x}, \mathbf{0}^{A'}) \mid \boldsymbol{x} \in MIV_s(\varphi_A) \ \}.$$

We notice that $MIV_s(\varphi_A)$ consists of only one element, that is, the minimum element of $V_s(\varphi_A)$, since the system φ_A is a coherent series system.

Proof. For an element $\boldsymbol{x} \in MIV_s(\varphi)$ ($s \neq 0$), there exists uniquely a subset A of C such that $|A| = k$, $\boldsymbol{x} = (\boldsymbol{x}^A, \mathbf{0}^{A'})$, $\boldsymbol{x}^A > \mathbf{0}$ and $\boldsymbol{x}^A \in MIV_s(\varphi_A)$, by Lemma 2.1 and that each φ_A is a coherent series system.
Let

$$A \subset C, \ |A| = k, \quad B \subset C, \ |B| = k, \quad A \neq B,$$
$$\boldsymbol{x} \in MIV_s(\varphi_A), \quad \boldsymbol{y} \in MIV_t(\varphi_B), \quad s \neq t, s \neq 0, t \neq 0.$$

We have $(A \backslash B) \cup (B \backslash A) \neq \phi$, $\boldsymbol{x} > \mathbf{0}$, $\boldsymbol{y} > \mathbf{0}$, and then, $(\boldsymbol{x}, \mathbf{0}^{A'})^B \not\geq \boldsymbol{y}$. Hence, $\varphi_B(\boldsymbol{x}, \mathbf{0}^{A'})^B = 0$, since $\boldsymbol{y}$ is the minimum element of $V_t(\varphi_B)$. Finally, have $\varphi(\boldsymbol{x}, \mathbf{0}^{A'}) = \varphi_A(\boldsymbol{x}) = s$ and $(\boldsymbol{x}, \mathbf{0}^{A'}) \in MIV_s(\varphi)$. □

Theorem 3.1 tells us that the type of minimal elements of $V_s(\varphi)$ is restricted to be $(\boldsymbol{x}, \mathbf{0}^{A'})$, $|A| = k$, $\boldsymbol{x} \in \Omega_A$, for the case of k-out-of-n:G system. Noticing this theorem, the next example shows us that the dual system of a k-out-of-n:G system is not necessarily $n-k+1$-out-of-n:G system.

Example 3.1. Let $\Omega_i = \{0,1,2,3\}$, $S = \{0,1,2\}$. A 2-out-of-3:G system φ is defined by specifying the minimal elements as
$$MIV_0(\varphi) = \{(0,0,0)\},$$
$$MIV_1(\varphi) = \{(1,2,0),(2,0,1),(0,1,2)\},$$
$$MIV_2(\varphi) = \{(2,3,0),(3,0,2),(0,2,3)\},$$
$(2,2,2)$ is easily verified a maximal element of $V_1(\varphi)$, and then, the dual system of φ is not a $n-k+1$-out-of-n:G, in this example a 2-out-of-3:G system. □

Example 3.1 shows us that the dual system of a k-out-of-n:G system is not generally a $n-k+1$-out-of-n:G system. But, as a special case, the dual system of a coherent n-out-of-n:G (series) system is a coherent 1-out-of-n:G (parallel) system, and vice versa.

Following Proposition 2.2 which asserts the existence of the minimum and maximum coherent series systems, there exist the minimum and maximum coherent k-out-of-n:G systems for given state spaces Ω_i ($i \in C$) and S. The following Corollary 3.1 shows that the dual system of the minimum (maximum) coherent k-out-of-n:G system is the maximum (minimum) coherent $n-k+1$-out-of-n:G system.

Definition 3.2.

1) A coherent k-out-of-n:G system is said to be maximum (minimum) when the series systems of Definition 3.1 (1) are all the maximum (minimum).
2) A coherent k-out-of-n:F system is said to be maximum (minimum) when the parallel systems of Definition 3.1 (2) are all the maximum (minimum).

The following Theorem 3.2 shows us the pattern of the maximal elements of the minimum (maximum) coherent k-out-of-n:G systems, of which proof needs the following lemma:

Lemma 3.1. Let C be a set with cardinal number n. For a subset $B \subset C$,
$$\forall A \subset C \text{ such that } |A| = k,\ A \cap B \neq \phi \qquad (3.1)$$
iff
$$|B| \geq n - k + 1.$$
Then, the minimum number of the cardinal number of B satisfying (3.1) is $n - k + 1$.

Proof. If $|B| \leq n - k$, then

$$|C\backslash B| = |C| - |B| \geq n - (n - k) = k, \quad B \cup (C\backslash B) = \phi,$$

which contradicts to (3.1). Then, (3.1) implies $|B| \geq n - k + 1$.
Suppose $|B| \geq n - k + 1$. For $A \subset C$ such that $|A| = k$, if $A \cap B = \phi$,

$$A \subset C\backslash B, \quad |A| \leq |C\backslash B| \leq n - (n - k + 1) = k - 1,$$

which contradicts to $|A| = k$. Thus $A \cap B \neq \phi$. □

Theorem 3.2. (Maximal elements of the minimum (maximum) coherent k-out-of-n:G systems) Let (Ω_C, S, φ) be the minimum (maximum) coherent k-out-of-n:G system. Then, for every $s \in S\backslash\{N\}$ and every maximum element x of $V_s(\varphi)$, there exists $B \subset C$ such that $|B| = n - k + 1$ and

$$x_i = \begin{cases} \min_{A: i \in A, |A|=k, A \subset C} (\inf V_{s+1}(\varphi_A))_i - 1, & i \in B, \\ N_i, & i \notin B. \end{cases} \quad (3.2)$$

By contraries, every $x \in \Omega_C$ constructed by the above formulae (3.2) by using every $B \subset C$ such that $|B| = n - k + 1$ is a maximal element of $V_s(\varphi)$.

Proof. Let $A_i \subset C$, $|A_i| = k$ $(i = 1, 2)$, $A_1 \cap A_2 \neq \phi$. Noticing how to construct the maximum (minimum) series system of Proposition 2.2, we have for every $i \in A_1 \cap A_2$ and every $s \in S$, the i-th coordinate state of every minimal element of $V_s(\varphi_{A_1})$ is equal to the i-th coordinate state of every minimal element of $V_s(\varphi_{A_2})$. Then, using Lemma 3.1, this theorem is proved. □

The next corollary is evident from Theorem 3.2.

Corollary 3.1. The dual system of the minimum (maximum) coherent k-out-of-n:G system is the coherent maximum (minimum) $n-k+1$-out-of-n:G system.

For k-out-of-n:F systems, we also have theorem and corollary similar to Theorem 3.2 and Corollary 3.1.

4 Modules of Coherent Systems

In this section, we examine the concepts of modules of multistate systems which are practically important, since systems in real situations have a hierarchic structure and each layer of the hierarchy consists of modules, and each module also consists of modules of smaller size. In other words, a system is constructed of systems of smaller size. Most part of the theory of binary coherent systems is devoted to an examination of modules.

We start with a definition of a modular decomposition, and then, a definition of a module is presented. Our argument is in reverse order to that of [1], but presents a unified approach to a module and a modular decomposition.

Definition 4.1. A partition $\mathcal{A} = \{A_j, 1 \leq j \leq m\}$ of C is called a modular decomposition of a system (Ω_C, S, φ) iff there exist systems $(\Omega_{A_j}, S_j, \chi_j)$ $(1 \leq j \leq m)$ and $(\prod_{j=1}^{m} S_j, S, \psi)$ satisfying

$$\forall \boldsymbol{x} \in \Omega_C, \quad \varphi(\boldsymbol{x}) = \psi\left(\chi_1\left(\boldsymbol{x}^{A_1}\right), \cdots, \chi_m\left(\boldsymbol{x}^{A_m}\right)\right).$$

Each $A_j \in \mathcal{A}$ is called a module of the system φ.

If systems χ_j $(1 \leq j \leq m)$ and ψ are increasing (coherent), then $\mathcal{A}$ is called an increasing (a coherent) modular decomposition and each $A_j \in \mathcal{A}$ is called an increasing (a coherent) module.

We can easily prove that any partition of C of any coherent series (parallel) systems is a coherent modular decomposition.

Proposition 4.1. If $\mathcal{A}$ is a modular decomposition of a coherent system φ, then ψ is relevant.

The proof of the proposition is easy and then omitted.

Proposition 4.1 tells us that we only examine the properties of each element of A_j $(1 \leq j \leq m)$ of a modular decomposition $\mathcal{A}$ to characterize a coherent modular decomposition. With a characterization of increasing and coherent modular decomposition, we prove a theorem almost similar to Three Modules Theorem of [6].

We define a pseudo-order $\stackrel{\varphi}{\leq}$ on Ω_A $(A \subset C)$ as the following: For $\boldsymbol{x}$ and $\boldsymbol{y}$ in Ω_A, $\boldsymbol{x} \stackrel{\varphi}{\leq} \boldsymbol{y}$ means $\varphi(\boldsymbol{x}, \boldsymbol{z}) \leq \varphi(\boldsymbol{y}, \boldsymbol{z})$ for every $\boldsymbol{z} \in \Omega_{A'}$. When $\boldsymbol{x} \stackrel{\varphi}{\leq} \boldsymbol{y}$ and $\boldsymbol{y} \stackrel{\varphi}{\leq} \boldsymbol{x}$, we define $\boldsymbol{x} \stackrel{\varphi}{=} \boldsymbol{y}$. The relation $\stackrel{\varphi}{=}$ is an equivalent relation on Ω_A. $(\Omega_A, \stackrel{\varphi}{\leq})$ is generally a partially pseudo-ordered set, and if $(\Omega_A, \stackrel{\varphi}{\leq})$ is a totally pseudo-ordered set, then Ω_A can be partitioned by the equivalent relation $\stackrel{\varphi}{=}$.

Proposition 4.2.

(i) Let φ be an increasing system. Then, a partition $\mathcal{A} = \{A_j, 1 \leq j \leq m\}$ of C is an increasing modular decomposition iff each pseudo-ordered set $(\Omega_{A_j}, \stackrel{\varphi}{\leq})$ $(1 \leq j \leq m)$ is a totally pseudo-ordered set.

(ii) Let φ be a coherent system. A partition $\mathcal{A} = \{A_j, 1 \leq j \leq m\}$ of C is a coherent modular decomposition iff the following two conditions are hold:

(ii-i) Each $(\Omega_{A_j}, \overset{\varphi}{\leq})$ $(i \leq j \leq m)$ is a totally pseudo-ordered set.

(ii-ii) For x and y of Ω_{A_j} satisfying $x \overset{\varphi}{\leq} y$ and $x \overset{\varphi}{\neq} y$, there exist (k_i, z) and (l_i, z) of Ω_{A_j} which satisfy $(k_i, z) \overset{\varphi}{=} x$ and $(l_i, z) \overset{\varphi}{=} y$ for $i \in A_j$ and $j = 1, \cdots, m$

Proof. From Proposition 2.4, (ii) of this proposition is easily proved by using (i), and then, we prove only (i). The necessity of the condition is clear from the definition of an increasing modular decomposition. Thus, we prove the sufficiency of the condition.

From our assumption, the equivalent relation $\overset{\varphi}{=}$ determines a partition $\mathcal{V}_{A_j} = \{V_i^{A_j}, 1 \leq j \leq m_{A_j}\}$ of Ω_{A_j} for every j $(1 \leq j \leq m)$, where we may assume $x \overset{\varphi}{\leq} y$ if $x \in V_k^{A_j}$, $y \in V_l^{A_j}$ for $k < l$. Then for $x \in V_k^{A_j}$ and $y \in V_l^{A_j}$, where $k < l$, we have $y \not\leq x$ by the increasing property of φ, and then, systems $(\Omega_{A_j}, S_j, \mathcal{V}_{A_j})$ or $(\Omega_{A_j}, S_j, \chi_j)$ $(1 \leq j \leq m)$ are determined, where $S_j = \{1, \cdots, m_j\}$ and $\chi_j(x) = s_j$ for $x \in V_{s_j} \in \mathcal{V}_{A_j}$. We now construct a system $(\prod_{j=1}^m S_j, S, \psi)$ as $\psi(s_1, \cdots, s_m) = \varphi(x_1, \cdots, x_m)$, where x_j is any element of $V_j^{A_j} \in \mathcal{V}_{A_j}$ $(1 \leq j \leq m)$, which is an increasing system and satisfies $\varphi(x) = \psi(\chi_1(x^{A_i}), \cdots, \chi_m(x))$ for any $x \in \Omega_C$. Coherent property is clear. □

Example 4.1. $\mathcal{A} = \{\{i\}, i \in C\}$ and $\mathcal{A} = \{C\}$ are modular decomposition of a system φ, and an increasing (coherent) modular decomposition if the system φ is increasing (coherent). □

Corollary 4.1.

(i) Let φ be an increasing system. Then, a subset A of C is an increasing module of the system iff $(\Omega_A, \overset{\varphi}{\leq})$ is a totally pseudo-ordered set, *i.e.*, A is an element of some modular decomposition $\mathcal{A}$ of the system φ.

(ii) Let φ be a coherent system. Then, a subset A of C is coherent module of the system φ iff Ω_A satisfies the conditions of Proposition 4.2 (ii), *i.e.*, A is an element of some coherent modular decomposition $\mathcal{A}$ of the system φ.

Proof. Since the increasing property of φ means that $(\Omega_i, \overset{\varphi}{\leq})$ is a totally pseudo-ordered set and a system (Ω_i, Ω_i, I), where I is the identity mapping of Ω_i, i.e., evidently a coherent system, then the corollary is obvious by considering the partition $\mathcal{A} = \{A, \{i\}, \ i \in A'\}$ of C. □

Corollary 4.1 shows us that a subset A of C is an increasing (a coherent) module iff Ω_A satisfies the conditions of Proposition 4.2(i) (Proposition 4.2(ii)), and then, we may treat the increasing (coherent) modules of increasing (coherent) systems without being conscious of increasing (coherent) modular decomposition.

Theorem 4.1. Let φ be a coherent system. If A and B are coherent modules of the system φ such that $A\backslash B$, $B\backslash A$ and $A \cap B$ are non-empty, then $A\backslash B$, $B\backslash A$ and $A \cup B$ are increasing modules of the system φ.

Proof. (Proof of that $A\backslash B$ is an increasing module) Let $\boldsymbol{x}$ and $\boldsymbol{y}$ be arbitrarily given elements of $\Omega_{A\backslash B}$. If $\varphi(\boldsymbol{x}, \boldsymbol{z}) < \varphi(\boldsymbol{y}, \boldsymbol{z})$ holds for some $\boldsymbol{z} \in \Omega_{(A\backslash B)'}$, then we have $\varphi(\boldsymbol{x}, \boldsymbol{w}) \leq \varphi(\boldsymbol{y}, \boldsymbol{w})$ for every $\boldsymbol{w} \in \Omega_{(A\backslash B)'}$. Since B is a coherent module, for arbitrarily given $i \in B\backslash A$ there exist $(k_i, \boldsymbol{u})$ and $(l_i, \boldsymbol{u})$ of Ω_B such that $(k_i, \boldsymbol{u}) \overset{\varphi}{=} \boldsymbol{z}^B$ and $(l_i, \boldsymbol{u}) \overset{\varphi}{=} \boldsymbol{w}^B$.

Since A is an increasing module,
$$\varphi(\boldsymbol{x}, (k_i, \boldsymbol{u}), \boldsymbol{z}^{(A \cup B)'}) < \varphi(\boldsymbol{y}, (k_i, \boldsymbol{u}), \boldsymbol{z}^{(A \cup B)'}),$$
$$\varphi(\boldsymbol{x}, (l_i, \boldsymbol{u}), \boldsymbol{w}^{(A \cup B)'}) \leq \varphi(\boldsymbol{y}, (l_i, \boldsymbol{u}), \boldsymbol{w}^{(A \cup B)'}).$$
Noting that $(l_i, \boldsymbol{u}) \overset{\varphi}{=} \boldsymbol{w}^B$, $\varphi(\boldsymbol{x}, \boldsymbol{w}) \leq \varphi(\boldsymbol{y}, \boldsymbol{w})$ holds, and hence, $A\backslash B$ is an increasing module by Corollary 4.1. A similar examination shows us that $B\backslash A$ is an increasing module.

(Proof of that $A \cup B$ is an increasing module) Let $\boldsymbol{x}$ and $\boldsymbol{y}$ be arbitrarily given elements of $\Omega_{A \cup B}$. We assume $\varphi(\boldsymbol{x}, \boldsymbol{z}) < \varphi(\boldsymbol{y}, \boldsymbol{z})$ for some $\boldsymbol{z} \in \Omega_{(A \cup B)'}$. Since A is a coherent module, for any $i \in A \cap B$ there exist $(k_i, \boldsymbol{u})$ and $(l_i, \boldsymbol{u})$ of Ω_A satisfying $(k_i, \boldsymbol{u}) \overset{\varphi}{=} \boldsymbol{x}^A$ and $(l_i, \boldsymbol{u}) \overset{\varphi}{=} \boldsymbol{y}^A$, and then
$$\varphi((k_i, \boldsymbol{u}), \boldsymbol{x}^{B\backslash A}, \boldsymbol{z}) < \varphi((l_i, \boldsymbol{u}), \boldsymbol{y}^{B\backslash A}, \boldsymbol{z}).$$
Since B is an increasing module,
$$\forall \boldsymbol{w} \in \Omega_{(A \cup B)'}, \ \varphi((k_i, \boldsymbol{u}), \boldsymbol{x}^{B\backslash A}, \boldsymbol{w}) \leq \varphi((l_i, \boldsymbol{u}), \boldsymbol{y}^{B\backslash A}, \boldsymbol{w}).$$
Using $(k_i, \boldsymbol{u}) \overset{\varphi}{=} \boldsymbol{x}^A$ and $(l_i, \boldsymbol{u}) \overset{\varphi}{=} \boldsymbol{y}^A$, we have $\varphi(\boldsymbol{x}, \boldsymbol{w}) \leq \varphi(\boldsymbol{y}, \boldsymbol{w})$ for every $\boldsymbol{w} \in \Omega_{(A \cup B)'}$, and then the proof is completed. □

Remark 4.1. It is easy to construct an example to show that $A \cap B$ is not an increasing module, even if A and B are coherent modules of a coherent system satisfying the condition that $A\backslash B, B\backslash$ and $A \cap B$ are nonempty.

5 Probabilistic Aspect of Coherent Systems

In this section we consider probabilistic properties of coherent systems, particularly we discuss the comparison of probability measures, and IFRA and NBU closure theorems in a situation slightly different from [23].

Section 2 supposes that the state spaces of the components and the systems are finite totally ordered sets, but in this section we assume more generally that Ω_i ($i \in C$) and S are at most countable partially ordered sets. As basic measurable spaces we set $(\Omega_i, \mathcal{A}_i)$, where $\mathcal{A}_i$ is the power set of Ω_i. $(\prod_{i=1}^n \Omega_i, \prod_{i=1}^n \mathcal{A}_i) = (\Omega_C, \mathcal{A}_C)$ is the product measurable set of $(\Omega_i, \mathcal{A}_i)$ ($1 \leq i \leq n$). The order on Ω_C is defined similarly to that of Section 2. Throughout this section, we assume φ to be an increasing measurable function from $(\Omega_C, \mathcal{A}_C)$ to $(S, \mathcal{S})$, where $\mathcal{S}$ is the power set of S and the definition of an increasing function is similar to that of Section 2. Furthermore, we assume that Ω_i ($1 \leq i \leq n$) and S are endowed with discrete topology. Then, φ is a continuous function.

Generally a subset W of an ordered set Ω is called increasing iff $x \in W$ and $x \leq y$ imply $y \in W$. The concept of increasing set plays an important role in the sequel.

Before proving the IFRA and NBU closure theorem, we present some remarks:

Remark 5.1.

(i) If $P(W) \geq Q(W)$ holds for every increasing set $W \in \mathcal{A}_C$, then we have $P(\varphi \geq s) \geq Q(\varphi \geq s)$ for every $s \in S$, where P and Q are probability measures on $(\Omega_C, \mathcal{A}_C)$.

(ii) If Ω_i ($1 \leq i \leq n$) and S are finite totally ordered sets and $(s, \cdots, s) \in V_s(\varphi)$ for every $s \in S$, then by Proposition 2.3 $P\{\,x \mid x \geq (s, \cdots, s)\,\} \leq P\{\,\varphi \geq s\,\} \leq 1 - P\{\,x \mid x \leq (s-1, \cdots, s-1)\,\}$, where P is a probability measure on $(\Omega_C, \mathcal{A}_C)$.

(iii) Let $(\Omega_C, \mathcal{A}_C, P) = (\prod_{i=1}^n \Omega_i, \prod_{i=1}^n \mathcal{A}_i, \prod_{i=1}^n P_i)$ and $(\Omega_C, \mathcal{A}_C, Q) = (\prod_{i=1}^n \Omega_i, \prod_{i=1}^n \mathcal{A}_i, \prod_{i=1}^n Q_i)$, both of which imply that the performances of the components are stochastically independent.

If $P_i(W_i) \geq Q_i(W_i)$ for every increasing set $W_i \in \mathcal{A}_i$ ($1 \leq i \leq n$), then $P(W) \geq Q(W)$ for every increasing set $W \in \mathcal{A}_C$. This is easily proved by using the indicator function of W and Fubini's Theorem. Hence, it is shown that (i) and (ii) of this remark generalize Theorems 4.2 and 4.4 of [7], respectively.

From now on we focus on proving the IFRA and NBU closure theorems in a situation slightly different from that of Ross [23]. Though he assumed that Ω_i ($1 \leq i \leq n$) and S were subsets of $\mathbf{R}^+ = [0, +\infty)$, which means that they are totally ordered sets, our requirement for Ω_i ($1 \leq i \leq n$) and S is that they are at most countable partially ordered sets.

We will use the following Proposition 5.1 to prove the IFRA and NBU closure theorems.

Lemma 5.1. Let P_i, Q_i and U_i be probability measures on $(\Omega_i, \mathcal{A}_i)$.

(i) If $P_i(W_i) \geq [Q_i(W_i)]^\alpha$ holds for every increasing set $W_i \in \mathcal{A}_i$ and $0 < \alpha < 1$, then $\int_{\Omega_i} f^\alpha dP_i \geq \left[\int_{\Omega_i} f dQ_i\right]^\alpha$ holds for $0 < \alpha < 1$, where f is an increasing measurable function from $(\Omega_i, \mathcal{A}_i)$ to $(\mathbf{R}^+, \mathcal{B}^+)$, where $\mathcal{B}^+$ is the class of Borel subset of $\mathbf{R}^+$.

(ii) If $U_i(W_i) \leq P_i(W_i)Q_i(W_i)$ holds for every increasing set $W_i \in \mathcal{A}_i$, then $\int_{\Omega_i} fg dU_i \leq \int_{\Omega_i} f dP_i \int_{\Omega_i} g dQ_i$, where f and g are increasing measurable functions from $(\Omega_i, \mathcal{A}_i)$ to $(\mathbf{R}^+, \mathcal{B}^+)$.

Proof.

(i) From the assumption, f is approximated by a step function of the form $\sum_{j=1}^m x_j I_{S_j}$, where S_j is an increasing set of $\mathcal{A}_i$, $S_1 \supset \cdots \supset S_m$, $x_j \geq 0$ and I_{S_j} is the indicator function of S_j ($1 \leq j \leq m$). Then, using an argument similar to that of Lemma 1 of [23] and taking the limit, (i) is evident.

(ii) From the assumption, f and g are approximated by step functions of the form $\sum_{j=1}^m x_j I_{S_j}$ and $\sum_{k=1}^n y_k I_{V_k}$, respectively, where S_j (V_k) is an increasing set of $\mathcal{A}_i$, $S_1 \supset \cdots \supset S_m$ ($V_1 \supset \cdots \supset V_n$), $x_j \geq 0$ ($y_k \geq 0$) and I_{S_j} (I_{V_k}) is the indicator function of S_j (V_k) ($1 \leq j \leq m$, $1 \leq k \leq n$). Then, using the assumption,

$$\int_{\Omega_i} \left(\sum_{j=1}^m x_j I_{S_j}\right) \left(\sum_{k=1}^n y_k I_{V_k}\right) dU_i$$

$$\leq \int_{\Omega_i} \left(\sum_{j=1}^m x_j I_{S_j}\right) dP_i \int_{\Omega_i} \left(\sum_{k=1}^n y_k I_{V_k}\right) dQ_i,$$

noticing that the intersection of increasing sets is also an increasing set. Hence (ii) is clear by taking the limit. □

Proposition 5.1. Let P_i, Q_i and U_i be probability measures on $(\Omega_i, \mathcal{A}_i)$ ($1 \leq i \leq n$).

(i) If $P_i(W_i) \geq [Q_i(W_i)]^\alpha$ holds for every increasing set $W_i \in \mathcal{A}_i$ and $0 < \alpha < 1$ $(1 \leq i \leq n)$, then for every $0 < \alpha < 1$ and increasing set $W \in \mathcal{A}_C$,

$$\left(\prod_{i=1}^n P_i\right)(W) \geq \left[\left(\prod_{i=1}^n Q_i\right)(W)\right]^\alpha.$$

(ii) If $U_i(W_i) \leq P_i(W_i)Q_i(W_i)$ holds for every increasing set $W_i \in \mathcal{A}_i$ $(1 \leq i \leq n)$, then for every increasing set $W \in \mathcal{A}_C$,

$$\left(\prod_{i=1}^n U_i\right)(W) \leq \left[\left(\prod_{i=1}^n P_i\right)(W)\right]\left[\left(\prod_{i=1}^n Q_i\right)(W)\right].$$

Proof. Mathematical induction on n proves the proposition.

(i) When $n = 1$, (i) of this proposition is evident from the assumption. Suppose that (i) holds for $n = n$. Letting I_W be the indicator function of W, we have by Fubini's theorem,

$$\left(\prod_{i=1}^{n+1} P_i\right)(W) = \int_{\prod_{i=1}^{n+1}\Omega_i} I_W d\prod_{i=1}^{n+1} P_i$$

$$= \int_{\Omega_{n+1}} dP_{n+1} \int_{\prod_{i=1}^n \Omega_i} (I_W)_{x_{n+1}} d\prod_{i=1}^n P_i.$$

The section $(I_W)_{x_{n+1}}$ $(x_{n+1} \in \Omega_{n+1})$ which is an increasing binary function denotes an increasing set in $\prod_{i=1}^n \mathcal{A}_i$. Then, using the induction hypothesis, Lemma 5.1(i) and Fubini's theorem,

$$\left(\prod_{i=1}^{n+1} P_i\right)(W) \geq \int_{\Omega_{n+1}} dP_{n+1} \left[\int_{\prod_{i=1}^n \Omega_i} (I_W)_{x_{n+1}} d\prod_{i=1}^n Q_i\right]^\alpha$$

$$\geq \left[\int_{\prod_{i=1}^{n+1}\Omega_i} I_W d\prod_{i=1}^{n+1} Q_i\right]^\alpha$$

$$= \left[\left(\prod_{i=1}^{n+1} Q_i\right)(W)\right]^\alpha.$$

(ii) When $n = 1$, (ii) of this proposition is clearly holds by the assumption.

Suppose that (ii) holds for $n = n$. Then,

$$\left(\prod_{i=1}^{n+1} U_i\right)(W) = \int_{\Omega_{n+1}} dU_{n+1} \int_{\prod_{i=1}^{n} \Omega_i} (I_W)_{x_{n+1}} d\prod_{i=1}^{n} dU_i$$

$$\leq \int_{\Omega_{n+1}} dU_{n+1} \int_{\prod_{i=1}^{n} \Omega_i} (I_W)_{x_{n+1}} d\prod_{i=1}^{n} dP_i \int_{\prod_{i=1}^{n} \Omega_i} (I_W)_{x_{n+1}} d\prod_{i=1}^{n} dQ_i$$

$$\leq \int_{\prod_{i=1}^{n+1} \Omega_i} I_W d\prod_{i=1}^{n+1} dP_i \int_{\prod_{i=1}^{n+1} \Omega_i} I_W d\prod_{i=1}^{n+1} dQ_i$$

$$= \left[\left(\prod_{i=1}^{n+1} P_i\right)(W)\right]\left[\left(\prod_{i=1}^{n+1} Q_i\right)(W)\right],$$

where the first inequality comes from the inductive hypothesis, and the second inequality from Lemma 5.1(ii). □

Let $(\Omega, \mathcal{A}, P)$ be a given probability space and T be a subinterval of $\mathbf{R}^+ = [0, \infty)$. We suppose $X_i(t)$ ($t \in T$) to be a measurable function from $(\Omega, \mathcal{A})$ to $(\Omega_i, \mathcal{A}_i)$ ($1 \leq i \leq n$), which is a stochastic process denoting the state of the i-th component at time t, and then, $\boldsymbol{X}(t) = (X_1(t), \cdots, X_n(t))$ ($t \in T$) is a measurable function from $(\Omega, \mathcal{A})$ to $(\Omega_C, \mathcal{A}_C)$.

Let μ_t be the probability measure induced by $\boldsymbol{X}(t)$ from $(\Omega, \mathcal{A}, P)$ and $\mu_{i,t}$ be the restriction of μ_t to the measurable space $(\Omega_i, \mathcal{A}_i)$ which is also the probability measure induced by $X_i(t)$ from $(\Omega, \mathcal{A}, P)$. If $\{X_i(t), t \in T\}$ ($1 \leq i \leq n$) are mutually independent stochastic processes, then $\mu_t = \prod_{i=1}^{n} \mu_{i,t}$.

Definition 5.1. Let $T_W = \inf\{ t \mid X_i(t) \notin W \}$ ($W \in \mathcal{A}_i$).

1) A stochastic process $\{X_i(t), t \in T\}$ is called IFR iff T_W is an IFR random variable for any increasing set $W \in \mathcal{A}_i$.
2) A stochastic process $\{X_i(t), t \in T\}$ is called IFRA iff T_W is an IFRA random variable for any increasing set $W \in \mathcal{A}_i$.
3) A stochastic process $\{X_i(t), t \in T\}$ is called NBU iff T_W is an NBU random variable for any increasing set $W \in \mathcal{A}_i$.

For the definitions of IFR, IFRA and NBU random variables, see [1].

Theorem 5.1. Let $\{X_i(t), t \in T\}$ ($1 \leq i \leq n$) be decreasing and right continuous with probability 1, and be mutually independent.

(i) If $\{X_i(t), t \in T\}$ ($1 \leq i \leq n$) are IFRA processes, then $\{\varphi(\boldsymbol{X}(t)), t \in T\}$ is an IFRA process.

(ii) If $\{X_i(t), t \in T\}$ $(1 \leq i \leq n)$ are NBU processes, then $\{\varphi(\boldsymbol{X}(t)), t \in T\}$ is an NBU process.

Proof. For any increasing set $W \in \mathcal{S}$, $\{\boldsymbol{x} \mid \varphi(\boldsymbol{x}) \in W\}$ is an increasing set of $\mathcal{A}_C$. Then, since $\{X_i(t), t \in T\}$ $(1 \leq i \leq n)$ are decreasing and right continuous, it is sufficient to prove that for every increasing set $W \in \mathcal{A}_C$,

$$\mu_{\alpha t}(W) \geq [\mu_t(W)]^\alpha \ (0 < \alpha < 1) \quad (\ \mu_{s+t}(W) \leq \mu_s(W)\mu_t(W)\ ).$$

Since for every increasing set $W_i \in \mathcal{A}_i$ $(1 \leq i \leq n)$,

$$\mu_{i,\alpha t}(W_i) \geq [\mu_{i,t}(W_i)]^\alpha \ (0 < \alpha < 1) \quad (\ \mu_{i,s+t}(W_i) \leq \mu_{i,s}(W_i)\mu_{i,t}(W_i)\ ),$$

and $\{X_i(t), t \in T\}$ $(1 \leq i \leq n)$ are mutually independent, Proposition 5.1 proves the theorem. □

6 Hazard Transform of Multistate Coherent Systems

In the theory of binary state coherent systems, a hazard transform plays a useful role for the proof of IFRA and NBU closure theorems and also the proof of that the preservation of IFR property determines the structure of binary state coherent systems as series systems [9]. This section is about a generalization of the concept of hazard transform to multistate systems and shows the similar results. In this section, we again assume Ω_i $(1 \leq i \leq n)$ and S to be finite totally ordered sets. First, we prove several lemmas and propositions for the proof of our main theorems.

Lemma 6.1.

(i) Suppose that W is an increasing subset of $\Omega_1 \times \Omega_2$. Then we have $W = \cup_{j=1}^m (A_j \times B_j)$, where A_j $(1 \leq j \leq m)$ are nonempty subsets of Ω_1 such that $A_1 \subset \cdots \subset A_m$ and $A_i \neq A_j$ $(i \neq j)$ hold, and B_j $(1 \leq j \leq m)$ are nonempty subsets of Ω_2 such that $\cup_{k=j}^m B_k$ $(1 \leq j \leq m)$ are increasing subsets of Ω_2. Then, $W = (P_{\Omega_1} W) \times (P_{\Omega_2} W)$ holds iff $m = 1$ holds.
(ii) Suppose that W is an increasing subset of $\prod_{i=1}^n \Omega_i$. Then, $W = \prod_{i=1}^n (P_{\Omega_i} W)$ holds iff W has the minimal element.

Lemma 6.2.

(i) $a_0^\alpha + a_1^\alpha - b_1^\alpha > [a_0 + a_1 - b_1]^\alpha$ holds for $0 < \alpha < 1$, $a_0 \geq a_1 > b_1 > 0$.
(ii) $\sum_{i=1}^{n-1} a_i^\alpha (b_i^\alpha - b_{i+1}^\alpha) + a_n^\alpha b_n^\alpha > \left[\sum_{i=1}^{n-1} a_i(b_i - b_{i+1}) + a_n b_n\right]^\alpha$ holds for $0 < \alpha < 1$, $n \geq 2$, $0 < a_1 < \cdots < a_n$, $b_1 > \cdots > b_n > 0$.

Lemma 6.1 is obvious and for the proof of Lemma 6.2, see [5] and [23].

Proposition 6.1. Let P_i and Q_i be probability measures on $(\Omega_i, \mathcal{A}_i)$ $(1 \leq i \leq n)$ and $0 < \alpha < 1$. Suppose that $P_i(W_i) = [Q_i(W_i)]^\alpha$ holds for every increasing set $W_i \in \mathcal{A}_i$, and $P_i(W_i) > P_i(W_i')$ holds for every increasing sets W_i and W_i' in $\mathcal{A}_i$ such that $W_i' \subset W_i$ and $W_i \neq W_i'$. Then, for each increasing set $W \in \mathcal{A}_C$,

$$\left(\prod_{i=1}^{n} P_i\right)(W) = \left[\left(\prod_{i=1}^{n} Q_i\right)(W)\right]^\alpha \quad \text{holds iff} \quad W = \prod_{i=1}^{n} P_{\Omega_i} W \quad \text{holds.}$$

Proof. "if" part is obvious. We prove "only if" part by the mathematical induction on n.

(The case of $n = 2$) If $W = (P_{\Omega_1} W) \times (P_{\Omega_2} W)$ does not hold, then using the same symbols of Lemma 6.1, we have $W = \cup_{j=1}^{m}(A_j \times B_j)$ $(m \geq 2)$, where $A_j \times B_j$ $(1 \leq j \leq m)$ are assumed to be mutually exclusive without loss of generality. Noticing

$$\left(\prod_{i=1}^{2} P_i\right)(W) = \sum_{j=1}^{m} P_1(A_j) P_2(B_j)$$

$$= \sum_{j=1}^{m-1} P_1(A_j) \left\{ P_2\left(\bigcup_{k=j}^{m} B_k\right) - P_2\left(\bigcup_{k=j+1}^{m} B_k\right) \right\} + P_1(A_m) P_2(B_m),$$

we have by the assumption and Lemma 6.2,

$$\left(\prod_{i=1}^{2} P_i\right)(W) = \sum_{j=1}^{m-1} [Q_1(A_j)]^\alpha \left\{ \left[Q_2\left(\cup_{k=j}^{m} B_k\right)\right]^\alpha - \left[Q_2\left(\cup_{k=j+1}^{m} B_k\right)\right]^\alpha \right\}$$

$$+ [Q_1(A_m)]^\alpha [Q_2(B_m)]^\alpha$$

$$> \left[\left(\prod_{i=1}^{2} Q_i\right)(W)\right]^\alpha \quad \text{(by Lemma 6.2),}$$

and then, "only if" part is proved for the case of $n = 2$.

Now assuming "only if" part to hold for $n = n$, we prove the proposition for $n = n + 1$.

Let I_W be an indicator function of an increasing subset $W \subset \prod_{i=1}^{n+1} \Omega_i$. If $W = \prod_{i=1}^{n+1}(P_{\Omega_i} W)$ does not hold, then for some $x_j \in \Omega_j$ an increasing subset of $\prod_{i=1, i \neq j}^{n+1} \Omega_i$ defines by the section $(I_W)_{x_j}$ is not a product set of increasing subsets of Ω_i $(1 \leq i \leq n+1, i \neq j)$. Thus, the inductive hypothesis gives us

$$\int_{\prod_{i=1, i\neq j}^{n+1} \Omega_i} (I_W)_{x_j} d\left(\prod_{i=1, i\neq j}^{n+1} P_i\right) > \left[\int_{\prod_{i=1, i\neq j}^{n+1} \Omega_i} (I_W)_{x_j} d\left(\prod_{i=1, i\neq j}^{n+1} Q_i\right)\right]^\alpha.$$

From the assumption of P_j, we have $P_j(\{x_j\}) > 0$. Hence, using Lemma 5.1 and Fubini's theorem, $\left(\prod_{i=1}^{n+1} P_i\right)(W) > \left[\left(\prod_{i=1}^{n+1} Q_i\right)(W)\right]^\alpha$ holds. □

Lemma 6.3. For $0 < a_1 < \cdots < a_n$, $b_1 > \cdots > b_n > 0$, $0 < \alpha_1 < \cdots < \alpha_n$, $\beta_1 > \cdots > \beta_n$, $n \geq 2$, we have the following inequality:

$$\left\{\sum_{j=1}^{n-1} a_j(b_j - b_{j+1}) + a_n b_n\right\}\left\{\sum_{j=1}^{n-1} \alpha_j(\beta_j - \beta_{j+1}) + \alpha_n \beta_n\right\}$$
$$> \sum_{j=1}^{n-1} a_j \alpha_j (b_j \beta_j - b_{j+1} \beta_{j+1}) + a_n \alpha_n b_n \beta_n.$$

Proof. Mathematical induction on n proves the lemma. □

Proposition 6.2. Let P_i, Q_i and U_i be probability measures on $(\Omega_i, \mathcal{A}_i)$ $(1 \leq i \leq n)$. Suppose that $U_i(W_i) = P_i(W_i) Q_i(W_i)$ holds for every increasing set $W_i \in \mathcal{A}_i$, and $P_i(W_i) > P_i(W_i') > 0$ and $Q_i(W_i) > Q_i(W_i') > 0$ hold for every increasing sets W_i and W_i' of $\mathcal{A}_i$ such that $W_i' \subset W_i$ and $W_i' \neq W_i$ hold. Then, for every increasing set $W \in \mathcal{A}_C$,

$$\left(\prod_{i=1}^n U_i(W)\right)(W) = \left[\left(\prod_{i=1}^n P_i(W)\right)(W)\right]\left[\left(\prod_{i=1}^n Q_i(W)\right)(W)\right]$$

holds iff $W = \prod_{i=1}^n (P_{\Omega_i} W)$ holds.

Proof. "if" part is obvious. We prove "only if" part by the mathematical induction on n.

(The case of $n = 2$) If $W = (P_{\Omega_1} W) \times (P_{\Omega_2} W)$ does not hold, then using the same symbols of Lemma 6.1, $W = \cup_{j=1}^m (A_j \times B_j)$ $(m \geq 2)$, where $A_j \times B_j$ $(1 \leq j \leq m)$ are mutually exclusive. Then, setting $P = \prod_{i=1}^2 P_i$, $Q = \prod_{i=1}^2 Q_i$ and $U = \prod_{i=1}^2 U_i$ and by Lemma 6.3,

$$U(W) = \sum_{j=1}^m U_1(A_j) U_2(B_j)$$
$$= \sum_{j=1}^{m-1} U_1(A_j) \left\{U_2\left(\cup_{k=j}^m B_k\right) - U_2\left(\cup_{k=j+1}^m B_k\right)\right\} + U_1(A_m) U_2(B_m)$$

$$= \sum_{j=1}^{m-1} P_i(A_j)Q_1(A_j) \left\{ P_2\left(\cup_{k=j}^m B_k\right) Q_2 \left(\cup_{k=j}^m B_k\right) \right.$$
$$\left. - P_2\left(\cup_{k=j+1}^m B_k\right) Q_2\left(\cup_{k=j+1}^m B_k\right) \right\} + P_1(A_m)Q_1(A_m)P_2(B_m)Q_2(B_m)$$
$$< \left[\sum_{j=1}^{m-1} P_1(A_j) \left\{ P_2\left(\cup_{k=j}^m B_k\right) - P_2\left(\cup_{k=j+1}^m B_k\right) \right\} + P_1(A_m)P_2(B_m) \right]$$
$$\times \left[\sum_{j=1}^{m-1} Q_1(A_j) \left\{ Q_2\left(\cup_{k=j}^m B_k\right) - Q_2\left(\cup_{k=j+1}^m B_k\right) \right\} + Q_1(A_m)Q_2(B_m) \right]$$
$$= P(W)Q(W).$$

Now assuming "only if" part to hold for $n = n$, we prove the proposition for $n = n+1$. Let I_W be the indicator function of an increasing subset $W \subset \prod_{i=1}^{n+1} \Omega_i$. If $W = \prod_{i=1}^{n+1} P_{\Omega_i}(W)$ does not hold, then for some $x_j \in \Omega_j$, an increasing subset of $\prod_{i=1, i \neq j}^{n+1} \Omega_i$ defined by the section $(I_W)_{x_j}$ is not a product set of increasing subsets of Ω_i ($1 \leq i \leq n+1$, $i \neq j$). Thus, by the inductive hypothesis,

$$\int_{\prod_{i=1, i \neq j}^{n+1} \Omega_i} (I_W)_{x_j} d \prod_{i=1, i \neq j}^{n+1} U_i$$
$$< \left[\int_{\prod_{i=1, i \neq j}^{n+1} \Omega_i} (I_W)_{x_j} d \prod_{i=1, i \neq j}^{n+1} P_i \right] \times \left[\int_{\prod_{i=1, i \neq j}^{n+1} \Omega_i} (I_W)_{x_j} d \prod_{i=1, i \neq j}^{n+1} Q_i \right].$$

From the assumption on P_j and Q_j, we have $P_j(\{x_j\}) > 0$ and $Q_j(\{x_j\}) > 0$. Hence, using Lemma 7.1 and Fubini's theorem,

$$\left(\prod_{i=1}^{n+1} U_i \right)(W) < \left[\left(\prod_{i=1}^{n+1} P_i \right)(W) \right] \times \left[\left(\prod_{i=1}^{n+1} Q_i \right)(W) \right]. \quad \square$$

Now we define a hazard transform of a multistate system as a mapping from $\prod_{i=1}^n \overline{\mathbf{R}}_{\leq}^{N_i}$ to $\overline{\mathbf{R}}_{\leq}^N$, where

$$\overline{\mathbf{R}}_{\leq}^m = \{ (x_1, \cdots, x_m) \mid 0 \leq x_1 \leq \cdots \leq x_m \leq +\infty \}.$$

Our definition is a straight extension of hazard transforms of binary state case [9] to multistate case.

Definition 6.1. A hazard transform of a system $(\Omega_{i=1}^n \Omega_i, S, \mathcal{V})$ is a mapping η from $\prod_{i=1}^n \overline{\mathbf{R}}_{\leq}^{N_i}$ to $\overline{\mathbf{R}}_{\leq}^{N}$ defined by the following procedure, where

$\Omega_i = \{0, 1, \cdots, N_i\}$ $(1 \leq i \leq n)$,
$S = \{0, 1, \cdots, N\}$,
$\mathcal{V} = \{V_s,\ s \in S\}$,
$W_j^i = \{j, j+1, \cdots, N_i\}$ $(1 \leq j \leq N_i,\ 1 \leq i \leq n)$,
$W_j = \cup_{k=j}^N V_k$ $(1 \leq j \leq N)$.

(the first step) For every i $(1 \leq i \leq n)$ and any given $\boldsymbol{x}^i = (x_1^i, \cdots, x_{N_i}^i) \in \overline{\mathbf{R}}_{\leq}^{N_i}$, determine the probability measure P_i on $(\Omega_i, \mathcal{A}_i)$ such that

$$P_i(W_j^i) = e^{-x_j^i} \quad (1 \leq j \leq N_i),$$

which is uniquely determined.

(the second step) Determine a probability measure on $(\Omega_C, \mathcal{A}_C)$ as $P = \prod_{i=1}^n P_i$, and then

$$(-\log P(W_1), -\log P(W_2), \cdots, -\log P(W_N)) \in \overline{\mathbf{R}}_{\leq}^{N}$$

is determined.

From now on we use the following operation rules for vectors:

$\boldsymbol{x} + \boldsymbol{y} = (x_1, \cdots, x_m) + (y_1, \cdots, y_m) = (x_1 + y_1, \cdots, x_m + y_m)$,
$\alpha \boldsymbol{x} = \alpha(x_1, \cdots, x_m) = (\alpha x_1, \cdots, \alpha x_m)$ $(\alpha : \text{a real number})$,
$\boldsymbol{x} = (x_1, \cdots, x_m) \geq \boldsymbol{y} = (y_1, \cdots, y_m) \iff x_i \geq y_i$ $(1 \leq i \leq m)$,
$\boldsymbol{x} = (x_1, \cdots, x_m) = \boldsymbol{y} = (y_1, \cdots, y_m) \iff x_i = y_i$ $(1 \leq i \leq m)$,

Proposition 6.3.

(i) Let η be the hazard transform of an increasing system $(\prod_{i=1}^n \Omega_i, S, \mathcal{V})$, then for $\boldsymbol{x}^i,\ \boldsymbol{y}^i \in \overline{\mathbf{R}}_{\leq}^{N_i}$ $(1 \leq i \leq m)$,

$$\eta(\boldsymbol{x}^1 + \boldsymbol{y}^1, \cdots, \boldsymbol{x}^m + \boldsymbol{y}^m) \geq \eta(\boldsymbol{x}^1, \cdots, \boldsymbol{x}^m) + \eta(\boldsymbol{y}^1, \cdots, \boldsymbol{y}^m). \quad (6.1)$$

(ii) Suppose that a system $(\prod_{i=1}^n \Omega_i, S, \mathcal{V})$ is a coherent system. Then, the equality in (6.1) holds iff the system is a series system.

Proof. Equation (6.1) is obvious from Proposition 5.1(ii) and the definition of hazard transforms.

"if" part of (ii) is immediate. If the system $(\prod_{i=1}^n \Omega_i, S, \mathcal{V})$ is a series coherent system, then each W_j $(1 \leq j \leq N)$ has the minimal element.

Thus, Lemma 6.1(ii) provides $W_j = \prod_{i=1}^{n}(P_{\Omega_i} W_j)$ $(1 \leq j \leq N)$, and thus, the equality in (6.1) follows by the definition of hazard transforms.

"only if" part of (ii) is also immediate. If the equality in (6.1) holds, we have $W_j = \prod_{i=1}^{n} P_{\Omega_i} W_i$ $(1 \leq j \leq N)$ by Proposition 6.2 and the definition of hazard transforms. Thus, W_j $(1 \leq j \leq N)$ has the minimal element by Lemma 6.1, and so does each V_j $(1 \leq j \leq N)$. Hence, the system $(\prod_{i=1}^{n} \Omega_i, S, \mathcal{V})$ is a series system. $\square$

Proposition 6.4.

(i) Let η be the hazard transform of an increasing system $(\prod_{i=1}^{n} \Omega_i, S, \mathcal{V})$. Then, for $\boldsymbol{x}^i \in \overline{\mathbf{R}}_{\leq}^{N_i}$ $(1 \leq i \leq n)$ and $0 < \alpha < 1$,

$$\eta(\alpha \boldsymbol{x}^1, \cdots, \alpha \boldsymbol{x}_m) \leq \alpha \eta(\boldsymbol{x}^1, \cdots, \boldsymbol{x}_m). \tag{6.2}$$

(ii) Suppose that a system $(\prod_{i=1}^{n} \Omega_i, S, \mathcal{V})$ is a coherent system. Thus, the equality in (6.2) holds iff the system is a series system.

Proof. (i) is obvious from Proposition 5.1(i) and the definition of hazard transforms. (ii) is immediate from Lemma 6.1 and Proposition 6.1. $\square$

In the sequel of this section, we examine some stochastic aspects of a system by using the hazard transform of the system. We use the symbols same as those in Section 5. Letting

$$\boldsymbol{H}^i(t) = \left(-\log \mu_{i,t}(W_1^i), \cdots, -\log \mu_{i,t}(W_{N_i}^i)\right),$$

we have the following proposition immediately, and thus, the proof is omitted:

Proposition 6.5. Suppose that $\{X_i(t), t \geq 0\}$ is decreasing and right continuous with probability 1. Then,

(i) $\{X_i(t), t \geq 0\}$ is IFR iff $\alpha \boldsymbol{H}^i(t_1) + \beta \boldsymbol{H}^i(t_2) \geq \boldsymbol{H}^i(\alpha t_1 + \beta t_2)$ holds for $t_1 \geq 0$, $t_2 \geq 0$, $\alpha \geq 0$, $\beta \geq 0$, $\alpha + \beta = 1$.
(ii) $\{X_i(t), t \geq 0\}$ is IFRA iff $\alpha \boldsymbol{H}^i(t) \geq \boldsymbol{H}^i(\alpha t)$ holds for $t \geq 0$, $0 < \alpha < 1$.
(iii) $\{X_i(t), t \geq 0\}$ is NBU iff $\boldsymbol{H}^i(t_1 + t_2) \geq \boldsymbol{H}^i(t_1) + \boldsymbol{H}^i(t_2)$ holds for $t_1 \geq 0$, $t_2 \geq 0$.

Let η be the hazard transform of a system $(\prod_{i=1}^{n} \Omega_i, S, \mathcal{V})$. If $\{X_i(t), t \in T\}$ $(1 \leq i \leq n)$ are mutually independent, $\mu_t = \prod_{i=1}^{n} \mu_{i,t}$ holds. Then,

$$H(t) = \eta(\boldsymbol{H}^1(t), \cdots, \boldsymbol{H}^n(t)), \tag{6.3}$$

where $\boldsymbol{H}(t) = (-\log \mu_t(W_1), \cdots, -\log \mu_t(W_N))$. Using Propositions 6.3, 6.4 and 6.5, and (6.3), IFRA and NUB closure theorems are immediately obtained.

Next, we prove that the preservation of IFR property determines the structure of multistate coherent system as a series system. For the proof of this proposition, we need the following lemma which is easily proved:

Lemma 6.4. For $a > \beta > 0$, $\alpha > a > 0$ and $\alpha - b \neq a - \beta$,

$$f(t) = \log[\exp\{-(a+b)t\} + \exp\{-(\alpha+\beta)t\} - \exp\{-(\alpha+a)t\}]$$

is neither convex nor concave in t.

Proposition 6.6. Let $\{X_i(t), t \in T\}$ be mutually independent, decreasing and right continuous with probability one, and let a system $(\prod_{i=1}^n \Omega_i, S, \mathcal{V})$ (or equivalently a system $(\prod_{i=1}^n \Omega_i, S, \varphi)$) be a coherent system. Then, $\{\varphi(\boldsymbol{X}(t)), t \in T\}$ is IFR whenever $\{X_i(t), t \in T\}$ ($1 \leq i \leq n$) are IFR iff the system is a series system.

Proof. ("if" part) Since $W_j = \prod_{i=1}^n P_{\Omega_i} W_j$ ($1 \leq j \leq N$) hold, $\mu_t(W_j) = \prod_{i=1}^n \mu_{i,t}(P_{\Omega_i} W_j)$ ($1 \leq j \leq N$) follows, and then,

$$\boldsymbol{H}(t) = \left(\sum_{i=1}^n -\log \mu_{i,t}(P_{\Omega_i} W_1), \cdots, \sum_{i=1}^n -\log \mu_{i,t}(P_{\Omega_i} W_N) \right).$$

Using $\alpha \boldsymbol{H}^i(t_1) + \beta \boldsymbol{H}^i(t_2) \geq \boldsymbol{H}^i(\alpha t_1 + \beta t_2)$ ($\alpha \geq 0, \beta \geq 0, \alpha + \beta = 1, 1 \leq i \leq n$), we have $\alpha \boldsymbol{H}(t_1) + \beta \boldsymbol{H}(t_2) \geq \boldsymbol{H}(\alpha t_1 + \beta t_2)$.

("only if" part) Let us consider probability measures

$$\mu_{i,t}(W_j^i) = \exp\{-t\alpha_j^i\} \quad (1 \leq j \leq N_i, \ \alpha_j^i \leq \alpha_{j+1}^i, \ 1 \leq i \leq n).$$

If the system is not series, then there exist $W_j \left(= \cup_{k=j}^N V_k, \ V_k \in \mathcal{V} \right)$ and $(x_{i_1}, \cdots, x_{i_{n-2}}) \in \prod_{k=1, k \neq i_{n-1}, k \neq i_n}^n \Omega_k$ such that the increasing set defined by the section $(I_{W_j})_{(x_{i_1}, \cdots, x_{i_{n-2}})}$ of the indicator function I_{W_j} is not a product set. Letting $\alpha_j^{i_k} \to 0$ ($j < x_{i_k}$) and $\alpha_j^{i_k} \to \infty$ ($j \geq x_{i_k}$ ($k = 1, \cdots, n-2$),

$$\mu_t(W_j) \to \int_{\Omega_{i_{n-1}} \times \Omega_{i_{n-2}}} (I_{W_j})_{(x_{i_1}, \cdots, x_{i_{n-2}})} d(\mu_{i_{n-1},t} \times \mu_{i_{n-2},t}),$$

where the right hand side is log concave in t by the assumption. We may set $i_{n-1} = 1$ and $i_{n-2} = 2$ without loss of generality. Noticing that the increasing set defined by $(I_{W_j})_{(x_{i_1},\cdots,x_{i_{n-2}})}$ is expressed as $\cup_{j=1}^{m}(A_j \times B_j)$ ($m \geq 2$) from Lemma 6.1. We here use the symbols same as those of the Lemma.

$$\int_{\Omega_1 \times \Omega_2} (I_{W_j})_{(x_3,\cdots,x_n)} d(\mu_{1,t} \times \mu_{2,t})$$

$$= \sum_{i=1}^{m}\left[\mu_{1,t}(A_i)\left\{\mu_{2,t}\left(\bigcup_{k=i}^{m} B_k\right) - \mu_{2,t}\left(\bigcup_{k=i+1}^{m} B_k\right)\right\}\right] + \mu_{1,t}(A_m)\mu_{2,t}(B_m)$$

is log concave in t. We may set

$$\mu_{1,t}(A_i) = \exp\{-\alpha_i t\}, \quad \mu_{2,t}\left(\bigcup_{k=i}^{m} B_k\right) = \exp\{-\beta_i t\},$$

$$\alpha_1 > \cdots > \alpha_m, \quad \beta_1 < \cdots < \beta_m, \quad \beta_2 - \beta_1 \neq \alpha_1 - \alpha_2.$$

Letting $\beta_3 \to \infty$ in the above equation, we have the limit function

$$\exp\{-(\alpha_1 + \beta_1)t\} + \exp\{-(\alpha_2 + \beta_2)t\} - \exp\{-(\alpha_1 + \beta_2)t\}$$

which is to be log concave. On the other hand, this function is not log concave by Lemma 6.4, which is a contradiction. Therefore, the system is series. □

7 Concluding Remarks

We have, in this chapter, examined a generalization of the concepts of binary state coherent systems like k-out-of-n systems, modules, IFR, IFRA and NBU processes, and so on. These are concerned with relations among the operating performances of systems and their components.

Throughout this work, we may recognize that a basic theory of reliability systems should be about algebraic and stochastic relations between two ordered sets through an increasing mapping from the one to the other. Some works have been done by [16, 20], but not sufficient. This chapter has considered only totally ordered finite sets except for the probabilistic examination, and the thorough generalization to arbitral ordered sets is remained to be an open problem.

Yu, Koren and Guo [27] defined multistate monotone coherent systems using partially ordered sets as state spaces of systems and components. But they did not define basic concepts as series systems, parallel systems, k-out-of-n systems and modules, and stochastic concepts as IFR, IFRA and NBU, and so on.

There are several definitions of multistate systems which are summarized in [18,19]. One of the important classes of multistate systems is EBW systems, which is an extension of Barlow and Wu [2] 's multistate systems. These multistate coherent systems are very close to binary state coherent systems, since the definition is based on the minimal cut and path sets of binary state coherent systems. Then, many properties of binary state systems are taken over to multistate case. For example, when restricted to this class, three modules theorem is perfectly held. Precise examinations have been seen in [24, 25].

References

1. Barlow, R. E. and Proschan, F. (1975). *Statistical Theory of Reliability of Life Testing*, New York, Holt, Rinehart and Winston.
2. Barlow, R. E. and Wu, A. S. (1978). *Coherent systems with multistate components*, Mathematics of Operations Research, Vol. 3, pp. 275–281.
3. Z. W. Birnbaum and . D. Esary (1965), *Modules of coherent binary systems*, SIAM Journal on Applied Mathematics, Vol. 13, pp. 444-462.
4. Birnbaum, Z. W., Esary, J. D. and Saunder, S. C. (1961). *Multi-component systems and structures and their reliability*, Technometrics, Vol. 3, pp. 55–77.
5. Block, H. W. and Savits, T. H. (1976). *The IFR closure problem*, Annals of Probability, Vol. 4, pp.1030-1032.
6. Butterworth, R. W. (1972). *A set theoretic treatment of coherent system*, SIAM Journal on Applied Mathematics, Vol. 22, pp. 590–598.
7. El-Neweihi, E., Proschan, F., and Sethuraman, J. (1978). *Multistate coherent systems*, Journal of Applied Probability, Vol. 15, pp. 675–688.
8. Esary, J. D. and Proschan, F. (1963). *Coherent structures of non-identical components*, Technometrics, Vol. 5, pp. 191–209.
9. Esary, J. D., Marshall, A. W. and Proschan, F. (1970). *Some reliability application of hazard transform*, SIAM Journal on Applied Mathematics, Vol. 18, pp. 331–359.
10. Hirsch, W. M., Meisner, M. and Boll, C. (1968). *Cannibalization in multicomponent systems and the theory of reliability*, Naval Research Logistics Quarterly, Vol. 15, pp. 331–359.
11. Hochberg, M. (1973). *Generalized multistate systems under cannibalization*, Naval Research Logistics Quarterly, Vol. 20, pp. 585–605.
12. Huang, J., Zuo, M. J. and Fang, Z. (2003). *Multi-state consecutive-k-out-of-n systems*, IIE Transactions, Vol. 35, pp. 527–534.
13. Lehmann, E. L. (1955). *Ordered families of distributions*, Annals of Mathematical Statistics, Vol. 26, pp. 399–416.
14. Mine, H. (1959). *Reliability of physical system*, IRE, **CT-6** Special Supplement, pp. 138–151.
15. Neveu, J. (1965). *Mathematical Foundations of Calculus of Probability*, Holden-Day, San Francisco.

16. Ohi, F. and Nishida, T. (1982). *A Definition of NBU Probability Measures*, Journal of Japan Statistical Society, Vol. 12, pp. 141–151.
17. Ohi, F. and Nishida, T. (1983). *Generalized multistate coherent systems*, Journal of Japan Statistical Society, Vol. 13, pp. 165–181.
18. Ohi F. and Nishida, T. (1984). *On Multistate Coherent Systems*, IEEE Transactions on Reliability, **R-33**, pp. 284–288.
19. Ohi F. and Nishida, T. (1984). *Multistate Systems in Reliability Theory*, Stochastic Models in Reliability Theory, Lecture Notes in Economics and Mathematical Systems *235*, Springer-Verlag, pp. 12-22.
20. Ohi, F. Shinmori, S. and Nishida, T. (1989). *A Definition of Associated Probability Measures on Partially Ordered Sets*, Mathematical Japonica, Vol. 34, pp. 403–408.
21. Ohi, F. and Shinmori, S. (1998). *A definition of generalized k-out-of-n multistate systems and their structural and probabilistic properties*, Japan Journal of Industrial and Applied Mathematics, Vol. 15, pp. 263–277.
22. H. Pham (editor), (2003). *Handbook of Reliability Engineering*, Springer.
23. Ross, S. M. (1979) *Mutivalued state component systems*, Annals of Probability, Vol. 7, pp. 379–383.
24. Shinmori, S., Ohi, F., Hagihara, H. and Nishida, T. (1989). *Modules for Two Classes of Multi-State Systems*, The Transactions of the IEICE, **E72**, pp. 600–608.
25. Shinmori, S., Hagihara, H., Ohi, F. and Nishida, T. (1989). *On an Extention of Barlow-Wu Systems - Basic Properties*, Journal of Operations Research Society of Japan, Vol. 32, pp. 159–172.
26. Shinmori, S., Ohi, F. and Nishida, T. (1990). *Stochastic Bounds for Generalized Systems in Reliability Theory*, Journal of Operations Research Society of Japan, Vol. 33, pp. 103–118.
27. Koren, K. Yu, I. and Guo, Y. (1994). Generalized Multistate Monotone Coherent Systems, *IEEE Transactions on Reliability*, Vol. 43, pp. 242–250.
28. Zuo, M J., Huang, J. and Kuo, W. (2003). Multi-state k-out-of-n systems, in *Handbook of Reliability Engineering* edited by H. Pham, Springer, pp. 3–17.

Chapter 2

Cumulative Damage Models

TAKASHI SATOW

Department of Social Systems Engineering
Tottori University
4-101 Koyama-Minami, Tottori 680-8552, Japan
E-mail: zwsatow@sse.tottori-u.ac.jp

1 Introduction

Reliability theory becomes major concerns of designers and managers engaged in making high quality of products. Many researchers have investigated complex phenomena of real systems and have studied many problems to improve their reliabilities. We have to pay attention to the matter what are essential laws of governing systems. From the point of views, we should make a grasp of processes generously and try to formulate it simply, avoiding small points. In other words, it would be necessary to construct mathematical models which outline the observational and theoretical features of phenomena.

Reliability of a product is the probability that the product will perform its intended function for a specified time period when operated in normal environmental conditions [1]. We are mainly interested in the failure time. The time is measured usually by the continuous time such as the age, the calendar and operating times, or sometimes by the discrete time such as the number of uses, stresses, shocks and transmissions. As typical functions which represent the continuous distributions of failure times, the exponential, gamma, Weibull and normal distributions are well-known. Further, the time to failure is often expressed as *first-passage time* in stochastic processes. It is noted that we use the word of *item* which means unit, device, part, component, structure, equipment, machine, system and etc.

In this chapter, we think about stochastic models where the item suffers damage such as wear, fatigue, deterioration, crack corrosion and erosion by arrived shocks [2]. A failure of the item depends on the total damage which is accumulated by the shock. Such damage models generate a cumulative process in stochastic processes [3], and sometimes called a shock damage model. A basic assumption for the shock damage model throughout this chapter is given as:

1) The item is subject to shocks which occur at a nonhomogeneous Poisson process.
2) The item suffers a non-negative random damage with an identical distribution. Its damages are independent with each other and are also independent of the process of shocks in 1).
3) Each random damage is accumulated to the current damage level of the item.
4) The damage level remains unchanged between shocks.

In addition to the above assumptions, the following reliability quantities become objects of our interest.

(i) The distribution of the total damage at time t.
(ii) The mean total damage at time t.
(iii) The distribution of the time to failure.
(iv) The mean time to failure.

Failures of the item during actual operations are costly in many situations or sometimes dangerous. We should inspect and maintain preventively the item before failure by appropriate methods such as repair, replacement and overhaul. Most famous three replacement policies for the shock damage model are the following three policies:

1. **Age replacement:** The item is replaced immediately when its age reaches at time T, or at failure, whichever occurs first. This policy is more effective when its failure rate increases with age and is the most commonly used policy. Many researchers have already studied the policy theoretically and practically.
2. **Shock number replacement:** The item is replaced before failure immediately when the Nth number of shocks has occurs. This is mainly employed when its age or operating time can not be known or be recorded.

This policy is one kind of discrete replacement policy when the time to failure is measured by the cycles to failure such as the number of inspections, repairs, faults, uses, flights and communications.
3. **Damage level(control limit) replacement**: The item is replaced before failure when the amount of damage has exceeded a threshold level Z at some shock. This policy is called a control limit policy. It would be better to use the level of damage as a replacement indicator for the case where the age or the operating time can not be known. However, it is necessary to be able to measure total damage.

Cox [4] defined the cumulative process as follows: Shocks occur at a renewal process and cause a random damage to the item. These damages have an additional property. A-Hameed and Proschan [5] considered that shocks occur at a nonhomogeneous Poisson process and used a gamma variable which represents an amount of damage caused by each shock. Esary, Marshall and Proschan [6] investigated various properties of the reliability function of this model, when shocks occur at a Poisson process. Taylor [7] and Zuckerman [8] also treated the model with a Poisson process. Feldman [9, 10] considered the case where the damage process forms a semi-Markov process. The above papers assumed that the damage level between shocks is constant. Posner and Zuckerman [11] considered the model where the damage level decreases between shocks. When multiple items are subject to shocks and suffer damage by shocks, this model yields multivariable distribution [12] and was studied by Marshall and Shaked [13].

Replacement policies for cumulative damage models have been proposed by many researchers: Zuckerman [8] considered the scheduled time policy for two net income problems when the damage process is the same as Taylor's model. The one is maximum problem of total long-run average net income per unit time, and the other is the maximum problem of total expected discount net income. A-Hameed and Shimi [14] considered the replacement policy under the condition that the breakdown of the item occurs only at the occurrence of a shock and it is replaced only at shock time. Zuckerman [15] treated the replacement model with the cost function of accumulated damage. Boland and Proschan [16] considered the periodic replacement model and gave sufficient conditions for the existence of a finite time period as an optimal policy. Puri and Singh [17] dealt with similar shock models.

Another important replacement policy is the control limit policy where the item is replaced at damage level. Taylor [7] denoted the cumulative damage by a terminating Markov process, and adopted two costs for the replacement of the item. Feldman [9] generalized Taylor's model by using a semi-Markov process and considered the replacement policy under the same cost structure. Nakagawa [18] derived the necessary and sufficient conditions for the optimal policy which minimizes the expected cost. Posner and Zuckerman [11] and Nakagawa and Kijima [19] proposed some modified models. Nakagawa [20] considered a discrete replacement policy where the item is replaced preventively before failure at the number of shocks. Kijima and Nakagawa [21] applied a sequential imperfect preventive maintenance(PM) policy to a cumulative damage model. Each preventive maintenance reduces the total damage. As a result, the probability of the item failure is decreased. If the item fails between PMs, it undergoes only minimal repair. Nakagawa and Murthy [22] dealt with discrete replacement policies for the system with two items: Item 1 causes damage to item 2 when item 1 fails. The damage is additive and item 2 fails when the total damage exceeds a failure level. The system is replaced at failure of item 2 or Nth failure of item 1. They obtained the expected cost rate and derived the optimal number N^* which minimizes it.

Murthy and Iskandar [23] proposed a different type of shock damage model: It relaxes occurring conditions of item failure. The item failure need not coincide with shock occurrence. Based on the shock damage model, Murthy and Iskandar [24] considered optimal age and shock number replacement policies. Chelbi and Ait-Kadi [25] derived the item time stationary availability for a typical shock damage model. Satow et. al. [26] applied the cumulative damage model to the garbage collection policy. We can see many findings in some books and surveys [2, 27–30].

2 Standard Cumulative Damage Model

2.1 Shock Arrival

Shocks occur at random times and each shock causes a random amount of damage to the item. A sequence of shock arrival is approximated by stochastic processes. One of the most powerful and flexible processes is a nonhomogeneous Poisson process(NHPP). It gives a definition of NHPP as follows [3]: For an integer-valued stochastic process $N(t)$ ($t \geq 0$) and a sequence of time points $0 \leq t_0 < t_1 < \cdots < t_n$,

1) $N(t_1)-N(t_0), N(t_2)-N(t_1), \cdots, N(t_n)-N(t_{n-1})$ are independent random variables. It has independent increment.

2) For arbitrary $t \geq 0$, $h > 0$ and $j = 0, 1, \cdots$, the random variable $N(t+h) - N(t)$ has the Poisson distribution

$$\Pr\{N(t+h) - N(t) = j\} = \frac{[R(t+h) - R(t)]^j}{j!} e^{-[R(t+h) - R(t)]}.$$

3) $N(0) = 0$.

The function $R(t)$ is a mean-value function of $N(t)$, and $R(t) \equiv \int_0^t \lambda(x) dx$. The function $\lambda(t)$ is an intensity function of $N(t)$.

Let random variables X_i ($i = 1, 2, \cdots$) denote a sequence of inter-arrival times between successive events. Random variables S_j ($j = 0, 1, 2, \cdots$) denote successive event times, where $S_j \equiv \sum_{i=0}^{j} X_i (j = 0, 1, 2, \cdots)$ and $S_0 = 0$. Then, the probability that at least j events occur in time duration $(0, t]$ is

$$\Pr\{S_j \leq t\} = \sum_{i=j}^{\infty} \frac{R(t)^i}{i!} e^{-R(t)} \quad (j = 0, 1, 2, \cdots). \tag{1}$$

Further, exactly j events occur in time duration $(0, t]$ is

$$\Pr\{N(t) = j\} = \Pr\{S_j \leq t\} - \Pr\{S_{j+1} \leq t\}$$
$$= \frac{R(t)^j}{j!} e^{-R(t)} \quad (j = 0, 1, 2, \cdots). \tag{2}$$

It should note that inter-arrival times between successive events are not always independent in case of NHPP.

2.2 Cumulative Damage

Damage due to shocks accumulates additively. The item fails when the total amount of damage has exceeded a certain level K. The level K is called a failure level and is assumed to constant. Such a stochastic model generates a cumulative process studied [2, 4, 6].

Some useful reliability measures for the cumulative damage model are summarized: The item is subject to shocks and suffers damage due to shocks. Let random variables W_i ($i = 1, 2, \cdots$) denote the damage due to ith shock. It is assumed that random variables W_i has an identical probability distribution function $G(x) \equiv \Pr\{W_i \leq x\}$ with finite mean $1/\mu$, independent of the number of shocks and nonnegative. The total damage up to the jth shock is $Z_j \equiv \sum_{i=1}^{j} W_i$, where $Z_0 \equiv 0$. Therefore, a distribution function of the total damage up to jth shock is

$$\Pr\{Z_j \leq x\} = G^{(j)}(x), \tag{3}$$

where $G^{(j)}(x)$ is the j-fold convolution of $G(x)$ with itself, and

$$G^{(0)}(x) \equiv \begin{cases} 0 & (x < 0), \\ 1 & (x \geq 0). \end{cases} \qquad (4)$$

Let $N(t)$ denote a counting process which denotes the total number of shocks received until $(0, t]$. The function $N(t)$ is a nonnegative function. Then, a distribution function of the total damage at time t is

$$Z(t) \equiv \begin{cases} \sum_{i=1}^{N(t)} W_i & (N(t) = 1, 2, \cdots), \\ 0 & (\text{otherwise}). \end{cases} \qquad (5)$$

2.3 Cumulative Damage Model

From a renewal theory [4], a probability distribution function that j shocks occur until time t is

$$\Pr\{N(t) = j\} = F_j(t) - F_{j+1}(t), \qquad (6)$$

since

$$\Pr\{S_j \leq t\} = F_j(t), \qquad (7)$$

where $F_0(t) \equiv 1$ $(t \geq 0)$. A probability distribution function that the total damage is less than or equal to x at time t is

$$\Pr\{Z(t) \leq x\} = \sum_{j=0}^{\infty} \Pr\{N(t) = j, Z_j \leq x\}$$

$$= \sum_{j=0}^{\infty} [F_j(t) - F_{j+1}(t)] G^{(j)}(x). \qquad (8)$$

A survival distribution function is

$$\Pr\{Z(t) > x\} = \sum_{j=0}^{\infty} [G^{(j)}(x) - G^{(j+1)}(x)] F_{j+1}(t). \qquad (9)$$

The total expected damage at time t is

$$E\{Z(t)\} = \sum_{j=1}^{\infty} F_j(t) \int_0^{\infty} \overline{G}(x) \, dx, \qquad (10)$$

where $\overline{G}(x) \equiv 1 - G(x)$. Since the distribution function $G(x)$ has a mean $1/\mu$, then (10) is

$$E\{Z(t)\} = \frac{1}{\mu} \sum_{j=1}^{\infty} F_j(t). \qquad (11)$$

Let a random variable Y be the time to item failure. The first-passage time distribution to item failure is

$$\Pr\{Y \leq t\} = \Pr\{Z(t) > K\} = \sum_{j=0}^{\infty} [G^{(j)}(K) - G^{(j+1)}(K)] F_{j+1}(t). \quad (12)$$

From (12), the mean time to item failure is

$$E\{Y\} = \sum_{j=0}^{\infty} G^{(j)}(K) \int_0^{\infty} [F_j(t) - F_{j+1}(t)] \, dt. \quad (13)$$

The failure rate of the item is

$$r(t)\Delta t \equiv \frac{\Pr\{t < Y \leq t + \Delta t\}}{\Pr\{Y > t\}}$$

$$= \frac{\sum_{j=0}^{\infty} [G^{(j)}(K) - G^{(j+1)}(K)] f_{j+1}(t) \Delta t}{\sum_{j=0}^{\infty} G^{(j)}(K)[F_j(t) - F_{j+1}(t)]}, \quad (14)$$

where $f(t)$ is a density function of $F(t)$, i.e., $dF(t)/dt = f(t)$. The function $f_j(t)$ is the j-th fold convolution of $f(t)$ with itself.

3 Failure Interaction Models

As an application of the shock damage model, it introduces a failure interaction model. We consider a system composed of two items denoted as items 1 & 2. Item 1 is repairable and it undergoes minimal repair at failure. The time to repair is small so that it can be ignored. Therefore, item 1 failures occur according to a nonhomogeneous Poisson process with an intensity function $\lambda(t)$ and a mean-value function $R(t)$, i.e., $R(t) \equiv \int_0^t \lambda(x) dx$. Let $N(t)$ be the total number of item 1 failures by time t. The probability that exactly j failures occur until time t is given by $H_j(t) = \{[R(t)]^j\}/j!\} e^{-R(t)}$ ($j = 0, 1, 2, \cdots$). As a result, the probability that j or more item 1 failures occur in $(0, t]$ is given by $F^{(j)}(t) = \sum_{i=j}^{\infty} H_i(t)$ ($j = 0, 1, 2, \cdots$). It causes a random amount of damage W_j ($j = 1, 2, \cdots$) with distribution $G(x)$ to item 2 when jth item 1 failure occurs. A function $G^{(j)}(x)$ is the j-fold Stieltjes convolution of $G(x)$ with itself. Item 2 fails whenever the total damage exceeds a failure level K. A system failure occurs whenever item 2 fails because both items fail simultaneously. We assume that item 2 is not repairable, and as a result, the failed system needs to be replaced by a new one. It calls corrective replacement.

The system failure, in general, results in a high cost. One way of reducing this cost is to replace the system preventively. From a cost point of view, preventive replacement would be cheaper than failure replacement. However, the preventive replacement implies discarding some useful life of the system. Hence, the preventive replacement needs to be executed in a manner which achieves a suitable trade off between this loss versus the risk of failure.

The system is replaced through the failure replacement when item 2 fails, *i.e.*, the damage of item 2 exceed K, or earlier through the preventive replacement when one of the following conditions occurs:

1) The system reaches an age T.
2) Nth failure of item 1 occurs.
3) The total damage to item 2 exceeds a level k ($< K$).

The notations T, N, k are called preventive parameters. It proposes three kinds of preventive replacement models by which two preventive parameters among T, N, k are combined. There are (T,k), (T,N) and (N,k) models. Three kinds of expected maintenance cost rates are derived as criteria functions.

The followings are assumed for the simplification:

1) The failures of item 1 and 2 are detected immediately.
2) The damage to item 2 is measured after each failure of item 1.
3) The times to repair item 1 and replace the system are small, so that, they can be approximated as being zero. In other words, the repair or replacements are instantaneous.
4) The cost of each minimal repair for item 1 is c_m. The cost of each failure (preventive) replacement for the system is c_1 (c_p) with $c_1 > c_p > c_m$.

3.1   Age and Damage Limit Model

The system is replaced at time T, damage limit k or system failure, whichever occurs first. The probability $\alpha_1(T,k)$ that the system is replaced at failure of item 2 due to total damage exceeding K is

$$\alpha_1(T,k) = \sum_{j=0}^{\infty} F_{j+1}(T) \int_0^k \overline{G}(K-x)\,\mathrm{d}G^{(j)}(x), \qquad (15)$$

the probability $\beta_1(T,k)$ that it is replaced preventively at age T is

$$\beta_1(T,k) = \sum_{j=0}^{\infty} H_j(T) G^{(j)}(k), \qquad (16)$$

and the probability $\delta_1(T,k)$ that it is replaced preventively when the total damage of item 2 exceeds k and is less than or equal to K is

$$\delta_1(T,k) = \sum_{j=0}^{\infty} F_{j+1}(T) \int_0^k [G(K-x) - G(k-x)] \, dG^{(j)}(x), \quad (17)$$

where it is proved that $\alpha_1(T,k) + \beta_1(T,k) + \delta_1(T,k) = 1$.

The expected number of minimal repairs over a cycle $\epsilon_1(T,k)$ is

$$\epsilon_1(T,k) = \sum_{j=1}^{\infty} F_j(T) G^{(j)}(k). \quad (18)$$

The expected one cycle length is

$$\zeta_1(T,k) = \sum_{j=0}^{\infty} G^{(j)}(k) \int_0^T H_j(t) \, dt. \quad (19)$$

As a result, from (15)–(19), the expected cost rate [30] is

$$C_1(T,k) = \frac{c_1 \alpha_1(T,k) + c_p[\beta_1(T,k) + \delta_1(T,k)] + c_m \epsilon_1(T,k)}{\zeta_1(T,k)}. \quad (20)$$

An optimal T^* and k^* are values which minimize $C_1(T,k)$ in (20). Necessary conditions of optimal T^* and k^* can be obtained from the first order conditions, $i.e.$, setting the derivatives of $C_1(T,k)$ with respect to T and k to zero. Differentiating $C_1(T,k)$ with respect to T and setting it equal to zero,

$$\sum_{j=0}^{\infty} F_{j+1}(T) B_j(k) - \frac{\sum_{j=0}^{\infty} H_j(T) B_j(k)}{\sum_{j=0}^{\infty} H_j(T) G^{(j)}(k)} \sum_{j=0}^{\infty} F_{j+1}(T) G^{(j)}(k) = c_p, \quad (21)$$

where

$$B_j(k) \equiv \int_0^k [(c_1 - c_p)G(K-x) - c_m G(k-x)] \, dG^{(j)}(x). \quad (22)$$

Denote the left-hand side of (21) by $J(T,k)$.

Differentiating $C_1(T,k)$ with respect to k and setting it equal to zero,

$$c_m \left\{ \frac{\sum_{j=1}^{\infty} F_j(T) g^{(j)}(k)}{\sum_{j=1}^{\infty} F_{j+1}(T) g^{(j)}(k)} \sum_{j=0}^{\infty} F_{j+1}(T) G^{(j)}(k) - \sum_{j=1}^{\infty} F_j(T) G^{(j)}(k) \right\}$$

$$+ (c_1 - c_p) \sum_{j=0}^{\infty} F_{j+1}(T) \int_0^k [G(K-x) - G(K-k)] \, dG^{(j)}(x) = c_p. \quad (23)$$

Denote the left-hand side of (23) by $Q(k,T)$.

On comparing $J(T,k)$ with $Q(k,T)$, we see that $Q(k,T)$ is always greater than $J(T,k)$ for T $(0 < T < \infty)$ and k, $(0 < k \leq K)$, because

$$J(T,k) - Q(k,T) = (c_f - c_p) \sum_{j=0}^{\infty} F_{j+1}(T) G^{(j)}(k)$$

$$\times \left\{ \frac{\sum_{j=0}^{\infty} H_j(T) \int_0^k [G(K-k) - G(K-x)] \, \mathrm{d}G^{(j)}(x)}{\sum_{j=0}^{\infty} H_j(T) G^{(j)}(k)} \right\}$$

$$+ c_m \sum_{j=0}^{\infty} F_{j+1}(T) G^{(j)}(k)$$

$$\times \left[\frac{\sum_{j=0}^{\infty} H_j(T) G^{(j+1)}(k)}{\sum_{j=0}^{\infty} H_j(T) G^{(j)}(k)} - \frac{\sum_{j=1}^{\infty} F_j(T) g^{(j)}(k)}{\sum_{j=1}^{\infty} F_{j+1}(T) g^{(j)}(k)} \right] < 0. \quad (24)$$

This implies that there does not exist (T^*, k^*) which satisfies (21) and (23) simultaneously. As a result, the optimal solution is given by either $k^* = K$ and $T^* < \infty$ or $T^* \to \infty$ and $k^* < K$.

3.2 Numerical Examples

It shows the expected cost rate $C_1(T,k)$ at preventive replacement age T and damage limit k: The item 1 failure occurs according to a homogeneous Poisson process with rate $\lambda = 1$. Damage caused by each failure is assumed to be an exponential distribution with mean 1. The failure level of item 2 damage K is 10. Maintenance costs c_1, c_p and c_m are 2, 1 and 0.1, respectively. Figure 1 shows the expected cost rate $C_1(T,k)$.

The optimal k^*, which below the failure level K, exists to the fixed age T. Sensitivity of k^* to T is small. The optimal T^* to the fixed damage is decreasing in k. However, sensitivity of T^* to k is large. For instance, the optimal k^* to $T = 12$ is 7.92, and $C_1(12, k^*) = 0.235$. The optimal T^* to $k = 9.6$ is 14.65, and $C_1(T^*, 9.6) = 0.256$. It can be shown that the function $C_1(T, k^*)$ is a decreasing function in T. On the other hand, $C_1(T^*, k)$ is an increasing function in k. From the above result, T should be long as much as possible to minimize $C_1(T, k)$ at this situation.

Figure 2 shows $C_1(T, k)$ to cost change when T is fixed. At both cases of $T = 10, 13$, the minimized expected cost rate $C_1(T, k^*)$ is sensitive to the minimal repair cost c_m. In comparison to the c_m change, a change of $C_1(T, k^*)$ to the replacement cost c_1 is small. On the other hand, the optimal k^* is sensitive to c_1 change.

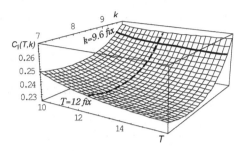

Fig. 1 Optimal age T^* and damage limit k^*

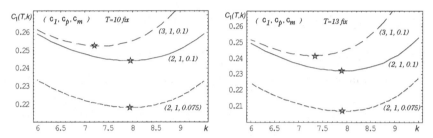

Fig. 2 Optimal damage limit k^* to cost change when $T=10(l.h.s)$, $T=13(r.h.s)$

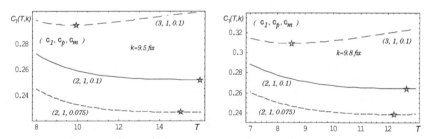

Fig. 3 Optimal age T^* to cost change when $k=9.5(l.h.s)$, $k=9.8(r.h.s)$

Figure 3 shows $C_1(T,k)$ to cost change when k is fixed. At both cases of $T=9.5, 9.8$, the optimal T^* are greatly decreasing in cost c_1. However, the optimal T^* are increasing in cost c_m.

3.3 Shock Number and Damage Limit (N,k) Model

The system is replaced at shock number N, damage limit k or the system failure, whichever occurs first. The probability $\alpha_2(N,k)$ that the system is replaced at failure of item 2 due to total damage exceeding K is

$$\alpha_2(N,k) = \sum_{j=0}^{N-1} \int_0^k \overline{G}(K-x)\,dG^{(j)}(x), \qquad (25)$$

the probability $\beta_2(N,k)$ that it is replaced preventively at shock number N is

$$\beta_2(N,k) = G^{(N)}(k), \qquad (26)$$

and the probability $\delta_2(N,k)$ that it is replaced preventively when the total damage of item 2 exceeds k and is less than or equal to K is

$$\delta_2(N,k) = \sum_{j=0}^{N-1} \int_0^k [G(K-x) - G(k-x)]\,dG^{(j)}(x), \qquad (27)$$

where it is proved that $\alpha_2(N,k) + \beta_2(N,k) + \delta_2(N,k) = 1$.

The expected number of minimal repairs over a cycle $\epsilon_2(N,k)$ is

$$\epsilon_2(N,k) = \sum_{j=1}^{N-1} G^{(j)}(k). \qquad (28)$$

The expected one cycle length is

$$\zeta_2(N,k) = \sum_{j=0}^{N-1} G^{(j)}(k) \int_0^\infty H_j(t)\,dt. \qquad (29)$$

Then, from (25)–(29), the expected cost rate is

$$C_2(N,k) = \frac{c_1\alpha_2(N,k) + c_p[\beta_2(N,k) + \delta_2(N,k)] + c_m\epsilon_2(N,k)}{\zeta_2(N,k)}. \qquad (30)$$

We first seek an optimal $k^*(N)$ which minimizes $C_2(N,k)$ for a fixed N. Differentiating $C_2(N,k)$ with respect to k and setting it equal to zero,

$$\sum_{j=0}^{N-1} \int_0^k [G(K-x) - G(K-k)]\,dG^{(j)}(x) = \frac{c_p - c_m}{c_1 - c_p}. \qquad (31)$$

Denote the left-hand side of (31) by $L(k,N)$. It is easily shown that

$$L(0,N) = 0 < \frac{c_p - c_m}{c_1 - c_p}, \quad L(K,N) = \sum_{j=1}^{N} G^{(j)}(K). \qquad (32)$$

$$\frac{dL(k,N)}{dk} = g(K-k)\sum_{j=0}^{N} G^{(j)}(K) > 0. \qquad (33)$$

Therefore, if $L(K,N) > (c_p - c_m)/(c_1 - c_p)$ then there exists a unique $k^*(N)$ ($< K$) which satisfies (31), and in this case, the optimal expected cost rate is

$$C_2(N, k^*(N)) = \lambda \left[(c_1 - c_r)\overline{G}(K - k^*(N)) + c_m\right]. \tag{34}$$

Define the set $\mathbf{N}_1$ as follows:

$$\mathbf{N}_1 \equiv \left\{ N \mid L(K,N) > \frac{c_p - c_m}{c_1 - c_p}, \ N > 0 \right\}. \tag{35}$$

The optimal control limit $k^*(N)$ is a strictly decreasing function in $N \in \mathbf{N}_1$, and $C_2(N, k^*(N))$ is a strictly increasing function in $k^*(N)$. As a result, if $\mathbf{N}_1$ is not a null set ($\mathbf{N}_1 \neq \emptyset$), then $C_2(N, k^*(N))$ will be a minimum as $N \to \infty$. In other words, the optimal (N^*, k^*) is $(\infty, k^*(\infty))$. If $\mathbf{N}_1$ is a null set ($\mathbf{N}_1 = \emptyset$), i.e., $L(K, \infty) \leq (c_p - c_m)/(c_1 - c_p)$, $C_2(N, k)$ is a decreasing in k for any N. As a result, the optimal k^* for any N is K.

3.4 Numerical Examples

It shows the expected cost rate $C_2(N, k)$ at preventive item 1 failure(minimal repair) numbers N and damage limit k: Item 1 failure occurs according to a homogeneous Poisson process with rate $\lambda = 1$. Damage caused by each item 1 failure is assumed to be an exponential distribution with mean 1. The failure level of item 2 damage K is 10. Maintenance costs c_1, c_p and c_m are 2, 1 and 0.1, respectively.

Figure 4 shows the expected cost rate $C_2(N, k)$. The optimal $k^*(N)$, which below the failure level K, exists to the fixed number N. Sensitivity

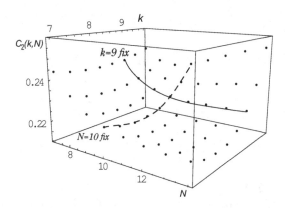

Fig. 4 Optimal shock number N^* and damage limit k^*

of k^* to the number N is very small. The optimal $N^*(k)$ to the fixed damage is decreasing in k. However, sensitivity of N^* to the damage k is large. For instance, the optimal $k^*(N)$ to the number $N = 10$ is 7.88, and $C_2(10, k^*) = 0.220$. The optimal $N^*(k)$ to the damage $k = 9$ is 12, and $C_2(N^*, 9) = 0.225$. It can be shown that $C_2(N, k^*)$ is a decreasing function in N. On the other hand, $C_2(N^*, k)$ is an increasing function in k. From the above result, the number N should be long as much as possible to minimize $C_2(N, k)$ in this situation.

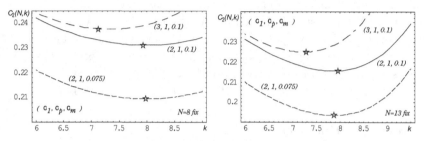

Fig. 5 Optimal damage limit k^* to cost change when $N = 8(l.h.s)$ and $N = 13(r.h.s)$

Figure 5 shows $C_2(N, k)$ to cost change when N is fixed. At both cases of $N = 8, 13$, the $C_2(N, k^*)$ is sensitive to the minimal repair cost c_m. In comparison to the c_m change, a change to the replacement cost c_1 is small. On the other hand, the optimal k^* is sensitive to c_1 change. This phenomenon is the same to the (T, k) model when T is replaced with N.

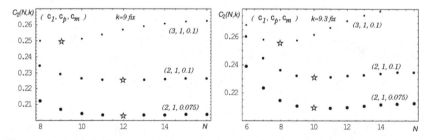

Fig. 6 Optimal shock number N^* to cost change when $k = 9(l.h.s)$ and $k = 9.3(r.h.s)$

Figure 6 shows $C_2(N, k)$ to cost change when k is fixed. At both cases of $k = 9, 9.3$, the optimal number N^* are decreasing in cost c_1. However, the optimal k^* do not change in cost c_m.

3.5 Age and Shock Number Model

The system is replaced at age T, shock number N or system failure, whichever occurs first. The probability $\alpha_3(T, N)$ that the system is replaced at failure of item 2 due to total damage exceeding K is

$$\alpha_3(T, N) = \sum_{j=0}^{N-1} F_{j+1}(T) \int_0^K \overline{G}(K - x)\, \mathrm{d}G^{(j)}(x), \tag{36}$$

the probability $\beta_3(T, N)$ that it is replaced preventively at age T is

$$\beta_3(T, N) = \sum_{j=0}^{N-1} H_j(T) G^{(j)}(K), \tag{37}$$

and the probability $\delta_3(T, N)$ that it is replaced preventively at shock number N is

$$\delta_3(T, N) = F_N(T) G^{(N)}(K), \tag{38}$$

where it is proved that $\alpha_3(T, N) + \beta_3(T, N) + \delta_3(T, N) = 1$.

The expected number of minimal repairs over a cycle $\epsilon_3(T, N)$ is

$$\epsilon_3(T, N) = \sum_{j=1}^{N-1} F_j(T) G^{(j)}(K). \tag{39}$$

Then, the expected one cycle length is given by

$$\zeta_3(T, N) = \sum_{j=0}^{N-1} G^{(j)}(K) \int_0^T H_j(t)\, \mathrm{d}t. \tag{40}$$

Then, from (36)–(40), the expected cost rate is

$$C_3(T, N) = \frac{c_1 \alpha_3(T, N) + c_p[\beta_3(T, N) + \delta_3(T, N)] + c_m \epsilon_3(T, N)}{\zeta_3(T, N)}. \tag{41}$$

This equation is the same as the extended model of [22]. It is difficult to derive the optimal (T^*, N^*) which minimizes $C_3(T, N)$ analytically. We need to use a numerical method to find the optimal parameters.

For the purpose of deriving the optimal (T^*, N^*) numerically, it use the following inequalities which is followed from the necessary condition for the optimal $N^*(T)$ which minimizes $C_3(T, N)$ for a fixed T.

Note that $N^*(T)$ to be optimal requires $C_3(T, N+1) > C_3(T, N)$ and $C_3(T, N) \leq C_3(T, N-1)$,

$$W(N, T) > c_p \text{ and } c_p \geq W(N - 1, T). \tag{42}$$

where

$$W(N,T) \equiv (c_1 - c_p)\left[\sum_{j=1}^{N} F_j(T)G^{(j)}(K) - \sum_{j=0}^{N-1} F_{j+1}(T)G^{(j)}(K)\frac{G^{(N+1)}(K)}{G^{(N)}(K)}\right]$$
$$+ c_m\left[\sum_{j=0}^{N-1} F_{j+1}(T)G^{(j)}(K)\frac{F_N(T)}{F_{N+1}(T)} - \sum_{j=1}^{N-1} F_j(T)G^{(j)}(K)\right]. \quad (43)$$

If $G^{(j+1)}(K)/G^{(j)}(K)$ is strictly decreasing in j, then $W(N,T)$ is strictly increasing in N since

$$W(N+1,T) - W(N,T) = \left\{(c_1 - c_p)\left[\frac{G^{(N+1)}(K)}{G^{(N)}(K)} - \frac{G^{(N+2)}(K)}{G^{(N+1)}(K)}\right]\right.$$
$$\left.+ c_m\left[\frac{F_{N+1}(T)}{F_N(T)} - \frac{F_{N+2}(T)}{F_{N+1}(T)}\right]\right\}\sum_{j=0}^{N} F_{j+1}(T)G^{(j)}(K) > 0. \quad (44)$$

Furthermore, $\lim_{N\to\infty} W(N,T) = \infty$. As a result, a finite $N^*(T)$ which satisfies (42) exists and is unique or two at most. The optimal (T^*, N^*), which minimize $C_3(T, N)$, are identical to $(T^*, N^*(T^*))$ which minimizes $C_3(T, N^*(T))$. Accordingly, T^* and N^* are derived numerically with finding the minimum $C_3(T, N^*(T))$.

3.6 Numerical examples

It shows the expected cost rate $C_3(T, N)$ at preventive item 1 failure *i.e.*, minimal repair, numbers N and age T: Item 1 failure occurs according to a homogeneous Poisson process with rate $\lambda = 1$. Damage caused by each item 1 failure is assumed to be an exponential distribution with mean 1. The failure level of item 2 damage K is 10. Maintenance costs c_1, c_p and c_m are 2, 1 and 0.1, respectively.

Figure 7 shows the expected cost rate $C_3(T, N)$. The optimal $N^*(T)$ exists to the fixed age T. Sensitivity of N^* to the age T is very small. The optimal $T^*(N)$ to the fixed number is decreasing in N. In comparison to the sensitivity of N^*, sensitivity of T^* to the number N is large. For instance, $T^*(N)$ to age $N = 12$ is 14.41, and $C_3(T^*, 12) = 0.256$. The optimal $N^*(T)$ to damage $T = 10$ is 8, and $C_3(10, N^*) = 0.251$. It shows that $C_3(T, N^*)$ is a decreasing function in T. On the other hand, $C_3(T^*, N)$ is an increasing function in N. From the above result, the number T should be long as much as possible to minimize $C_3(T, N)$ in this situation.

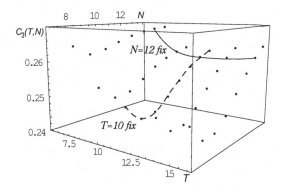

Fig. 7 Optimal replacement age T^* and shock number N^*

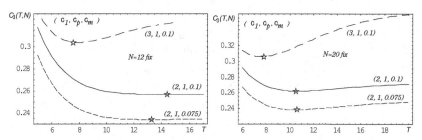

Fig. 8 Optimal replacement age T^* to cost change when $N = 12(l.h.s)$ and $N = 20(r.h.s)$

Figure 8 shows $C_3(T,N)$ to cost change when N is fixed. A sensitivity of $C_3(T,N1)$ to the age T is larger than a sensitivity of $C_3(T,N2)$ for $N1 < N2$ to the age T. The optimal T^* is greatly influenced by c_1 change under small N. However, $C_3(T^*(N1), N1)$ is lower than the $C_3(T^*(N2), N2)$. Therefore, we should decide the number N carefully, and set it as low as possible.

Figure 9 shows $C_3(T,N)$ to cost change when T is fixed. As described before, $C_3(T, N^*(T))$ has tendency to take low values when T is large.

3.7 Conclusions

It has applied three preventive parameters T, N and k to the shock damage interaction model. Three combination models, which are formulated with two preventive parameters, have been proposed. The expected costs rates are derived as the criterion for the decision making of the system

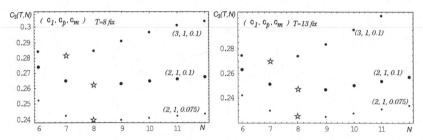

Fig. 9 Optimal shock number N^* to cost change when $T = 8(l.h.s)$ and $T = 13(r.h.s)$

maintenance. In each model, the optimal replacement policies to minimize the expected cost rates are discussed.

At the age and damage limit (T, k) model, the optimal finite solution (T^*, k^*) does not exist. The optimal solution is given by either $k^* = K$ and $T^* < \infty$ or $T^* \to \infty$ and $k^* < K$. We can confirm the phenomenon by Figure 1. At the number and damage limit (N, k) model, the optimal solution of finite N and $k(< K)$ does not exist. The optimal solution is given by either $k^* = K$ and any N ($\mathbf{N}_1 = \emptyset$) or $N^* \to \infty$ ($\mathbf{N}_1 \neq \emptyset$) and $k^* < K$. We can confirm the phenomenon by Figure 4. At the age and number (T, N) model, there is a possibility of the existence of the finite optimal solution (T^*, N^*).

From the results of numerical examples, it is confirmed that the damage limit k greatly contributes to the expected cost minimization. If we can control the damage limit k, then the k policy should be positively adopted as the control variable.

4 Oppotunistic Replacement Model

An opportunistic maintenance model is one of useful methods for a system administrator. In general, many maintenance models have assumed that the item is replaced at scheduled time or failure. However, the administrator is sometimes forced to execute maintenance at other timing. In order to achieve such requirements, the opportunistic maintenance model was proposed [31–33]. In this section, it proposes opportunistic replacement models for the shock damage model.

An amount W_j of damage due to the jth shock has an identical distribution $G(x) \equiv \Pr\{W_j \leq x\}$. The other assumptions are the same as in Sections 2 and 3.

It assumes that the maintenance opportunity arises at only shock time. If the opportunity arises then the item is replaced. The opportunity arises with probability $p(t)$ when the shock occurs at time t. With probability $q(t) \equiv 1 - p(t)$, the item suffers damage due to the shock. We calls this situation an age dependent opportunity for the shock damage model.

A probability that the maintenance opportunity does not occur in time $(0, t]$ is

$$\overline{F}_p(t) = e^{-\int_0^t p(x)\lambda(x)dx}. \tag{45}$$

A probability that j shocks occur with no maintenance opportunity in time $(0, t]$ is

$$\frac{[\int_0^t q(x)\lambda(x)dx]^j}{j!} e^{-\int_0^t q(x)\lambda(x)dx}. \tag{46}$$

4.1 Age Model

We have three kinds of replacements: If the item does not failed and no maintenance opportunity arise in time $(0, T]$, then the item is replaced preventively at time T. It calls preventive replacement. If the item fails before time T, i.e., the total damage exceeds the failure level K, then the item is replaced at failure when no opportunity has occurred. It calls failure replacement. The item is replaced with the maintenance opportunity before time T and the item failure. It calls opportunistic replacement.

A probability that it executes the preventive replacement at time T is

$$A_1(T) = \sum_{j=1}^{\infty} G^{(j)}(K) V_j(T) \overline{F}(T), \tag{47}$$

the probability that it executes failure replacement is

$$B_1(T) = \sum_{j=0}^{\infty} \int_0^K \overline{G}(K-x) \, dG^{(j)}(x) \int_0^T V_j(t) \, dF(t), \tag{48}$$

and the probability that it executes the opportunistic replacement before time T and the item failure is

$$D_1(T) = \sum_{j=1}^{\infty} G^{(j)}(K) \int_0^T p(t) V_{j-1}(t) \, dF(t), \tag{49}$$

where $\overline{F}(t) \equiv H_0(t)$ and

$$V_j(t) \equiv \frac{[\int_0^t q(x)\lambda(x) \, dx]^j}{j!}. \tag{50}$$

An expected renewal cycle, i.e., the expected replacement time is

$$E_1(T) = \sum_{j=0}^{\infty} G^{(j)}(K) \int_0^T V_j(t)\overline{F}(t)\,\mathrm{d}t. \tag{51}$$

Therefore, the expected cost rate is

$$C_A(T) = \frac{c_p A_1(T) + c_1[B_1(T) + D_1(T)]}{E_1(T)}, \tag{52}$$

where c_p is the preventive replacement cost and c_1 is an unplanned replacement cost. The unplanned replacement cost is defined as the cost of failure and opportunistic replacement.

We seek an optimal age T^* which minimizes $C_A(T)$ in (52). Differentiating $C_A(T)$ with respect to T and setting it equal to zero,

$$W_1(T) = \frac{c_p}{c_1 - c_p}, \tag{53}$$

where

$$W_1(T) \equiv \lambda(T)\left[1 - \frac{q(T)\sum_{j=0}^{\infty} G^{(j+1)}(K)V_j(T)}{\sum_{j=0}^{\infty} G^{(j)}(K)V_j(T)}\right]$$

$$\times \sum_{j=0}^{\infty} G^{(j)}(K) \int_0^T V_j(t)\overline{F}(t)\,\mathrm{d}t$$

$$- \sum_{j=0}^{\infty} \int_0^K \overline{G}(K-x)\,\mathrm{d}G^{(j)}(x) \int_0^T V_j(t)\,\mathrm{d}F(t)$$

$$- \sum_{j=1}^{\infty} G^{(j)}(K) \int_0^T p(t)V_{j-1}(t)\,\mathrm{d}F(t). \tag{54}$$

Further, differentiating $W_1(T)$ with respect to T,

$$\frac{\mathrm{d}\left\{\lambda(T)\left[1 - \frac{q(T)\sum_{j=0}^{\infty} G^{(j+1)}(K)V_j(T)}{\sum_{j=0}^{\infty} G^{(j)}(K)V_j(T)}\right]\right\}}{\mathrm{d}T} \sum_{j=0}^{\infty} G^{(j)}(K) \int_0^T V_j(t)\overline{F}(t)\,\mathrm{d}t. \tag{55}$$

The function $W_1(t)$ takes value 0 at $T \to 0$ to the following value at $T \to \infty$.

$$\lambda(\infty) \sum_{j=0}^{\infty} G^{(j)}(K) \int_0^{\infty} V_j(x)\overline{F}(x)\,\mathrm{d}x - 1, \tag{56}$$

where $\lambda(\infty) \equiv \lim_{T\to\infty} \lambda(T)$.

It is difficult to discuss an optimal policy to minimize the expected cost rate in (52) under previous general functions. We give the following assumptions to relax the difficulty:

1) the probability $q(t)$ is constant to time, i.e., $q(t) = q$ and $p(t) = p$,
2) the intensity function $\lambda(t)$ is not a decreasing function in time t,
3) a function $G^{(j+1)}(x)/G^{(j)}(x)$ is a decreasing function in j.

If these assumptions are satisfied then (54) is a increasing function in T, i.e., the sign of (55) is positive. Therefore, it gives an optimal replacement policy as follows:

(i) If the following condition is satisfied, then a finite and unique optimal replacement time T^* to minimize $C_A(T)$ exists:

$$\lambda(\infty) \sum_{j=0}^{\infty} G^{(j)}(K) \int_0^{\infty} V_j(x)\overline{F}(x)\,\mathrm{d}x > \frac{c_1}{c_1 - c_f}. \tag{57}$$

Then, the optimum expected cost rate $C_A(T^*)$ is

$$C_A(T^*) = (c_1 - c_p)\lambda(T^*)\left[1 - \frac{q(T^*)\sum_{j=0}^{\infty} G^{(j+1)}(K)V_j(T^*)}{\sum_{j=0}^{\infty} G^{(j)}(K)V_j(T^*)}\right]. \tag{58}$$

(ii) If the condition (57) is not satisfied, i.e.,

$$\lambda(\infty) \sum_{j=0}^{\infty} G^{(j)}(K) \int_0^{\infty} V_j(x)\overline{F}(x)\,\mathrm{d}x \leq \frac{c_1}{c_1 - c_f}, \tag{59}$$

then $T^* \to \infty$. In other words, we should not execute any preventive age replacement. The optimum expected cost rate is

$$C_A(\infty) = \frac{c_1}{\sum_{j=0}^{\infty} G^{(j)}(K) \int_0^{\infty} V_j(t)\overline{F}(t)\,\mathrm{d}t}. \tag{60}$$

4.2 Numerical examples

It shows the expected cost rate $C_A(T)$ with the replacement age T. As long as there is no attention, the following parameters are assumed to be a basic set: The shock occurs according to a homogeneous Poisson process with rate $\lambda = 1$, i.e., an expected time interval between sequential shock arrival is a unit time. Damage caused by each shock is assumed to be an exponential distribution with mean $1/\mu = 1$. The failure damage level K is assumed to be 10. Maintenance costs c_1 and c_p are 2 and 1, respectively. The probability of maintenance opportunity $p(t)$ is assumed to be an exponential distribution with mean 10, i.e., $p(t) = 1 - \mathrm{e}^{-0.1t}$.

Figure 10 shows the expected cost rate $C_A(T)$. An optimal age T^* to minimize $C_A(T)$ exists as a finite value. In case of Figure 10, the failure

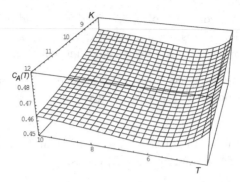

Fig. 10 Optimal age T^* with exponential opportunity

level K does not have influence to the optimal T^*, but it has somewhat higher influence on $C_A(T^*)$.

Figure 11 also shows $C_A(T)$ to some parameters change. Under the same expected failure time, the optimal T^* decreases with the parameter s, but the optimal $C_A(T^*)$ behaves just like increases with s. The optimal T^* increases with a ratio which divides c_p by c_1, and $C_A(T^*)$ increases with the above ratio.

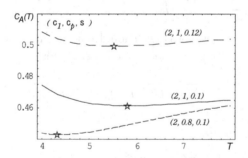

Fig. 11 Optimal age T^* to cost and opportunity rate

As a special case of this model, the probability $q(t)$ and $p(t)$ are assumed to be constant value q and $p \equiv 1 - q$, respectively. In this special case, the failure replacement cost c_1 is assumed to be 5. In comparison with two cases of $p = 0.05$, 0.15 in Figure 12, the movement of the optimal T^* to each cost and parameters is almost same. However, $C_A(T)$ takes quite different value. It presumes the reason as follows: If the probability of replacement opportunity p is high, then the opportunistic replacement is frequently generated in a young age. It causes the high expected cost rate.

To avoid such the frequent replacement, the optimal T^* becomes long for high probability p.

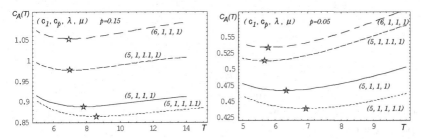

Fig. 12 Optimal age T^* with constant opportunity when $p = 0.15(l.h.s)$ and $p = 0.05(r.h.s)$

4.3 Damage Limit Model

A damage limit model pays attention to the amount of accumulated damage. A decision making, which should replace or not replace the item, is only done at each shock arrival. The item is replaced preventively at damage x $(0 \leq k \leq x < K)$ when the maintenance opportunity have not occurred. It calls preventive replacement. If the total damage has exceeded K, then the item is replaced as failure replacement. The item is replaced at the maintenance opportunity before preventive and failure replacements. It calls opportunistic replacement. At other cases of $\{Z(t) < k\}$, the item is left alone.

Probability of preventive replacement is

$$A_2(k) = \sum_{j=1}^{\infty} \int_0^k \int_{k-x}^{K-x} dG(y) \, dG^{(j-1)}(x) \int_0^{\infty} q(t) V_{j-1}(t) \, dF(t), \quad (61)$$

the probability of failure replacement is

$$B_2(k) = \overline{G}(K) + \sum_{j=1}^{\infty} \int_0^k \overline{G}(K-x) \, dG^{(j)}(x) \int_0^{\infty} q(t) V_{j-1}(t) \, dF(t), \quad (62)$$

and the probability of opportunistic replacement is

$$D_2(k) = \sum_{j=1}^{\infty} \int_0^k G(K-x) \, dG^{(j-1)}(x) \int_0^{\infty} p(t) V_{j-1}(t) \, dF(t), \quad (63)$$

where

$$V_j(t) \equiv \frac{[\int_0^t q(x)\lambda(x) \, dx]^j}{j!}. \quad (64)$$

An expected renewal cycle, i.e., the expected replacement time is

$$E_2(k) = \sum_{j=0}^{\infty} G^{(j)}(k) \int_0^{\infty} V_j(t)\overline{F}(t)\,dt. \tag{65}$$

Therefore, an expected cost rate is

$$C_B(k) = \frac{c_p A_2(k) + c_1[B_2(k) + D_2(k)]}{E_2(k)}, \tag{66}$$

where c_p is the preventive replacement cost and c_1 is an unplanned replacement cost. The unplanned replacement cost is defined as the cost of failure and opportunistic replacement.

We seek an optimal age k^* which minimizes $C_B(k)$ in (66). Differentiating $C_B(k)$ with respect to k and setting it equal to zero,

$$W_2(k) \equiv \frac{c_1}{c_1 - c_p}, \tag{67}$$

where

$$W_2(k) \equiv \sum_{j=1}^{\infty} \int_0^k \int_{k-x}^{K-x} dG(y)\,dG^{(j-1)}(x) \int_0^{\infty} V_j(x)\,dF(x)$$

$$- \left\{ \frac{\sum_{j=1}^{\infty} \left[G(K-k)g^{(j-1)}(k) - g^{(j)}(k) \right] \int_0^{\infty} V_j(x)\,dF(x)}{\sum_{j=0}^{\infty} g^{(j)}(k) \int_0^{\infty} V_j(t)\overline{F}(t)\,dt} \right\}$$

$$\times \sum_{j=0}^{\infty} G^{(j)}(k) \int_0^{\infty} V_j(t)\overline{F}(t)\,dt, \tag{68}$$

where the function $g^{(j)}(x)$ is a density function of $G^{(j)}(x)$.

The function (68) takes value 0 at $k \to 0$, and

$$\lim_{k \to K} W_2(k) = \frac{\sum_{j=1}^{\infty} g^{(j)}(K) \int_0^{\infty} V_j(t)\,dF(t)}{\sum_{j=1}^{\infty} g^{(j)}(K) \int_0^{\infty} V_j(t)\overline{F}(t)\,dt}$$

$$\times \sum_{j=0}^{\infty} G^{(j)}(K) \int_0^{\infty} V_j(t)\overline{F}(t)\,dt. \tag{69}$$

It is also difficult to discuss an optimal policy to minimize the expected cost rate $C_B(k)$ in (66) under previous general functions. In order to relax the difficulty, it assumes that $q(x) = q$, and $p(x) = p$. Under this conditions, the function $W_2(k)$ is a increasing function in k since

$$\frac{dW_2(k)}{dk} = g(K-k) \sum_{j=0}^{\infty} q^{j+1} G^{(j)}(k) > 0. \tag{70}$$

Therefore, we have the following optimal policy:

(i) If the following condition is satisfied, then a finite and unique optimal replacement time k^* to minimize the expected cost rate $C_B(k)$ exists:

$$\lim_{k \to K} W_2(k) = \sum_{j=0}^{\infty} q^j G^{(j)}(K) > \frac{c_1}{c_1 - c_p}. \tag{71}$$

Then, the optimum expected cost rate $C_B(k^*)$ is

$$C_B(k^*) = (c_1 - c_p) \left\{ \frac{\sum_{j=1}^{\infty} q^j [g^{(j)}(k^*) - G(K-k)g^{(j-1)}(k^*)]}{\sum_{j=1}^{\infty} q^j g^{(j)}(k^*) \int_0^{\infty} H_j(t) dt} \right\}. \tag{72}$$

(ii) If the condition (71) is not satisfied, i.e.,

$$\lim_{k \to K} W_2(k) = \sum_{j=0}^{\infty} q^j G^{(j)}(K) \leq \frac{c_1}{c_1 - c_p}, \tag{73}$$

then $k^* \to K$. In other words, we should not make any preventive replacement. The optimum expected cost rate is

$$C_B(\infty) = \frac{c_1}{\sum_{j=1}^{\infty} q^j G^{(j)}(K) \int_0^{\infty} H_j(t) dt}. \tag{74}$$

4.4 Numerical Examples

It shows the expected cost rate $C_B(k)$ with damage limit k. A basic set for this numerical example is the same as the basic set of the age model.

Figure 13 shows the expected cost rate $C_B(k)$. An optimal k^* to minimize $C_B(k)$ exists under failure level K. The optimal k^* increases with K. In comparison to T^*, a sensitivity of k^* to K is quite large. There is a possibility that rough k^* can be presumed from K.

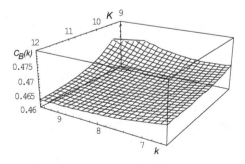

Fig. 13 Optimal damage limit k^* with exponential opportunity

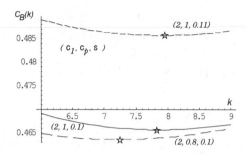

Fig. 14   Optimal damage limit k^* to cost and opportunity rate

Figure 14 also shows $C_B(k)$ to some parameters change. The optimal $C_B(k^*)$ behaves just like an increasing function in s. The optimal k^* increases with a ratio c_p/c_1.

In Figure 15, the probabilities $q(t)$ and $p(t)$ are constant, i.e., $q(t) = q$ and $p(t) = 1 - q$, and $c_1 = 5$. In comparison to $C_B(k)$ at $p = 0.2$, a sensitivity of $C_B(k)$ at $p = 0.1$ is large. The expected cost rate $C_B(k)$ at $p = 0.2$ takes considerably higher value than $C_B(k)$ at $p = 0.1$.

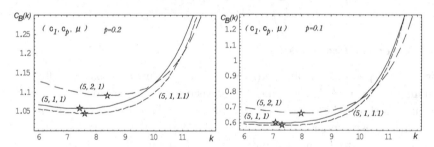

Fig. 15   Optimal damage limit k^* with constant opportunity when $p = 0.2(l.h.s)$ and $p = 0.1(r.h.s)$

4.5   Conclusions

It is difficult to predict the occurrence of maintenance opportunity generated by an internal or external causes. The occurrence of opportunity is assumed as a probability event. It proposes two maintenance models that are age T and damage limit k models. At formulation of the expected cost rate, the opportunity arises with probability $p(t)$ when the shock occurs at time t. The necessary condition to minimize the expected cost rate is derived under probability $p(t)$. Under some additional assumptions, the optimal policy is discussed. From the result of numerical examples, the mean time to the occurrence of opportunity has great influence on the optimal cost. It is not permitted to ignore the opportunity. In other words, we have to replace the item when the opportunity occurs even if the item is so young. It would be one of reasons for the great influence. If it is possible to select opportunities then the optimal cost can be reduced.

References

1. Blischke, W. R and Murthy, D. N. P. (2000). *Reliability Modeling, Prediction and Optimization*, John Wiley & Sons, New York.
2. Nakagawa, T. (2007) *Shock and Damage Models in Reliability Theory*, Springer-Verlag, London.
3. Osaki, S. (1992) *Applied Stochastic System Modeling*, Springer-Verlag, Berlin.
4. Cox, D. R. (1962). *Renewal Theory*, Methuen, London.
5. A-Hameed, M. S. and Proschan, F. (1973). *Nonstationary Shock Models*, Stochastic Processes and Their Applications, **1**, pp. 383-404.
6. Esary, J. D., Marshall, A. W. and Proschan, F. (1973). *Shock models and Wear Processes*, Annals of Probability, **1**, pp. 627–649.
7. Taylor, H. M. (1975). *Optimal replacement under additive damage and other failure models*, Naval Research Logistic Quarterly, **22**, pp. 1–18.
8. Zuckerman, D. (1977). *Replacement models under additive damage*, Naval Research Logistic Quarterly, **24**, pp. 549–558.
9. Feldman, R. M. (1976). *Optimal replacement with semi-Markov shock models*, Journal of Applied Probability, **13**, pp. 108–117.
10. Feldman, R. M. (1977). *Optimal replacement with semi-Markov shock models using discounted cost*, Mathematics of Operations Research, **2**, pp. 78–90.
11. Posner, M. J. M. and Zuckerman, D.(1984). *A replacement model for an additive damage model with restoration*, Operations Research Letters, **3**, pp. 141–148.
12. Barlow, R. E. and Proschan, F. (1965). *Statistical Theory of Reliability and Life Testing: Probability Models*, Holt, Rinehart and Winston, New York.
13. Marshall, A. W. and Shaked, M. (1979). *Multivariate shock models for distributions with increasing hazard rate average*, Annals of Probability, **7**, pp. 343–358.
14. A-Hameed, M. S. and Shimi, I. N. (1978). *Optimal replacement of damage devices*, Journal of Applied Probability, **15**, pp. 153–161.
15. Zuckerman, D. (1980). *A note on the optimal replacement time of damage devices*, Naval Research Logistic Quarterly, **27**, pp. 521–524.
16. Boland P. J. and Proschan, F. (1983). *Optimal replacement of a system subject to shocks*, Operations Research, **31**, pp. 679–704.
17. Puri, P. S. and Singh, H.(1986). *Optimal replacement of a system subject to shocks: a mathematical lemma*, Operations Research, **34**, pp. 782–789.

18. Nakagawa, T. (1976). *On a replacement problem of a cumulative damage model*, Operational Research Quarterly, **27**, pp. 895–900.
19. Nakagawa, T. and M. Kijima, M. (1989). *Replacement policies for a cumulative damage model with minimal repair at failure*, IEEE Transactions on Reliability, **38**, pp. 176–182.
20. Nakagawa, T. (1984). *A summary of discrete age replacement policies*, European Journal of Operational Research, **17**, pp. 382–392.
21. Kijima, M. and Nakagawa, T. (1992). *Replacement policies of a shock model with imperfect preventive replacement*, European Journal of Operational Research, **57**, pp. 100–110.
22. Nakagawa, T. and Murthy, D. N. P. (1993). *Optimal replacement policies for a two-unit system with failure interactions*, Operations Research, **27**, pp. 427–438.
23. Murthy, D. N. P. and Iskandar, B.P. (1991a). *A new shock damage model: part I- Model formulation and analysis*, Reliability Engineering and System Safety, **31**, pp. 191–208.
24. Murthy, D. N. P. and Iskandar, B.P. (1991b). *A new shock damage model: part II- Optimal maintenance policies*, Reliability Engineering and System Safety, **31**, pp. 211–231.
25. Chelbi, A. and Ait-Kadi, D. (2000). *Generalized inspection strategy for randomly failing systems subjected to random shocks*, International Journal of Production Economics, **64**, pp. 379–384.
26. Satow, T., Yasui, K. and Nakagawa, T. (1996). *Optimal garbage collection policies for a database in a computer system*, RAIRO Operations Research, **30**, pp. 359–372.
27. Barlow, R. E. and Proschan, F. (1965). *Mathematical Theory of Reliability*, John Wiley & Sons, New York.
28. Cho, D. I. and Parlar, M. (1991). *A survey of maintenance models for multiunit systems*, European Journal of Operational Research, **51**, pp. 1–23.
29. Wang, H. (2002). *A survey of maintenance policies of deteriorating systems*, European Journal of Operational Research, **139**, pp. 469–489.
30. Nakagawa, T. (2005). *Maintenance Theory and Reliability*, Springer-Verlag, London.
31. Dekker, R. and Smeitink, E. (1991). *Opportunity-based block replacement*, European Journal of Operational Research, **53**, pp. 46–63.
32. Dekker, R. and Dijkstra, M. (1992). *Opportunity-based age replacement: Exponentially distributed times between opportunities*, Naval Research Logistics, **39**, pp. 175–190.
33. Dekker, R. and Smeitink, E. (1994). *Preventive maintenance at opportunities of restricted duration*, Naval Research Logistics, **41**, pp. 335–353.

Chapter 3

Extended Inspection Models

SATOSHI MIZUTANI

Department of Media Informations
Aichi University of Technology
50-2 Manori, Nishihazama-cho, Gamagori 443-0047, Japan
E-mail: mizutani@aut.ac.jp

1 Introduction

In recent years, a lot of systems such as digital circuit or other devices for information processing are widely used. Therefore, it is necessary to check the systems at suitable times to detect its failure as soon as possible. However, if the inspection is done so frequently, then the system might be incurred much loss cost and work. Therefore, we should determine an optimal schedule of inspection for making a trade-off between the cost of failure and inspection.

In this chapter, we derive the mean time of one cycle and the expected cost rate, where we define *one cycle* as the time from the beginning of system operation to the detection of the failure. Further, we discuss optimal inspection policies which minimize the expected cost rates.

Barlow and Proschan [1] summarized optimal inspections policies which minimize the expected cost. Ross [2] and Osaki [3] explained plainly the stochastic processes and applied them to typical reliability models. Kaio and Osaki [37-39] compared with some inspection policies. Ben-Daya and Duffuaa [4], and Gertsbakh [5], Osaki [32], Pham [33], Nakagawa [6] overviewed many maintenance policies. The modified models were considered where checking times are nonnegligible [7], a unit is inoperative during checking times [8], checking hastens failure [9], and when failure's symptoms are delayed [10]. The imperfect inspection model, where there exist some failures which can not be detected upon inspection, was first treated in [11–13]. Periodic testing models to detect intermittent faults are discussed [14–18]

Inspection models have been recently applied to many actual systems: Christer *et al.* [19–21] reported the inspection maintenances of building, industrial plant and underwater structure. Sim *et al.* [22–24] analyzed the periodic test of combustion turbine units and standby equipments in dormant systems and nuclear generating stations. Ito and Nakagawa [25] discussed optimal policies for FADEC (Full-Authority Digital Engine Control) which is a control device of gas turbine engines and mainly consists of a two-unit system.

In Section 2, we consider the inspection policy for a two-unit system. First, the system operates as a two-unit system and is checked periodically. When one of the unit fails, the system operates as a single-unit system. We introduce two costs of one check for a two-unit system and a single-unit system.

In Section 3, we consider the inspection policy for a system which is checked by two types of inspection: The cost of type-1 inspection is lower than that of type-2 inspection. Therefore, the system is checked by type-1 inspection more frequently than type-2 inspection. However, there exist some failures which can not be detected by type-1 inspection, although they can be detected by type-2 inspection. That is, we assume that the failures are classified into two cases with certain probability: One of failures can be detected by type-1 inspection, and the other can be detected only by type-2 inspection.

In Section 4, we consider an inspection policy for a system with self-testing. The failure of system with self-testing can be detected during it is operating without external inspection. However, the detection by self-testing may has some latency. On the other word, the failures may not be detected rapidly. Therefore, to achieve a high reliability, the external inspection should check the system at scheduled times. Thus, when the system fails, the failure is detected by self-testing or at the next periodic inspection, whichever occur first.

In Section 5, we consider the maintenance and inspection policies when a system has to operate for a finite interval. In actual fields, most of the systems have a finite span of use. Using the partition method [26, 36], a finite interval is divided into some parts of inspection. Concretely, we

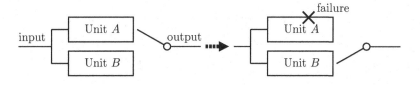

Fig. 1 Two-unit system

describe two models: Periodic inspection and sequential inspection policies. The unit of each model has to be operating for a finite interval $[0, S]$. We discuss an optimal number of inspection which minimizes the expected cost rate.

2 Periodic Policy for a Two-Unit System

2.1 *Model and Assumptions*

We consider the inspection policy for a two-unit system such as a digital control device for aircraft engine [25, 27]. The system is configured with two units, where we call the units respectively unit A and unit B. At first, unit A is connected with the part of output, and unit B operates as a hot standby unit. Further, we assume each unit is checked periodically to detect early the failure. When, one of units fails and the failure is detected by periodic inspection, the system operates as a single-unit system (Figure 1). An example of inspection method is comparing the output codes of each unit. The other examples are watch-dog timer, inputing test pattern codes and checking its output codes, checking the parameters such as voltages, resistance, impedance, and so on [28].

For simplify, we assume that the intervals of inspection for a two-unit and single-unit systems are the same. Then, the mean time and the expected cost rate are derived. An optimal interval of inspection which minimizes the expected cost is analytically derived. Finally, numerical examples are given when the failure time distribution is exponential.

For this model, we define the following assumptions:

1) After detecting the failure of unit A, if unit B has not failed then unit B is connected with part of output. The cost to detach the failed unit are negligible.
2) Both intervals of inspection for the two-unit system and the single-unit

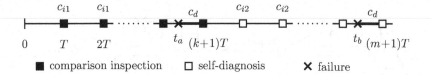

Fig. 2 Case 1 of periodic comparison-checking

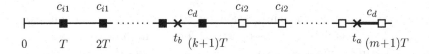

Fig. 3 Case 2 of periodic comparison-checking

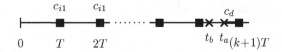

Fig. 4 Case 3 of periodic comparison-checking

system are the same. That is, the system is checked always at periodic times kT ($k = 1, 2, \ldots, 0 < T < \infty$). The failure of each unit is detected only by the periodic inspection.

3) We assume that unit A fails at time t_a ($0 < t_a < \infty$) and unit B fails at time t_b ($0 < t_b < \infty$), where the failure time of each unit has an independent and identical general distribution $F(t)$ with finite mean $1/\lambda$, where $\overline{F}(t) \equiv 1 - F(t)$.

4) A cost c_{i1} is the loss cost for one check of the two-unit system, and c_{i2} is the cost for one check of the single-unit system. A cost c_d is the loss cost per unit of time for the time elapsed between a failure of a unit which is connected with the part of output and its detection at the next time of inspection. Further, c_r is the constant cost for maintenance or replacement, when the second failure is detected.

The total expected cost of one cycle is classified into the following three cases:

(a) $kT < t_a \leq (k+1)T \leq mT < t_b \leq (m+1)T$

Suppose that unit A fails during $(kT, (k+1)T]$ ($k = 0, 1, \cdots$) before unit B fails, and after that, unit B fails during $(mT, (m+1)T]$ ($m = k+1, k+2, \cdots$)

(Figure 2). Then, the expected cost of one cycle is

$$\sum_{m=1}^{\infty}\int_{mT}^{(m+1)T}\Big[\sum_{k=0}^{m-1}\int_{kT}^{(k+1)T}\{c_{i1}(k+1)+c_{i2}(m-k)$$
$$+c_d\big[(k+1)T-t_a+(m+1)T-t_b\big]\}dF(t_a)\Big]dF(t_b). \qquad (1)$$

(b) $kT < t_b \leq (k+1)T \leq mT < t_a \leq (m+1)T$

Suppose that unit B fails during $(kT, (k+1)T]$ $(k = 0, 1, \cdots)$ before unit A fails. Thereafter, unit A fails during $(mT, (m+1)T]$ $(m = k+1, k+2, \cdots)$ (Figure 3). Then, the expected cost of one cycle is

$$\sum_{m=1}^{\infty}\int_{mT}^{(m+1)T}\Big\{\sum_{k=0}^{m-1}\int_{kT}^{(k+1)T}\big[c_{i1}(k+1)+c_{i2}(m-k)$$
$$+c_d\big((m+1)T-t_a\big)\big]dF(t_b)\Big\}dF(t_a). \qquad (2)$$

(c) $kT < t_a, t_b \leq (k+1)T$

Suppose that both units A and B fail during $(kT, (k+1)T]$ $(k = 0, 1, 2, \cdots)$, and their failures are detected by the next inspection (Figure 4). Then, the expected cost of one cycle is

$$\sum_{k=0}^{\infty}\int_{kT}^{(k+1)T}\big[c_{i1}(k+1)+c_d\big((k+1)T-t_a\big)\big]\big[F((k+1)T)-F(kT)\big]dF(t_a). \qquad (3)$$

Thus, the total expected cost of one cycle is obtained by summing (1), (2) and (3), and the maintenance cost c_r as follows:

$$c_r+(c_{i1}-c_{i2})\sum_{m=0}^{\infty}\overline{F}(mT)^2+c_{i2}\sum_{m=0}^{\infty}\big[1-F(mT)^2\big]$$
$$+c_d\sum_{m=0}^{\infty}\big[1+F(mT)\big]\int_{mT}^{(m+1)T}\big[F(t)-F(mT)\big]dt. \qquad (4)$$

Similarly, the mean time of one cycle is

$$2\sum_{m=1}^{\infty}\int_{mT}^{(m+1)T}\Big[\sum_{k=0}^{m-1}\int_{kT}^{(k+1)T}(m+1)TdF(t_b)\Big]dF(t_a)$$
$$+\sum_{k=0}^{\infty}\int_{kT}^{(k+1)T}(k+1)T\big[F((k+1)T)-F(kT)\big]dF(t_a)$$
$$=T\Big[2\sum_{m=0}^{\infty}\overline{F}(mT)-\sum_{m=0}^{\infty}\overline{F}(mT)^2\Big]=T\sum_{m=0}^{\infty}\big[1-F(mT)^2\big]. \qquad (5)$$

Thus, the expected cost rate is, from (4) and (5),

$$C(T) = \frac{c_r + (c_{i1}-c_{i2})\sum_{m=0}^{\infty}\overline{F}(mT)^2 + c_{i2}\sum_{m=0}^{\infty}[1-F(mT)^2] + c_d\sum_{m=0}^{\infty}[1+F(mT)]\int_{mT}^{(m+1)T}[F(t)-F(mT)]dt}{T\sum_{m=0}^{\infty}[1-F(mT)^2]}. \quad (6)$$

Obviously,

$$C(0) \equiv \lim_{T\to 0} C(T) = \infty, \qquad C(\infty) \equiv \lim_{T\to\infty} C(T) = c_d. \quad (7)$$

Therefore, there exists an optimal inspection interval T^* $(0 < T^* \le \infty)$ which minimizes $C(T)$.

2.2 Optimal Policy

For simplify of analysis, we assume that the failure time of each unit has an exponential distribution $F(t) = 1 - e^{-\lambda t}$. Then, the expected cost rate in (6) is

$$C(T) = \frac{c_{i1} + 2c_{i2}e^{-\lambda T} + c_r(1 - e^{-2\lambda T}) - c_d(1 + e^{-\lambda T} - 2e^{-2\lambda T})/\lambda}{T(1 + 2e^{-\lambda T})} + c_d. \quad (8)$$

Differentiating $C(T)$ with respect to T and putting it equal to 0,

$$4c_{i1}(1 - e^{-\lambda T}) + (4c_{i2} + 2c_r\lambda T e^{-\lambda T})(1 - e^{-2\lambda T})$$
$$+ \left[2(c_{i2}-c_{i1}) + \frac{c_d}{\lambda}(1 + 2e^{-\lambda T})^2\right][1 - (1 + \lambda T)e^{-\lambda T}]$$
$$- c_r(1 + 2e^{-\lambda T})[1 - (1 + 2\lambda T)e^{-2\lambda T}] = 6c_{i2} + 3c_{i1}. \quad (9)$$

Letting denote the left-hand side of (9) by $Q(T)$,

$$Q(0) = 0, \qquad Q(\infty) = 6c_{i2} + 2c_{i1} + \frac{1}{\lambda}c_d - c_r. \quad (10)$$

Thus, if $c_d/\lambda > c_{i1} + c_r$ then there exists exists a finite T^* $(0 < T^* < \infty)$ which satisfies (9).

2.3 Numerical Examples

We calculate the optimal interval T^* which minimizes the expected cost rate $C(T)/c_{i2}$ in (8) when $F(t) = 1 - e^{-\lambda t}$. All costs are normalized to c_{i2} as a unit cost, i.e., they are divided by c_{i2}.

Table 1 gives $\lambda T^* \times 10^5$ which minimizes the expected cost rate $C(T)/c_{i2}$ for $c_d/(\lambda c_{i2}) \times 10^{-7} = 1\text{-}10$ and $c_r/c_{i2} \times 10^{-5} = 1, 5, 10$ when $c_{i1}/c_{i2} =$

Table 1 Optimal interval $\lambda T^* \times 10^5$ when $c_{i1}/c_{i2} = 0.1$

$c_d/(\lambda c_{i2}) \times 10^{-7}$	$c_r/c_{i2} \times 10^{-5}$		
	1	5	10
1	37.26	36.62	35.86
2	26.40	26.17	25.90
3	21.57	21.45	21.30
4	18.69	18.61	18.50
5	16.72	16.66	16.59
6	15.26	15.22	15.16
7	14.13	14.10	14.05
8	13.22	13.19	13.16
9	12.47	12.44	12.41
10	11.83	11.81	11.78

Table 2 Optimal interval $\lambda T^* \times 10^5$ when $c_r/c_{i2} = 5 \times 10^5$

c_{i1}/c_{i2}	$c_d/(\lambda c_{i2}) \times 10^{-7}$		
	1	5	10
0.01	35.82	16.30	11.55
0.05	36.18	16.46	11.67
0.1	36.62	16.66	11.81
0.5	39.95	18.18	12.88

0.1. It is shown that λT^* decreases as $c_d/(\lambda c_{i2})$ or c_r/c_{i2} increases. This indicates that when the loss cost c_d and the replacement cost c_r is large, it is better to shorten the interval of inspection to detect failures as soon as possible. For example, when $c_d/(\lambda c_{i2}) \times 10^{-7} = 5$ and $c_r/c_{i2} \times 10^{-5} = 10$, $\lambda T^* \times 10^5 = 16.59$. That is, when the mean time to failure of each unit is $1/\lambda = 3 \times 10^4$ hours (approximately 3.5 years), $c_{i1}/c_{i2} = 1/3 \times 10$, $c_d/c_{i2} = 1/3 \times 10^3$ and $c_r/c_{i2} = 10^5$, the optimal interval T^* is about 4.977 hours.

Table 2 presents $\lambda T^* \times 10^5$ which minimizes $C(T)/c_{i2}$ for $c_{i1}/c_{i2} = 0.01$, 0.05, 0.1, 0.5 and $c_d/(\lambda c_{i2}) \times 10^{-7} = 1, 5, 10$ when $c_r/c_{i2} = 5 \times 10^5$. This indicates that λT^* increases as c_{i1}/c_{i2} increases. Thus, if the cost c_{i1}/c_{i2} of inspection is higher, the interval of inspection should be larger.

3 Periodic Policy for a System with Two Types of Inspection

This section considers a system which is checked periodically by type-1 and type-2 inspections. The cost of type-1 inspection is lower than that

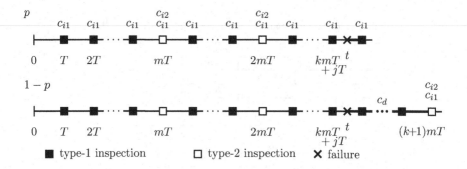

Fig. 5 Two types of inspection

of type-2 inspection. Therefore, type-1 inspection checks the system more frequently than type-2 inspection. On the other hand, we assume that type-2 inspection can detect any failures which can not be detected by type-1 inspection. That is, it is assumed that type-2 inspection can detect all failures of the system.

A typical example of such a inspection policy is electronic control device checked periodically by self-diagnosis and external inspection with tester [29]. Coverage of external inspection is larger than self-diagnosis, although the cost performance of external inspection is more worse than that of self-diagnosis. Therefore, we also classify the failure into two cases such that the high-cost inspection and low-cost one, where intervals of high-cost inspection are larger than those of low-cost self-diagnosis.

The mean time of one cycle and the expected cost rate are derived. Further, it is discussed analytically an optimal number of type-1 inspection between type-2 inspection which minimizes the expected cost. Finally, a numerical example is given when the failure time is exponential.

3.1 Model and Assumptions

Consider a system which is checked periodically by two-types of inspection. For this model, we define the following assumptions:

1) Type-1 inspection checks the system at periodic times jT ($j = 1, 2, \ldots,$ $j \neq km$ ($k = 1, 2, \ldots$)), and type-2 inspection checks the system at periodic times kmT ($k = 1, 2, \ldots$) for m ($m = 1, 2, \ldots$) and some specified T ($0 < T < \infty$), i.e., type-2 inspection is done at every m times of type-1 inspection.

2) Type-1 and type-2 inspections can detect the failure which occurred with probability p $(0 < p \leq 1)$, although type-1 inspection cannot detect the failure which occurred with $q \equiv 1-p$. That is, the failure which occurred with q can be detected only by type-2 inspection.
3) The failure time has a general distribution $F(t)$ with finite mean $1/\lambda$, where $\overline{\Phi}(t) \equiv 1 - \Phi(t)$ for any function $\Phi(t)$. When the failure occurs, it is classified according to 2).
4) Let c_{i1} be the cost of one check by type-1 inspection, and $c_{i1} + c_{i2}$ be the cost of one check by type-2 inspection. Let c_d be the loss cost per unit of time for the time elapsed between a failure and its detection. Further, c_r be the replacement cost, and we assume $c_d/\lambda > c_{i2} + c_r$, i.e., if the system failed while the MTTF then the cost is over one of check by type-2 inpsection and replacement. The system is replaced when its failure is detected by inspection. Any failure does not occur between the first failure and its detection. If the failure is detected, then the system is maintained and is as good as new.

Figure 5 shows the processes of the system with two types of inspection: The horizontal axis represents time of the process. Upper side shows the case that the system fails at time t $(kmT + jT < t \leq kmT + (j+1)T)$ and the failure which occurred with probability p is detected by type-1 inspection at time $kmT + (j+1)T$. The lower side shows that the failure which occurred with probability $1-p$ is detected only by type-2 inspection at time $(k+1)mT$.

Then, the mean time of one cycle is

$$p \sum_{k=0}^{\infty} \sum_{j=0}^{m-1} \int_{kmT+jT}^{kmT+(j+1)T} [kmT + (j+1)T]\,\mathrm{d}F(t) + q \sum_{k=0}^{\infty} \int_{kmT}^{(k+1)mT} (k+1)mT\,\mathrm{d}F(t)$$

$$= pT \sum_{k=0}^{\infty} \overline{F}(kT) + qmT \sum_{k=0}^{\infty} \overline{F}(kmT) \qquad (m = 1, 2, \dots). \tag{11}$$

Further, the total expected cost of one cycle is

$$p \sum_{k=0}^{\infty} \left\{ \sum_{j=0}^{m-1} \int_{kmT+jT}^{kmT+(j+1)T} \{c_{i1}(km+j+1) + c_d[kmT+(j+1)T-t]\}\,\mathrm{d}F(t) \right.$$

$$+ c_{i2} \left[\sum_{j=0}^{m-2} \int_{kmT+jT}^{kmT+(j+1)T} k\,\mathrm{d}F(t) + \int_{(k+1)mT-T}^{(k+1)mT} (k+1)\,\mathrm{d}F(t) \right] \Bigg\}$$

$$+ q \sum_{k=0}^{\infty} \int_{kmT}^{(k+1)mT} \{(c_{i1}m + c_{i2})(k+1) + c_d[(k+1)mT - t]\}\,\mathrm{d}F(t)$$

$$= (c_{i1} + c_d T)\left[p\sum_{k=0}^{\infty}\overline{F}(kT) + qm\sum_{k=0}^{\infty}\overline{F}(kmT)\right]$$

$$+ c_{i2}\left[p\sum_{k=1}^{\infty}\overline{F}((km-1)T) + q\sum_{k=0}^{\infty}\overline{F}(kmT)\right] - \frac{c_d}{\lambda}. \tag{12}$$

Thus, the expected cost rate is,

$$C(m;T) = \frac{c_{i2}\left[p\sum_{k=1}^{\infty}\overline{F}((km-1)T) + q\sum_{k=0}^{\infty}\overline{F}(kmT)\right] - \frac{c_d}{\lambda} + c_r}{pT\sum_{k=0}^{\infty}\overline{F}(kT) + qmT\sum_{k=0}^{\infty}\overline{F}(kmT)}$$

$$+ c_d + \frac{c_{i1}}{T} \qquad (m = 1, 2, \ldots). \tag{13}$$

3.2 Optimal Policy

We assume that the failure distribution is exponential, *i.e.*, $F(t) = 1 - e^{-\lambda t}$. Then, the total expected cost rate in (13) can be rewritten as

$$C(m;T) = \frac{c_{i2}[1 - p(1-e^{-\lambda(m-1)T})] - (\frac{c_d}{\lambda} - c_r)(1 - e^{-\lambda mT})}{qm(1 - e^{-\lambda T}) + p(1 - e^{-\lambda mT})}\left(\frac{1 - e^{-\lambda T}}{T}\right)$$

$$+ c_d + \frac{c_{i1}}{T} \qquad (m = 1, 2, \ldots). \tag{14}$$

Clearly,

$$C(1;T) = c_d + \frac{c_{i1}}{T} + \frac{c_{i2} - (c_d/\lambda - c_r)(1 - e^{-\lambda T})}{T},$$

$$C(\infty;T) = c_d + \frac{c_{i1}}{T}.$$

Therefore, if $c_d/\lambda - c_r > c_{i2}/(1 - e^{-\lambda T})$ then there exists a finite m^* ($1 \leq m^* < \infty$).

Letting $C(m+1;T) \geq C(m;T)$,

$$\frac{\sum_{k=1}^{m}(e^{\lambda kT} - 1)}{p + qe^{\lambda mT}} \geq \frac{c_{i2}\left(\frac{1}{1-e^{-\lambda T}} + pe^{\lambda T}\right)}{q\left(\frac{c_d}{\lambda} - c_r + pc_{i2}e^{\lambda T}\right)} \qquad (m = 1, 2, \ldots). \tag{15}$$

Letting denote the left-hand of (15) by $L(m)$,

$$L(1) = \frac{1 - e^{-\lambda T}}{pe^{-\lambda T} + q}, \qquad L(\infty) = \frac{1}{q(1 - e^{-\lambda T})},$$

$$L(m+1) - L(m) = \frac{[p(e^{\lambda(m+1)T} - 1) + qe^{\lambda mT}(m+1)(e^{\lambda T} - 1)]}{(p + qe^{\lambda(m+1)T})(p + qe^{\lambda mT})} > 0.$$

Therefore, from assumption 6 and $c_d/\lambda - c_r > qc_{i2}$, there exists an optimal number m^* ($1 \leq m^* < \infty$) which satisfies (15).

Table 3 Optimal number m^* which minimize $C(m;T)/c_{i1}$ when $c_{i2}/c_{i1} = 10$ and $c_r/c_{i1} = 1000$

p	$1/(\lambda T) = 300$		$1/(\lambda T) = 600$	
	$c_d T/c_{i1}$			
	100	1000	100	1000
0.50	11	4	16	5
0.55	12	4	16	5
0.60	12	4	17	6
0.65	13	4	19	6
0.70	14	5	20	6
0.75	16	5	22	7
0.80	17	6	25	8
0.85	20	6	28	9
0.90	25	8	35	11
0.95	35	11	49	15
1.00	∞	∞	∞	∞

3.3 Numerical Example

We compute numerically the optimal inspection number m^* which minimizes the expected cost rate $C(m;T)/c_{i1}$ when $F(t) = 1 - e^{-\lambda t}$. All costs are normalized to c_{i1} as a unit cost, i.e., they are divided by c_{i1}.

Table 3 gives the optimal number m^* which minimizes $C(m;T)/c_{i1}$ for $1/(\lambda T) = 300, 600$, $c_d T/c_{i1} = 100, 1000$ and $p = 0.50, 0.55, 0.60, 0.65, 0.70, 0.75, 0.80, 0.85, 0.90, 0.95, 1.00$ when $c_{i2}/c_{i1} = 10$ and $c_r/c_{i1} = 1000$. This indicates that m^* increase as p and $1/(\lambda T)$ increase, and $c_d T/c_{i1}$ decrease. Especially, when p goes to 1, m^* goes to infinity. This results means that it would be better not to done type-2 inspection when type-1 inspection can detect any failures.

4 Periodic Policy for a System with Self-Testing

We consider a system which can detect the failure with a delay time distribution. That is, the system can detect without external inspection, although the time to detect the failure has a delay time. Therefore, it is necessary to check periodically by external inspection for early detection of the failure.

A typical example is digital circuit which has a property of self-testing: When the system has a failure which is in an assumed failure set, if the system has at least one input code for output a code which are not in a correct output code set, then the system is called that it has a property

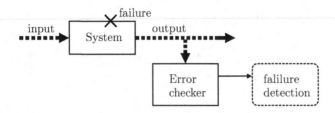

Fig. 6 System with self-testing

of *self-testing*. Therefore, such a failure can be detected without periodic inspection, *i.e.*, the system with self-testing can detect any failure during the normal operation [28] (Figure 6).

However, even if the system has input codes to detect some failures, they may not be inputed rapidly to the system. Therefore, some failures may not be detected rapidly by self-testing. Hence, for detection of the failures early and surely, it would be necessary to check by inspection such as inputting a set of test codes at periodic times. In this case, if the system fails, then its failure is detected by self-testing or the next periodic inspection, whichever occur first. However, checking periodic inspection so frequently will incur much loss cost.

In this model, we obtain the expected cost rate. Optimal intervals of periodic inspection which minimize the total expected cost and the expected cost rate are analytically derived, Further, we consider the case where there exist some failures which cannot be detected by self-testing with probability p. Finally, numerical examples are given when both times of failure and its detection by self-testing are exponential.

4.1 Model and Assumptions

Consider a system which can detect the failure with a delay time distribution. For this model, we define the following assumptions:

1) The system is checked at times kT ($k = 1, 2, \ldots$) by periodic inspection.
2) The failure time distribution has a general distribution $F(t)$ with finite mean $1/\lambda$, where $\overline{F}(t) \equiv 1 - F(t)$.
3) The time from a failure to its detection by self-testing is a general distribution $G(x)$ with finite mean $1/\mu$ ($\mu > \lambda$). Thus, when the system fails, its failure will be detected by self-testing or at the next periodic inspection, whichever occurs first.

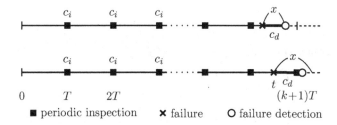

Fig. 7 Processes of system with self-testing

4) Let $d(t) \equiv g(t)/\overline{G}(t)$, where $g(t)$ is a density of $G(t)$ and $\overline{G}(t) \equiv 1 - G(t)$, where $d(t)\mathrm{d}t$ represents the probability that the failure of system is detected during $(t, t+\mathrm{d}t)$ by *self-testing*. We call $d(t)$ *self-detection rate*. It would be practically estimated that $d(t)$ is a decreasing function.

5) A cost c_i is the cost for one check by periodic inspection, and c_d is the loss cost per unit of time for the time elapsed between a failure and its detection by self-testing or periodic inspection, whichever occurs first. A cost c_r is the replacement or maintenance cost. Especially, we assume $c_d > \lambda c_r + \mu c_i$.

Figure 7 shows the processes of the system with self-testing: The horizontal axises presents the process of time, and the system fails at time t $(kT < t \le (k+1)T)$. The upper side shows the case the failure is detected at time $t+x$ $(< (k+1)T)$ by self-testing, and the lower side shows the case where its failure is detected at time $(k+1)T$ by periodic inspection.

Then, the mean time of one cycle is

$$\sum_{k=0}^{\infty} \int_{kT}^{(k+1)T} \left[\int_0^{(k+1)T-t} (t+x)\,\mathrm{d}G(x) + (k+1)T\,\overline{G}((k+1)T - t) \right] \mathrm{d}F(t)$$

$$= \frac{1}{\lambda} + \sum_{k=0}^{\infty} \int_0^T \left[\overline{F}(kT) - \overline{F}((k+1)T - x)\right] \overline{G}(x)\,\mathrm{d}x. \tag{16}$$

In a similar way, the total expected cost of one cycle is

$$\sum_{k=0}^{\infty} \int_{kT}^{(k+1)T} \left[\int_0^{(k+1)T-t} (c_i k + c_d x)\,\mathrm{d}G(x) \right.$$

$$\left. + \{c_i(k+1) + c_d[(k+1)T - t]\}\overline{G}((k+1)T - t) \right] \mathrm{d}F(t) + c_r$$

$$= c_i \sum_{k=0}^{\infty} \left\{ \overline{F}(kT) - \int_0^T \left[\overline{F}(kT) - \overline{F}((k+1)T - x)\right] \mathrm{d}G(x) \right\}$$

$$+ c_d \sum_{k=0}^{\infty} \int_0^T \left[\overline{F}(kT) - \overline{F}((k+1)T - x)\right] \overline{G}(x)\,\mathrm{d}x + c_r. \tag{17}$$

Therefore, the expected rate is, from (16) and (17),

$$C(T) = \frac{c_i \sum_{k=0}^{\infty} \left\{\overline{F}(kT) - \int_0^T [\overline{F}(kT) - \overline{F}((k+1)T - x)]\,\mathrm{d}G(x)\right\} - c_d/\lambda + c_r}{\sum_{k=0}^{\infty} \int_0^T [\overline{F}(kT) - \overline{F}((k+1)T - x)]\overline{G}(x)\,\mathrm{d}x + 1/\lambda} + c_d. \tag{18}$$

Evidently,

$$C(0) \equiv \lim_{T \to 0} C(T) = \infty, \qquad C(\infty) \equiv \lim_{T \to \infty} C(T) = \frac{c_d/\mu + c_r}{1/\lambda + 1/\mu}.$$

Thus, there exists an optimal time T^* $(0 < T^* \le \infty)$ which minimizes $C(T)$.

4.2 Optimal Policy

Consider the optimization problem of minimizing $C(T)$ in (18). In particular, when $F(t) = 1 - e^{-\lambda t}$, (18) is

$$C(T) = \frac{c_i \left[1 - \int_0^T (1 - e^{-\lambda(T-x)})\,\mathrm{d}G(x)\right] - \left(\frac{c_d}{\lambda} - c_r\right)(1 - e^{-\lambda T})}{\int_0^T (1 - e^{-\lambda(T-x)})\overline{G}(x)\,\mathrm{d}x + \frac{1}{\lambda}(1 - e^{-\lambda T})} + c_d. \tag{19}$$

Differentiating $C(T)$ with respect to T and putting it equal to zero,

$$(c_d - \lambda c_r)\int_0^T (e^{\lambda x} - 1)\overline{G}(x)\,\mathrm{d}x - c_i\left[\int_0^T (e^{\lambda x} - 1)\,\mathrm{d}G(x) + \int_0^T \lambda e^{\lambda x}\overline{G}(x)\,\mathrm{d}x\right.$$

$$+ \int_0^T \lambda e^{\lambda x}\,\mathrm{d}G(x)\int_0^T (1 - e^{-\lambda(T-x)})\overline{G}(x)\,\mathrm{d}x$$

$$\left. - \int_0^T \lambda e^{\lambda x}\overline{G}(x)\,\mathrm{d}x \int_0^T (1 - e^{-\lambda(T-x)})\,\mathrm{d}G(x)\right] = c_i. \tag{20}$$

Letting denote the left-hand side of $Q(T)$,

$$Q(0) \equiv \lim_{T \to 0} Q(T) = 0,$$

$$Q(\infty) \equiv \lim_{T \to \infty} Q(T) = (c_d - \lambda c_r)\int_0^{\infty} (e^{\lambda x} - 1)\overline{G}(x)\,\mathrm{d}x$$

$$- c_i\left[\int_0^{\infty} (e^{\lambda x} - 1)\,\mathrm{d}G(x) + \frac{\lambda}{\mu}\int_0^{\infty} e^{\lambda x}\,\mathrm{d}G(x)\right],$$

$$Q'(T) = (e^{\lambda T} - 1)\overline{G}(T)c_i \left[\frac{c_d - \lambda c_r}{c_i} - d(T) - \frac{\lambda}{1 - e^{-\lambda T}}\right]$$
$$+ \lambda e^{\lambda T}\overline{G}(T)c_i \int_0^T (1 - e^{-\lambda(T-x)})\overline{G}(x)\,dx \left[\frac{\int_0^T (1 - e^{-\lambda(T-x)})\,dG(x)}{\int_0^T (1 - e^{-\lambda(T-x)})\overline{G}(x)\,dx} - d(T)\right].$$

Because we assumed that the self-detection rate $d(t)$ is decreasing, *i.e.*, $d(T) \leq g(x)/\overline{G}(x)$ for $0 \leq x \leq T$, it follows that
$$\frac{\int_0^T (1 - e^{-\lambda(T-x)})\,dG(x)}{\int_0^T (1 - e^{-\lambda(T-x)})\overline{G}(x)\,dx} \geq \frac{\int_0^T (1 - e^{-\lambda(T-x)})\,dG(x)}{\int_0^T (1 - e^{-\lambda(T-x)})g(x)/d(T)\,dx} = d(T). \quad (21)$$

Further, $\lim_{T \to 0} Q'(T) = -\lambda c_i$. Thus, $Q(T)$ decreases at first, and after that, increases to $Q(\infty)$. Therefore, if $Q(\infty) > c_i$, then there exists a finite and unique T^* ($0 < T^* < \infty$) which minimizes $C(T)$.

Case 1. We assume $G(x) = 1 - e^{-\mu x}$. Then, (20) is
$$(c_d - \lambda c_r - \mu c_i)\left(\frac{1 - e^{-(\mu-\lambda)T}}{\mu - \lambda} - \frac{1 - e^{-\mu T}}{\mu}\right) - c_i \frac{\lambda(1 - e^{-(\mu-\lambda)T})}{\mu - \lambda} = c_i. \quad (22)$$

Let denote the left-hand side of (22) by $Q(T)$. Then,
$$Q(\infty) \equiv \lim_{T \to \infty} Q(T) = (c_d - \lambda c_r - \mu c_i)\frac{\lambda}{\mu(\mu - \lambda)},$$
$$Q'(T) = e^{-(\mu-\lambda)T}[(c_d - \lambda c_r - \mu c_i)(1 - e^{-\lambda T}) - \lambda c_i].$$

Thus, $Q(T)$ starts from 0 and decreases for a while, and increases strictly to $Q(\infty)$.

Therefore, we have the optimal policy:

(i) If $(c_d - \lambda c_r - \mu c_i)\lambda/[\mu(\mu - \lambda)] > c_i$, then there exists a finite and unique T^* which satisfies (22).
(ii) If $(c_d - \lambda c_r - \mu c_i)\lambda/[\mu(\mu - \lambda)] \leq c_i$, then $T^* = \infty$.

Case 2. We assume $G(x) = p(1 - e^{-\mu x})$, where it is assumed that p ($0 < p < 1$) is probability that the failure can be detected by self-testing. On the other hand, $q \equiv 1 - p$ is the probability that the failure can not be detected by self-testing, *i.e.*, it can be detected only by periodic inspection. That is, there exist some failures which can not be detected by self-testing. In the design of a complex system, it would be difficult to design a system which can detect any failure. Then, the self-detection rate is
$$d(x) = \frac{p\mu e^{-\mu x}}{q + pe^{-\mu x}},$$

which is a decreasing function from $p\mu$ to 0. If $p = 1$, then this model corresponds to the case where $G(x)$ is a standard exponential distribution, and if $p = 0$, then this model corresponds to simple periodic inspection model [1].

Then, the expected cost rate $C(T)$ in (18) is

$$\frac{c_i\left[1 - \mu p\left(\frac{1-e^{-\mu T}}{\mu} - \frac{e^{-\lambda T} - e^{-(\mu-\lambda)T}}{\mu-\lambda}\right)\right] - \left(\frac{c_d}{\lambda} - c_r\right)(1 - e^{-\lambda T})}{p\left(\frac{1-e^{-\mu T}}{\mu} + \frac{1-e^{-\lambda T}}{\lambda} - \frac{e^{-\lambda T} - e^{-(\mu-\lambda)T}}{\mu-\lambda}\right) + qT} + c_d. \quad (23)$$

Equation (20) is

$$p\left[(c_d - \lambda c_r - \mu c_i)\left(\frac{1 - e^{-(\mu-\lambda)T}}{\mu-\lambda} - \frac{1 - e^{-\mu T}}{\mu}\right) - \frac{c_i\lambda}{\mu-\lambda}(1 - e^{-(\mu-\lambda)T})\right]$$

$$+ q\left[\frac{c_d - \lambda c_r}{\lambda}(e^{\lambda T} - \lambda T - 1) - c_i(e^{\lambda T} - 1)\right]$$

$$- pq\left[\frac{\lambda\mu T}{\mu-\lambda}(1 - e^{-(\mu-\lambda)T}) - (e^{\lambda T} - 1)(1 - e^{-\mu T})\right] = c_i. \quad (24)$$

Letting denote the left-hand side of (24) by $Q(T)$,

$$Q(\infty) \equiv \lim_{T\to\infty} Q(T)$$

$$= p\left[(c_d - \lambda c_r - \mu c_i)\left(\frac{1}{\mu-\lambda} - \frac{1}{\mu}\right) - \frac{\lambda c_i}{\mu-\lambda}\right]$$

$$+ \lim_{T\to\infty} q\left\{\left[\frac{c_d - \lambda(c_r + qc_i)}{\lambda}\right](e^{\lambda T} - 1) - \left(c_d - \lambda c_r + \frac{\lambda\mu pc_i}{\mu-\lambda}\right)T\right\}.$$

Therefore, if $c_d - \lambda(c_r + qc_i) > 0$, then $Q(\infty) = \infty$. Thus, there exists an optimal interval T^* ($0 < T^* < \infty$) which minimizes $C(T)$. Obviously, if $\lambda(c_d - \lambda c_r - \mu c_i)/\mu^2 > c_i$, then $c_d - \lambda(c_r + qc_i) > 0$.

4.3 Numerical Examples

We compute numerical examples for these models when $F(t) = 1 - e^{-\lambda t}$. First, we calculate an optimal interval T^* which minimizes the expected cost rate $C(T)$ in (19) when $G(x) = 1 - e^{-\mu x}$. Second, we calculate T^* which minimizes the expected cost $C(T)$ in (23) when $G(x) = p(1 - e^{-\mu x})$. The cost c_d is normalized to c_i as a unit cost, i.e., it is divided by c_i.

Table 4 gives the optimal T^* which minimizes $C(T)$ in (19) for $1/\mu = 20, 30, 40, 50, 60, 70, 80, 90, 100, \infty$, and $c_d/c_i = 100, 250, 500$ when $G(x) = 1 - e^{-\mu x}$, $1/\lambda = 3 \times 10^5$, $c_r/c_i = 10^4$. From the optimal policy, if $1/\mu \leq 2/[-\lambda + \sqrt{\lambda^2 + 4\lambda(c_d - \lambda c_r)/c_i}]$, then $T^* = \infty$, i.e., it should not check the system periodically.

Table 4 Optimal interval T^* to minimize $C(T)$ when $G(x) = 1 - e^{-\mu x}$, $1/\lambda = 3 \times 10^5$, $c_r/c_i = 10^4$

$1/\mu$	c_d/c_i		
	100	250	500
20	∞	∞	∞
30	∞	∞	68.69
40	∞	107.75	52.21
50	∞	80.79	46.71
60	194.32	71.34	43.87
70	144.89	66.33	42.12
80	126.50	63.18	40.93
90	116.34	61.02	40.07
100	109.78	59.43	39.42
∞	77.48	49.00	34.64

Table 5 Optimal interval T^* to minimize $C(T)$ when $G(x) = p(1-e^{-\mu x})$, $1/\lambda = 3\times 10^5$, $c_r/c_i = 10^4$ and $c_d/c_i = 100$

$1/\mu$	p			
	0.0	0.2	0.5	0.9
20	77.48	85.55	105.98	229.82
30	77.48	84.59	102.31	209.62
40	77.48	83.72	98.72	181.48
50	77.48	82.98	95.68	153.43
60	77.48	82.38	93.24	133.49
70	77.48	81.88	91.29	120.87
80	77.48	81.47	89.73	112.65
90	77.48	81.12	88.47	106.98
100	77.48	80.83	87.43	102.85

Table 5 shows the optimal T^* which minimizes $C(T)$ for $1/\mu = 20$–100 and $p = 0.0, 0.2, 0.5, 0.9$ when $G(x) = p(1 - e^{-\mu x})$, $1/\lambda = 3 \times 10^5$, $c_d/c_i = 100$ and $c_r/c_i = 10^4$. For example, if $1/\mu = 40$ and $p = 0.5$ then $T^* = 98.72$. This indicates that T^* increases as p increases, where note that if $p = 0$ then the system can not detect its failure by self-testing.

5 Optimal Policies for a Finite Interval

This section considers optimal maintenance and inspection policies for a unit which has to operate for a finite interval. Practically, the working times of most units are finite in actual fields. A typical example of the finite intervals is a lease term.

There have little papers treated with replacements for a finite time span. Barlow and Proschan [1] derived the optimal sequential policy for age replacement for a finite interval. Christer [30] and Ansell et al. [34] gave the asymptotic costs of age replacement for a finite interval. Visicolani [40] suggested a checking schedules with finite horizon. Nakagawa et al. [35, 36] considered the inspection model for a finite working time and gave the optimal policy by partitioning the working time into equal parts.

This section describes modified optimal policies which convert three usual models for a finite interval: (1) Periodic inspection policy and (2) sequential inspection policy.

5.1 Periodic Inspection Policy

A unit has to be operating for a finite interval $[0, S]$ and fails according to a general distribution $F(t)$. To detect failures, the unit is checked at periodic time kT ($k = 1, 2, \ldots, N$). Let c_1 be the cost for one check, c_2 be the cost per unit of time for the time elapsed between a failure and its detection at the next check, and c_3 be the replacement cost. Then, the total expected cost until failure detection or time S is

$$C(N) = \sum_{k=0}^{N-1} \int_{kT}^{(k+1)T} \{c_1(k+1) + c_2[(k+1)T - t]\} \, dF(t) + c_1 N \overline{F}(NT) + c_3$$

$$= \left(c_1 + \frac{c_2 S}{N}\right) \sum_{k=0}^{N-1} \overline{F}\left(\frac{kS}{N}\right) - c_2 \int_0^S \overline{F}(t) \, dt + c_3 \qquad (N = 1, 2, \ldots). \tag{25}$$

It is evident that

$$C(1) = c_1 + c_2 \int_0^S F(t) \, dt + c_3, \qquad C(\infty) \equiv \lim_{N \to \infty} C(N) = \infty.$$

Thus, there exists a finite number N^* ($1 \leq N^* < \infty$) that minimizes $C(N)$.

Table 6 presents the optimum number N^* and the total expected cost

$$\widetilde{C}(N^*) \equiv C(N^*) + c_2 \int_0^S \overline{F}(t) \, dt - c_3,$$

when $F(t) = 1 - e^{-\lambda t^2}$ and $S = 100$. In this case, we set the mean failure time equal to S, i.e., $\lambda S^2 = \pi/4$. The optimal N^* decreases as the check cost c_1 increases.

Table 6 Optimum number N^* and expected cost $\widetilde{C}(N^*)/c_2$ when $F(t) = 1 - e^{-\lambda t^2}$, $S = 100$ and $\lambda S^2 = \pi/4$

c_1/c_2	N^*	$\widetilde{C}(N^*)/c_2$
2	4	92.25
3	3	95.26
5	2	100.19
10	2	109.30
20	1	120.00
30	1	130.00

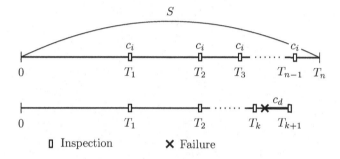

Fig. 8 Sequential inspection policy for a finite interval

5.2 Sequential Inspection Policy

Nakagawa et al. [35, 36] have considered the periodic inspection model for a finite interval $(0, S]$. In this section, we extend this model to a sequential inspection policy as follows:

1) An operating unit is checked at successive times $0 < T_1 < T_2 < \cdots < T_n$ (Figure 8), where $T_0 \equiv 0$ and $T_n \equiv S$.
2) The failure time has a general distribution $F(t)$ with finite mean $1/\lambda$, where $\overline{F}(t) \equiv 1 - F(t)$.
3) A cost c_i is the cost of one check and c_d is the cost per unit of time for the time elapsed between a failure and its detection at the next check.

Then, the total expected cost until the detection of failure or time S is

$$\mathbf{C}(n) = \sum_{k=0}^{n-1} \int_{T_k}^{T_{k+1}} [c_i(k+1) + c_d(T_{k+1} - t)] \, dF(t) + c_i n \overline{F}(S) \quad (n = 1, 2, \ldots). \tag{26}$$

Putting that $\partial \mathbf{C}/\partial T_k = 0$,

$$T_{k+1} - T_k = \frac{F(T_k) - F(T_{k-1})}{f(T_k)} - \frac{c_i}{c_d} \qquad (k = 1, 2, \ldots, n-1), \qquad (27)$$

and the resulting expected cost is

$$\mathbf{C}(n) + c_d \int_0^S \overline{F}(t)\,dt = \sum_{k=0}^{n-1}[c_i + c_d(T_{k+1} - T_k)]\overline{F}(T_k) \qquad (n = 1, 2, \ldots). \qquad (28)$$

For example, when $n = 3$, the checking times T_1 and T_2 are given by the solutions of equations

$$S - T_2 = \frac{F(T_2) - F(T_1)}{f(T_2)} - \frac{c_i}{c_d},$$

$$T_2 - T_1 = \frac{F(T_1)}{f(T_1)} - \frac{c_i}{c_d},$$

and the total expected cost is

$$\mathbf{C}(3) + c_d \int_0^S \overline{F}(t)\,dt$$
$$= c_i + c_d T_1 + [c_i + c_d(T_2 - T_1)]\overline{F}(T_1) + [c_i + c_d(S - T_2)]\overline{F}(T_2).$$

Therefore, we compute optimal T_k ($k = 1, 2, \ldots, n-1$) which satisfy (27), and substituting them into (28), we obtain the total expected cost $\mathbf{C}(n)$. Next, comparing $\mathbf{C}(n)$ for all $n \geq 1$, we can get an optimal checking number n^* and checking times T_k^* ($k = 1, 2, \ldots, n^*$).

5.3 Numerical Examples

We compute numerically optimal policies for each model. Table 7 shows optimal n^* for periodic replacement with minimal repair for $1/\lambda = 10$–100, and $c_p = 5, 6, 7, 8, 9, 10, 15, 20, 25$ when $S = 100$, $c_m = 1$ and $F(t) = 1 - e^{-\lambda t^2}$. This indicates that n^* decreases as $1/\lambda$ or c_p increases.

Table 8 shows the optimal n^* for simple replacement for $1/\lambda = 10$–100, and $c_p = 0.5, 0.6, 0.7, 0.8, 0.9, 1, 2, 4, 6$ when $S = 100$, $c_d = 1$ and $F(t) = 1 - e^{-\lambda t}$. This shows similar results with Table 7, i.e., n^* decreases as $1/\lambda$ or c_p increases.

Table 9 gives the checking time T_k ($k = 1, 2, \ldots, n$) and the expected cost $\widetilde{\mathbf{C}}(n) \equiv \mathbf{C}(n)/c_d + \int_0^S \overline{F}(t)dt$ when $S = 100$, $c_i/c_d = 2$ and $F(t) = 1 - e^{-\lambda t^2}$. In this case, we set that the mean failure time is equal to S, i.e.,

$$\int_0^\infty e^{-\lambda t^2}\,dt = \frac{1}{2}\sqrt{\frac{\pi}{\lambda}} = S.$$

Table 7 Optimal n^* for periodic replacement when $S = 100$, $c_m = 1$ and $F(t) = 1 - e^{-\lambda t^2}$

$1/\lambda$	c_p							
	5	6	7	8	9	10	15	20	25
---	---	---	---	---	---	---	---	---	---
10	14	13	12	11	11	10	8	7	6
20	10	9	8	8	7	7	6	5	4
30	8	7	7	6	6	6	5	4	4
40	7	6	6	6	5	5	4	4	3
50	6	6	5	5	5	4	4	3	3
60	6	5	5	5	4	4	3	3	3
70	5	5	5	4	4	4	3	3	2
80	5	5	4	4	4	4	3	3	2
90	5	4	4	4	4	3	3	2	2
100	4	4	4	4	3	3	3	2	2

Table 8 Optimal n^* for simple replacement when $S = 100$, $c_d = 1$ and $F(t) = 1 - e^{-\lambda t}$

$1/\lambda$	c_p							
	0.5	0.6	0.7	0.8	0.9	1	2	4	6
---	---	---	---	---	---	---	---	---	---
10	28	25	23	21	20	19	12	7	5
20	21	19	17	16	15	14	9	6	5
30	17	16	14	13	12	12	8	5	4
40	15	14	13	12	11	10	7	5	4
50	13	12	11	11	10	9	6	4	3
60	12	11	10	10	9	9	6	4	3
70	11	10	10	9	8	8	6	4	3
80	11	10	9	8	8	7	5	4	3
90	10	9	9	8	7	7	5	3	3
100	10	9	8	8	7	7	5	3	3

Comparing $\widetilde{C}(n)$ for $n = 1, 2, \ldots, 9$, the expected cost is minimum at $n = 4$. That is, the optimal checking number is $n^* = 4$ and checking times are 44.1, 66.0, 84.0, 100.

6 Conclusions

This chapter has showed the optimal inspection and maintenance policies for high reliable systems. We have suggested several useful models where systems such as digital control devices are checked by inspection. We have obtained the optimal policies analytically by making a trade-off between the loss cost of failures and the cost of inspection. Using the reliability theory,

Table 9 Checking times T_k and expected cost $\widetilde{C}(n) \equiv C(n)/c_d + \int_0^S \overline{F}(t)dt$ when $S = 100$, $c_i/c_d = 2$ and $F(t) = 1 - e^{-\lambda t^2}$

n	1	2	3	4	5	6	7	8	9
T_1	100	64,1	50.9	44.1	40.3	38.1	36.8	36.3	36.1
T_2		100	77.1	66.0	50.8	56.2	54.3	53.3	53.1
T_3			100	84.0	75.4	70.5	67.8	66.6	66.3
T_4				100	88.6	82.3	78.9	77.3	77.0
T_5					100	92.1	87.9	85.9	85.5
T_6						100	94.9	92.5	92.0
T_7							100	97.2	96.6
T_8								100	99.3
T_9									100
$\widetilde{C}(n)$	10.20	93.44	91.52	91.16	91.47	92.11	92.91	93.79	94.70

we have obtained the mean time and the expected cost rates. Further, we have discussed analytically the optimal inspection and maintenance schedules which minimize these expected costs, have given numerical examples of each model, and have evaluated them.

References

1. Barlow, R. E. and Proschan, F. (1965). *Mathematical Theory of Reliability*, New York, John Wiley & Sons.
2. Ross, S. M. (1970). *Applied Probability Models with Optimization Applications*, Holden-Day, San Francisco.
3. Osaki, S. (1992). *Applied Stochastic System Modeling*, Springer-Verlag, Berlin.
4. Ben-Daya, M. and Duffuaa, S. O. (2000). *Overview of maintenance modeling areas*, Maintenance, Modeling and Optimization, pp. 3–35.
5. Gertsbakh, I. (2000). *Reliability Theory with Applications to Preventive Maintenance*, Springer-Verlag, Berlin.
6. Nakagawa, T. (2005). *Maintenance Theory of Reliability*, Springer-Verlag, London.
7. Luss, H. and Kander, Z. (1974). *Inspection policies when duration of checking is non-negligible*, Operations Research, Vol. Q25, pp. 299–309.
8. Luss, H. (1976). *Inspection policies for a system which is inoperative during inspection periods*, AIIE Transaction, Vol. 9, pp. 189–194.
9. Wattanapanom, N. and Shaw, L. (1979). *Optimal inspection schedules for failure detection in a model where tests hasten failures*, Operations Research, Vol. 27, pp. 303–317.
10. Sengupta, B. (1980). *Inspection procedures when failure symptoms are delayed*, Operations Research, Vol. 28, pp. 768–776.

11. Weiss, G. H. (1962). *A problem in equipment maintenance*, Management Science, Vol. 8, pp. 266–277.
12. Coleman, J. J and Abrams, I. J. (1962). *Mathematical model for operational readiness*, Operations Research, Vol. 10, pp. 126–138.
13. Morey, R. C. (1967). *A criterion for the economic application of imperfect inspections*, Operations Research, Vol. 15, pp. 695–698.
14. Su, S. Y. H., Koren, I. and Malaiya, Y. K. (1978). *A continuous-parameter Markov model and detection procedures for intermittent faults*, IEEE Transactions on Computer, Vol. C-27, No. 6, pp. 567–570.
15. Koren, I. (1979). *Analysis of signal reliability measure and an evaluation procedure*, IEEE Transactions on Computer, Vol. C-28, No. 3, pp. 224–249.
16. Koren, I. and Su, S. H. Y. (1979). *Reliability analysis of N-modular redundancy systems with intermittent and permanent faults*, IEEE Transactions on Computer, Vol. C-28, No. 7, pp. 514–520.
17. Nakagawa, T. and Yasui, K. (1989). *Optimal testing-policies for intermittent faults*, IEEE Transactions on Reliability, Vol. 38, No. 5, pp. 577–580.
18. Nakagawa, T., Motoori, M. and Yasui, K. (1990). *Optimal testing policy for a computer system with intermittent faults*, Reliability Engineering and System Safety, Vol. 27, No. 2, pp. 213–218.
19. Christer, A. H. (1982). *Modelling inspection policies for building maintenance*, Journal of Operational Research Society Vol. 33, pp. 723–732.
20. Christer, A. H and Waller, W. M. (1984). *Delay time models of industrial inspection maintenance problems*, Journal of Operational Research Society, Vol. 35, pp. 401–406.
21. Christer, A. H., MacCallum, K. L., Kobbacy, K., Bolland, J. and Hessett, C. (1989). *A system model of underwater inspection operations*, Journal of Operational Research Society, Vol. 40, pp. 551–565.
22. Sim, S. H. (1984). *Availability model of periodically tested standby combustion turbine units*, IIE Transactions, Vol. 16, pp. 288–291.
23. Sim, S. H and Wang, L. (1984), *Reliability of repairable redundant systems in nuclear generating stations*, European Journal of Operational Research, Vol. 17, pp. 71–78.
24. Sim, S. H. (1985), *Unavailability analysis of periodically tested components of dormant systems*, IEEE Transactions on Reliability, Vol. R-34, No. 1, pp. 88–91.
25. Ito, K. and Nakagawa, T. (2003). *Optimal self-diagnosis policy for FADEC of gas turbine engines*, Mathematical and Computer Modelling, Vol. 38, No. 11–13, pp. 1243–1248.
26. Nakagawa, T. (2008). *Advanced Reliability Models and Maintenance Policies*, Springer-Verlag, London.
27. Hjelmgren, K., Svensson, S. and Hannius, O. (1998). *Reliability Analysis of a Single-Engine Aircraft FADEC, 1998 Proceedings Annual Reliability and Maintainability Symposium*, pp. 401–407.
28. Lala, P. K. (2001). *Self-Checking and Fault Tolerant Digital Design*, Morgan Kaufman Pub., San Francisco.
29. O'Connor, P. (2001). *Test Engineering, John Wiley & Sons*, Chichester.

30. Christer, A. H. (1978). *Refined asymptotic costs for renewal reward processes*, Journal of Operational Research Society, Vol. 29, pp. 577–583.
31. Valdez-Flores, C., and Feldman, R. M. (1989). *A survey of preventive maintenance models for stochastic deteriorating single-unit systems*, Naval Logistics Quarterly, vol. 36. pp. 419–446.
32. Osaki, S. (eds.) (2002). *Stochastic Models in Reliability and Maintenance*, Springer-Verlag. Berlin.
33. Pham, H. (eds). (2003). *Handbook of Reliability Engineering*, Springer-Verlag. London.
34. Ansell, J., Bendell, A. and Humble, S. (1984) *Age replacement under alternative cost criteria*, Management Science, Vol. 30, pp. 358–367.
35. Nakagawa, T., Yasui, K. and Sandoh, H. (2004) *Note on optimal partition problems in reliability models*, Journal of Quality in Maintenance Engineering, Vol. 10. pp. 282–287.
36. Nakagawa, T. and Mizutani, S. (2009). *A summary of maintenance policies for a finite interval*, Reliability Engineering and System Safety, Vol. 94, pp. 89–96.
37. Kaio, N. and Osaki, S. (1984). *Some remarks on optimum inspection policies*, IEEE Transactions on Reliability, Vol. R. 33, pp. 277–279.
38. Kaio, N. and Osaki, S. (1988). *Inspection policies: Comparisons and modifications*, R. I. R. O. Operations Research, Vol. 22 pp. 387–400.
39. Kaio, N. and Osaki, S. (1989). *Comparison of inspection policies*, Journal of Operations Research Society, Vol. 40, pp. 499–503.
40. Visicolani, B. (1991). *A note on checking schedules with finite horizon*, RAIRO Operations Research, Vol. 25, pp. 203–208.

PART 2
COMPUTER SYSTEMS

Chapter 4

Stochastic Analyses for Hybrid State Saving and Its Experimental Validation

MAMORU OHARA*, MASAYUKI ARAI[†] and SATOSHI FUKUMOTO[‡]

Faculty of System Design
Tokyo Metropolitan University
6-6 Asahigaoka, Hino, Tokyo 191-0065, Japan
*E-mail: *ohara@iel.sd.tmu.ac.jp; [†]m_arai@tmu.ac.jp; [‡]s-fuku@tmu.ac.jp*

1 Introduction

This chapter discusses uncoordinated checkpointing with logging for practical applications running with limited resources. We focus on emerging applications using such uncoordinated techniques, in which it is required that they can roll back on an error to recovery points other than the latest checkpoint. A typical of such applications is an optimistic parallel discrete event simulation (PDES). For example, when we find our system was intruded by malicious attackers, we should roll it back not to the latest checkpoint but to earlier time point at which it was not invaded. Another example of such applications is a kind of in distributed speculative execution, which is used in, for example, concurrency control for distributed databases [DDB], distributed constraint satisfaction problem in distributed artificial intelligence systems, and distributed event simulations [3, 5]. In such applications, a process does not wait for slower processes even if it needs information held in the slower ones for improving performance; it guesses their results and continues processing. If it fails on the speculation, *i.e.*, it made a bad guess, it rolls back and retries in the manner as that mentioned in the following section. Uncoordinated checkpointing and its crossbred with other techniques are often used in such applications.

Especially, hybrid approaches using both uncoordinated checkpointing and logging are frequently used for reducing recovery overhead [4]. In these approaches, each process saves differences between before/after states as log data on every event processing in addition to uncoordinated periodic checkpointing. Using the log data, the hybrid approaches stop the domino effects and improve performance of recovery operation. Overhead for comparing before/after states and recording the differences is usually negligible compared to that for coordinated checkpointing in most applications.

We present a discrete time model evaluating the total expected overhead for a Hybrid State Saving (HSS), which was proposed by Soliman et al. [1] and frequently used in such applications, assuming that the number of available checkpoints each process can hold is finite. Soliman et al. introduced such a hybrid technique, called Hybrid State Saving (HSS) technique, and analyzed the total expected time to execute a finite-length task, generalizing Lin's analysis [2]. In their analysis, it was assumed that each process could hold an infinite number of checkpoints. Thus, it was possible to guarantee that there would be one or more checkpoints near any recovery point. Although this assumption is convenient to derive analytically the optimal checkpoint interval, it is not realistic in many actual applications. Their analysis also assumed that a process may roll back to any time point, which might be an infinitely long time ago. However, the rollback is usually bound within some finite interval in practical applications [3,5]. The upper bound of the rollback interval is given according to characteristics of applications. We evaluate the total overhead added for HSS per event where the number of available checkpoints that each process can hold is finite. In our model, the rollback distance is also bound to some finite interval, reflecting most situations in many actual applications. Therefore, the recovery overhead for the checkpointing scheme is described by using a truncated geometric distribution as the rollback distance distribution. Although it is difficult to derive analytically the optimal checkpoint interval, which minimizes the total expected overhead, substituting other simple probabilistic distributions instead of the truncated geometric distribution enables us to do this explicitly.

This chapter discusses hybrid state saving schemes with realistic restrictions. The number of checkpoints that can be held in a process is limited by finite storage size; the rollback distance is also bound to some finite interval determined by application characteristics, independently from the number of checkpoints. Soliman et al. assumed that the rollback distance obeys a geometric distribution. We generalize it to a truncated geometric distribution and develop evaluation models for the total expected overhead that is added to the execution of every ordinary event. Using a trapezoidal distribution and a triangular distribution instead of a truncated geometric distribution for the rollback distance distribution, we create a good approximation for the evaluation of overhead. We explicitly derive an optimal checkpoint interval by using these approximations.

We also examine the effectiveness in a concrete application of the proposed stochastic model. As one of the concrete applications, we implement a parallel distributed logic circuit simulator software using the Time Warp technique, which is frequently used for optimistic parallel distributed simulation. We first introduce the idea of optimistic parallel simulation using the Time Warp technique and then describe the architecture and behavior of the simulator. Next, numerical examples obtained by the simulator for some circuits are presented. Through these numerical examples, we discuss the impact of checkpoint intervals on simulation time. These numerical examples show that we can obtain the optimal checkpoint intervals for realistic application from the analytical model. Moreover, we find that per-process optimization realized by the proposed analytical model can reduce the total simulation time of the distributed application.

This chapter is organized as follows: Section 2 outlines the Time Warp technique and Hybrid State Saving to explain the background of our evaluation models. Section 3 briefly presents a discrete time evaluation model with a limited number of checkpoints and bound rollbacks, and analytically derives the optimal checkpoint interval. Section 4 introduces the implemented logic circuit simulator and numerical examples are presented in Section 5. Finally, Section 6 summarizes this chapter.

2 Time Warp Simulation and Hybrid State Saving

Time Warp technique improves the performance of parallel discrete event simulation in an optimistic manner [3]. This technique is an example of the application of HSS, and can be used to improve the performance of computer simulations of network systems, file systems [8], functionality verification of logic circuits, *etc.*, by optimistically parallelizing calculations [5]. The Time Warp technique is sometimes used for concurrency control for distributed database systems [6].

In PDES, a series of events for a discrete event system is processed in a predefined causal order. Events in discrete time simulation and serialized operations included in database transactions are typical examples of such causally ordered events. For example, in distributed discrete event simulation (PDES), it is natural that when an event occurs at a given time and another event is generated as a consequence of the first event, the causal event must be handled before the resultant event.

To ensure in-order execution of the events in distributed systems, pessimistic approaches have traditionally been used [5]. Some techniques employ centralized approaches, in which, a central controller determines when and which process should execute an event. In other pessimistic approaches, processes synchronize with each other in order to ensure the event execution order. Obviously, in large distributed systems with centralized approaches, the load is very unequally distributed between the controller and the other processes. In addition, synchronization overhead can have a significant impact on the performance of parallel applications.

Using the Time Warp technique, parallel distributed simulation PDES is performed optimistically. That is, no synchronization for ensuring the event execution order is employed during normal processing for improving parallelism. Without synchronization, the process may sometimes handle events in incorrect order. When the process detects an error caused by wrong-ordered event execution, it returns to an earlier time point at which the process can properly handle the out-of-order events, and the process re-executes the events in the correct order. Here, we can use *checkpointing and recovery* (C/R) techniques, which are traditionally used for dependable systems. While there are several approaches for realizing distributed discrete event systems in the Time Warp technique, uncoordinated checkpointing or a hybrid with logging are usually used.

In the Time Warp technique, a process can detect the wrong order of event execution by checking timestamp value of interprocess messages. A process notifies other processes of events to be handled by them through the messages. All messages are timestamped with a kind of logical time, a single time series over a distributed system, called *virtual time*. Events carried by these messages must be handled in the timestamp order. The timestamp value denotes the time of the event to be scheduled in simulations. A message delivering an event is constructed as shown in Figure 1. Here, `id` is the identifier of the message, and `virtual_send_time` is a time value when the event included in the message was generated. Also, `virtual_receive_time` is the time to process the event. The `event` field holds application-specific event information.

Time Warp Message
id
virtual_send_time
virtual_receive_time
event (see Figure 9)

Fig. 1   Message construction of Time Warp simulation

In addition, each process has its own local clock indicating the virtual time value of the event it is currently processing. When a process receives a message, it decides whether the event delivered by the message can be processed in the correct order by comparing `virtual_receive_time` the timestamp value of the message with value of its local clock. If the local clock value is larger than `virtual_receive_time`, the timestamp value, the event delivered is already out of order.

The process detecting the out-of-order events should perform a recovery operation in order to redo the event processing in the correct order. In the recovery operation, the process detecting the error rolls back to the state just before the time indicated by the timestamp of the received out-of-order message. This point in time to which the process should return is called the *recovery point* in recovery operation.

In recovery, the failed process sends special messages called anti-messages that cancel the effects of previous messages sent to other processes after the recovery point. As mentioned above, a process has a queue holding outgoing messages in order to generate anti-messages. An anti-message has the same contents as the corresponding normal message but its identifier has a negative value with an absolute value that is equal to that of the corresponding normal value. A process receiving an anti-message searches the corresponding normal message in the incoming message queue and annihilates the pair of normal and anti-messages. If the event in the corresponding message is already processed, the receiving process also needs to return to the state before it processes the event in the message. Therefore, a phenomena similar to rollback propagation in uncoordinated checkpointing can occur in Time Warp simulation. That is, additional anti-messages may be generated in the cascaded rollback, and they may invoke rollbacks in other processes.

The time distance of rollback is naturally limited to a finite interval by the global virtual time (GVT) in the Time Warp simulation. The GVT is a kind of global clock and is defined as the earliest virtual time value in the system. That is, candidates for the GVT are the local clock values of all processes and the virtual time values of all timestamps of in-transit messages, which have already been sent but have not yet been handled. Rollbacks with the Time Warp technique are caused by out-of-date timestamped messages. Thus, a process never definitely rolls back before the GVT.

Figure 2 illustrates an example of message exchange in Time Warp simulation. The local clocks are denoted as numbers surrounded by frames in Figure 2, which is advanced after processing each event. Events are generated internally or are delivered by messages from other processes. When a process receives a message, it compares the message's timestamp with its own local clock, and if the timestamp is older (smaller) than its own local clock, the process recognizes an error and invokes the recovery operation. For example, the middle process in Figure 2, the local clock of which is initially set at 3, executes a number of internally-generated events and advances its clock. The middle process also generates resulting events, some of which should be executed by other processes. The middle process sends messages in order to deliver the events. By sending the rightmost message in Figure 2, the middle process asks the upper process to execute an event at virtual time 8. The upper process receives and unpacks the message and tries to schedule the event at 8. However, its local clock has already advanced up to 9. The upper process detects an error in event execution order at this point and therefore invokes the recovery operation.

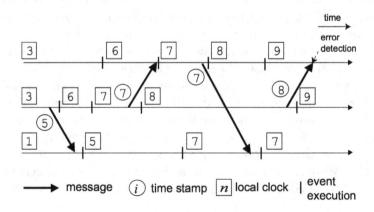

Fig. 2 Diagram of Time Warp technique

Before we discuss how to construct a model to evaluate the overhead from checkpointing, logging, and recovery operations in HSS schemes, let us outline how the operations checkpointing, logging, and recovery operations are invoked and performed. Several checkpoint/restart techniques have been used for the Time Warp simulation. Traditionally, uncoordinated checkpointing or logging was often used alone. Soliman et al. [1] proposed a hybrid technique of uncoordinated checkpointing and logging, called Hybrid State Saving (HSS), and showed that the HSS can reduce simulation time comparing the traditional techniques. HSS combines uncoordinated checkpointing with logging. In HSS, each process creates checkpoints independently every time a constant number of events are executed. No synchronization for checkpointing between processes is done to improve performance during normal processing. In addition to the periodic checkpointing, each process executes logging of differences in its state after every event execution. The logging in HSS saves the differences bi-directionally. This can help the recovery operation as mentioned below.

In this chapter, we discuss applications such as distributed speculation, in which recovery points are given probabilistically, *i.e.*, processes do not necessarily roll back to the latest checkpoints. Generally, a process involved in distributed computation often requires information maintained in other processes. In conventional approaches, the information-consumer process must wait for the information-maintainer process to generate required information if progress of task in the maintainer is slower than that of the consumer. It can be a bottleneck of system performance. In contrast, in distributed speculation, the consumer does not wait for the slower maintainer. Instead, it makes some assumptions about the required information and continues processing with the assumptions. After a while, the maintainer will generate the information and notify it to the consumer. Since the consumer will not have to roll back unless the notified (true) information would be inconsistent with the assumption, it is expected that the performance of the system is improved. However, if the notified information and the assumption would be inconsistent, the consumer process will roll back to the time point at which the wrong assumption was constructed and rerun with the true information. That is, processes in such applications can roll back by failures in the speculation in addition to process crashes.

Figures 3 and 4 show the recovery operation of HSS where the recovery point is in between two successive checkpoints. In both figures, the process detecting an error loads the closest checkpoint to the recovery point in the first phase of the recovery operation. It loads the checkpoint just before the

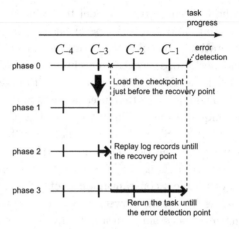

Fig. 3 Forward recovery operation of HSS with replaying log records

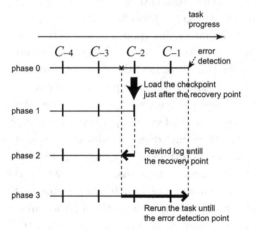

Fig. 4 Backward recovery operation of HSS with rewinding log records

recovery point in Figure 3 and one just after the recovery point in Figure 4. It then replays forward or rewinds backward the log records until it reaches the recovery point of the consistent state (Phase 2) and reruns the task up to the error detection point (Phase 3). HSS can efficiently roll back to the recovery point which is far from the checkpoint just before the recovery point by using the bi-directional log as shown in Figure 4.

The process can roll back to the recovery point with less overhead by using the log from the closest checkpoint.

Soliman et al. [1] presented an evaluation model assuming this type of recovery from both sides of the recovery point. They analyzed the total expected time to execute a finite-length task. We present a concrete example of such schemes in the Time Warp technique in Section 4. We evaluate the total overhead added for HSS on every ordinary event. In addition, we build two new constraints reflecting the characteristics of Time Warp simulation in our stochastic model, which were not considered in Soliman's model; the number of available checkpoints that each process can hold is finite, and the rollback distance is also bound to some finite interval, *i.e.*, the GVT.

3 Analytical Model

In the previous study [7], we quantitatively evaluated the overhead of HSS assuming that a process can only hold a limited number of checkpoints and construct a model that determines the optimal checkpoint interval. This section briefly outlines our stochastic model for HSS.

Figure 5 has examples of checkpointing executions when there are three available checkpoints. The vertical lines denote checkpoints and possible recovery points are marked with a ×. In Figure 5(a), the checkpoint interval seems to be too short. As rollback from checkpoint C_{-3} will be applied to return to almost all recovery points, the significance of checkpoints C_{-2} and C_{-1} is small. Conversely, the checkpoint interval in Figure 5(b) seems to be too long. Therefore, as rollback from C_{-1} or the error detection point, or rollforward from C_{-2} or C_{-1} will be used in most recoveries in the hybrid state saving, the effect of C_{-3} is not large. Intuitively, the execution example in Figure 5(c) would be the most effective. Of course, in order to examine the optimal checkpoint interval, we have to take account of the checkpointing overhead in normal operation as well as the recovery overhead.

Defining T as the constant checkpoint interval, we estimate the total expected overhead $H(T)$ added to every ordinary event execution, for which we aim to minimize. Let us assume the number M of effective checkpoints to be finite. This is naturally determined by size of the storage of the process.

(a) Example with too short checkpoint intervals

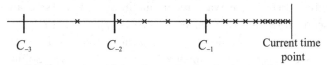

(b) Example with too long checkpoint intervals

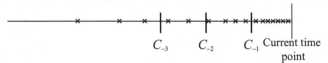

(c) Example with suitable checkpoint intervals

× : Recovery Points C_{-3}, C_{-2}, C_{-1} : Checkpoints

Fig. 5   Checkpointing execution for process with three available checkpoints

In this model, X expresses the time interval between the error detection point and the recovery point. As previously mentioned, X is bound to some finite interval in realistic applications. The probability function $f(x)$ for X is distributed in the interval $[0, L]$, where L is the rollback bound determined by each application's characteristic such as GVT in the Time Warp technique mentioned in the next section. Hence, the cumulative distribution function

$$F(x) \equiv \Pr\{X \leq x\} = \sum_{y=0}^{x} \Pr\{X = y\}$$

$$= \sum_{y=0}^{x} f(y) \qquad (x = 0, 1, 2, \cdots, L) \qquad (1)$$

properly satisfies $F(L) = 1$.

We also define C as the overhead for single checkpoint creating and loading, and denote the average overhead of saving and restoring data items updated by an event with the logging as δ. If $R(T)$ is the expected overhead for an error to roll back to the recovery point from the current time point and λ is the number of incidents causing rollbacks to occur in a process per event, the total expected overhead $H(T)$ that is added to every ordinary event execution can be obtained by

$$H(T) = \frac{C}{T} + \delta + \lambda \left[R(T) + \left(1 + \frac{C}{T} + \delta\right) E[X] \right]. \tag{2}$$

Apparently, the checkpointing overhead appended to every ordinary event execution is C/T. Hence, the expected event re-execution overhead is estimated as $E[X]$, and the retaking checkpointing and logging overhead is estimated as $(C/T + \delta)E[X]$.

The proposed analytical model was constructed in consideration of the relations between the number of effective checkpoints M and the rollback bound L. There are two situations:

1) M is relatively large compared with L that is, processes can efficiently locate one or more checkpoints near every possible recovery point.
2) L is relatively large so that process may sometimes roll back to much earlier time points than when the oldest checkpoint was created.

Since a process creates checkpoints periodically, the former situation is expressed as

$$L \leq MT, \tag{3}$$

where Soliman's model [1] with infinite checkpoints can be also used approximately. Let us first consider the model for case 1).

Roughly speaking, a process has both checkpoints before and after the recovery point in case 1). Therefore, we assume that replaying or rewinding the log is limited to the range of $[0, T/2]$. If we assume that distance between the recovery point and its nearest checkpoint would uniformly distributes in $[0, T/2]$, then the mean amount of the distance can be regarded as $T/4$. Thus, the expected overhead for an error to roll back to the recovery point from the current time point is

$$R(T) = C + \delta \frac{T}{4}, \tag{4}$$

where δ denotes average overhead of saving and restoring data items updated by an event with the logging.

Soliman et al. [1] have assumed that the rollback distance obeys a geometric distribution in $[0, \infty]$. Generalizing on this, we assume a random variable X has a geometric distribution truncated for $[0, L]$, that is,

$$f(x) = \frac{g(x)}{\sum_{y=0}^{L} g(y)}, \tag{5}$$

where $g(x) \equiv p(1-p)^x$. The average of $f(x)$ can be obtained by

$$E[X] = \frac{q}{p}\left[1 - \frac{(L+1)pq^L}{1-q^{L+1}}\right], \qquad (6)$$

where $q \equiv 1-p$. Note that if $L \to \infty$, $f(x)$ becomes the ordinary geometric distribution.

Thus, we can obtain the optimal checkpoint interval T^* which minimizes $H(T)$ by solving

$$h(T) = H(T+1) - H(T) = 0, \qquad (7)$$

that is,

$$T^* = \frac{-1 + \sqrt{1 + \frac{16}{\delta\lambda}C\left(1 + \lambda E[X]\right)}}{2}. \qquad (8)$$

If $L \leq MT^*$, where T^* is the optimal checkpoint interval obtained above, is satisfied, it is appropriate that we assume the situation to be as in case 1) expressed by (3). Otherwise, when L is greater than MT^*, we should use case 2) for the analytical model. The border value of L, by which we decide which case is to be applied, L_{border}, satisfies

$$L_{border} = M \cdot T^*|_{L=L_{border}}. \qquad (9)$$

Note that $E[X]$ in T^* includes L_{border}th order polynomials (see (6)). Solving the nonlinear equation (9) numerically, we can obtain L_{border}.

However, it is not convenient that we need to numerically solve the above equation in order to decide which case is suitable. Thus, we introduce an additional approximation to explicitly obtain L_{border}. As mentioned later, intuitively, the geometric distribution truncated by $[0, L]$ given by (5) is similar to a uniform distribution if L is small and $f(x)$ has similar values throughout $[0, L]$. In such case, we assume a uniform distribution as the probability function for the random variable X:

$$f(x) = \frac{1}{L+1}. \qquad (x = 0, 1, 2, \cdots, L) \qquad (10)$$

Simulation results in our previous work [7] have shown that this approximation works well. Using the uniform distribution, we can obtain L_{border} explicitly as follows:

$$L_{border} = \frac{M}{2}\left[\frac{2}{\delta}CM - 1 + \sqrt{\left(1 - \frac{2}{\delta}CM\right)^2 + \frac{16C}{\delta\lambda}}\right]. \qquad (11)$$

In case 2), processes may roll back to an earlier time point than that of the oldest checkpoint they hold. As with case 1), we first derive the recovery overhead. The overhead for a single recovery operation $r(X)$ is

$$r(X) = \begin{cases} \delta\frac{T}{8} & 0 \leq X < T/4 \\ C + \delta\frac{T}{8} & T/4 \leq X < T/2 \\ C + \delta\frac{T}{4} & T/2 \leq X < (M-\frac{1}{2})T \\ C + \delta\left[X - (M-\frac{1}{2})T\right] & (M-\frac{1}{2})T \leq X \leq L, \end{cases} \quad (12)$$

where we assume the error detection point and the recovery point would be at the middle of the checkpoint interval. Thus, the expected recovery overhead $R(T)$ is

$$R(T) = \sum_{x=0}^{L} r(x)f(x). \quad (13)$$

Substituting (5) into (13),

$$R(T) = \frac{1}{1-q^{L+1}}\left(\delta\frac{T}{8} + Cq^{\frac{T}{4}} + \delta\frac{T}{8}q^{\frac{T}{2}} + \delta\left[\frac{q}{p} - \frac{T}{4}\right]q^{(M-\frac{1}{2})T}\right.$$
$$\left. + \left\{\delta\left[\left(M-\frac{1}{2}\right)T - \frac{1}{p} - L\right] - C\right\}q^{L+1}\right). \quad (14)$$

As was the case with L_{border}, it is quite difficult to derive an explicit formula for T^* by solving the nonlinear equation $h(T) = 0$ obtained with (14), (2), and (7). Therefore, we introduce additional approximations. For X, we use a trapezoidal distribution or a triangular distribution chosen according to the relation between $E[X]$ and L. Intuitively, when the expected rollback distance $E[X]$ is relatively large compared with L, $f(L)$ of the truncated geometric distribution is not close to 0. It is expected that we can achieve a good approximation by using a trapezoidal distribution in such situations. There is a special case where $f(0) \approx f(L)$, which occurs when $E[X]$ approaches $L/2$. In such situations, the trapezoidal distribution approaches the uniform distribution. Conversely, when $E[X]$ is relatively small, $f(L)$ is negligible small compared with $f(0)$, and thus, it is better to use a triangular distribution rather than a trapezoidal distribution. Assuming that the average of the trapezoidal or triangular distributions is equivalent to that of the truncated geometric distribution, we can decide which distribution can be used as approximations according to the subcases as follows:

- When $L/3 \leq E[X] < L/2$, use a trapezoidal distribution,

- When $E[X] < L/3$, use a triangular distribution.

Figures 6 and 7 present numerical examples of these approximations. As seen in Figure 6 ($p = 0.0015$, $L = 800$, and $E[X] = 321.6$), the trapezoidal distribution can well approximate the truncated geometric distribution. On the other hand, when $E[X]$ is much smaller than L, for example in Figure 7 ($p = 0.015$, $L = 500$, and $E[X] = 65.7$), the geometric distribution no longer resembles the truncated geometric distribution and the triangular distribution is the better approximation.

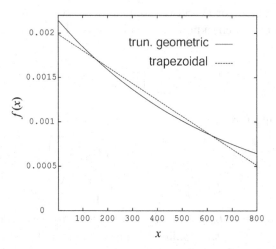

Fig. 6 Approximation of truncated geometric distribution with a trapezoidal distribution when $p = 0.0015, L = 800$, and $E[X] = 321.6$

We then use these approximations to derive explicitly the optimal checkpoint interval. Let us start with the trapezoidal distribution. The trapezoidal distribution can be expressed as
$$f(x) = b - ax \quad (x = 0, 1, 2, \cdots, L). \tag{15}$$
Obviously, value of the probability function $f(x)$ for any x is in $[0, 1]$, therefore, a and b are positive real values less than or equal to 1.

$E[X]$ is the barycentric position of the trapezoid. Thus, the coefficient a and intercept b are derived as functions of $E[X]$. Since
$$E[X] = (L+1)\left(b - \frac{aL}{2}\right) \tag{16}$$
is satisfied and the cumulative function is
$$F(L) = \frac{[f(0) + f(L)](L+1)}{2} = 1, \tag{17}$$

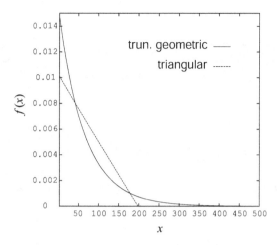

Fig. 7 Approximation of truncated geometric distribution with a triangular distribution when $p = 0.015, L = 500$, and $E[X] = 65.7$

we obtain

$$a = \frac{6(L - 2E[X])}{L(L+1)(L+2)}, \quad b = \frac{2(2L+1-3E[X])}{(L+1)(L+2)}. \quad (18)$$

Substituting (12) and (15) into (14),

$$R(T) = \frac{a\delta}{192}\left(1+24M^2-32M^3\right)T^3 + \frac{aC+\delta\left[16bM^2+4M(a-2b)-3a-2b\right]}{32}T^2$$
$$+ \left\{\frac{\delta}{12}(2M-1)\left[a-3b+3(a-2b)L+3aL^2\right] - \frac{C}{8}(a+2b)\right\}T$$
$$+C(L+1)\left(b-\frac{a}{2}L\right) + \frac{\delta L}{6}(L+1)(3b-a-2aL). \quad (19)$$

We are now ready to derive the optimal checkpoint interval by using the trapezoidal distribution. To obtain the optimal checkpoint interval T^*, we need to solve $h(T) = 0$ obtained by using (19), (2), and (7). That is, we obtain T^* by solving the following four-dimensional equation:

$$T^4 + A_3 T^3 + A_2 T^2 + A_1 T + A_0 = 0, \quad (20)$$

where

$$A_3 \equiv \frac{1}{32M^3 - 24M^2 - 1} \times$$
$$\left\{-\frac{4C}{\delta} + \frac{2}{a}\left[32aM^3 - 8(3a+4b)M^2 - 8(a-2b)M + 5a+4b\right]\right\}, \quad (21)$$

$$A_2 \equiv \frac{2}{3a\left(32M^3 - 24M^2 - 1\right)} \times$$
$$\left(\frac{3C(a+8b)}{\delta} + 64aM^3 - 48(a+3b)M^2 + 4\{-13a+30b\right.$$
$$\left. - 12L[a(L+1) - 2b]\}M + 3\{11a - 2b + 8L\left[a(L+1) - 2b\right]\}\right), \quad (22)$$

$$A_1 \equiv \frac{1}{3a\left(32M^3 - 24M^2 - 1\right)} \times$$
$$\left(\frac{6C(3a+8b)}{\delta} + 32aM^3 - 24(a+4b)M^2 + 8\{-7a+18b\right.$$
$$\left. - 12L[a(L+1) - 2b]\}M + 3\{11a - 12b + 16L\left[a(L+1) - 2b\right]\}\right), \quad (23)$$

$$A_0 \equiv \frac{64C(1 + \lambda E[X])}{a\delta\lambda(32M^3 - 24M^2 - 1)}. \quad (24)$$

Let P, Q, R, α, and β be calculated as follows:

$$P = -\frac{3A_3^2}{8} + A_2, \qquad Q = \frac{A_3^3}{8} - \frac{A_3 A_2}{2} + A_1, \quad (25)$$

$$R = -\frac{3A_3^4}{256} + \frac{A_3^2 A_2}{16} - \frac{A_3 A_1}{4} + A_0, \quad (26)$$

$$\alpha = \sqrt{2\gamma - P}, \qquad \beta = \sqrt{\gamma^2 - R}, \quad (27)$$

where γ is one of the solutions of the following three-dimensional equation:

$$Q^2 - 4(2\gamma - P)(\gamma^2 - R) = 0. \quad (28)$$

Four solutions of (20) can be obtained:

$$T^* = \left[\frac{-\alpha \pm \sqrt{\alpha^2 - 4(\gamma + \beta)}}{2} - \frac{A_3}{4}\right], \quad (29)$$

and

$$T^* = \left[\frac{\alpha \pm \sqrt{\alpha^2 - 4(\gamma - \beta)}}{2} - \frac{A_3}{4}\right]. \quad (30)$$

When there are two real-valued solutions of the four-dimensional equation, we choose the smaller positive one as the optimal checkpoint interval.

Next, we derive the optimal checkpoint interval by using the triangular distribution. The probability function of the triangular distribution is

$$f(x) = b - ax \quad (0 \leq x \leq 3E[X]). \tag{31}$$

For the triangular distribution, a and b are derived because $3E[X] = b/a$:

$$a = \frac{2}{3E[X](3E[X]+1)}, \quad b = \frac{2}{3E[X]}. \tag{32}$$

The triangular distribution gives $f(x) = 0$ for $x > 3E[X] (= b/a)$. Therefore, we should consider two subcases in the same way as in case 1). They are:

- $3E[X] \leq MT$,
- $MT < 3E[X]$.

Let us consider the former case: When $3E[X] \leq MT$, it can be regarded as that both checkpoints are near every recovery point, as in case 1). Hence, the rollback overhead for an error is $r(X) = C + \delta T/4$. Thus,

$$R(T) = \frac{b}{2}\left(1 + \frac{b}{a}\right)\left(C + \delta\frac{T}{4}\right), \tag{33}$$

where a and b are given by (32).

Then, we can derive the optimal checkpoint interval by solving the following equation using (33), (2), and (7):

$$\frac{b\delta\lambda}{8}\left(1+\frac{b}{a}\right) - \frac{C}{T(T+1)}(1+\lambda E[X]) = 0. \tag{34}$$

That is,

$$T^* = \frac{1}{2}\left[-1 + \sqrt{1 + \frac{32C(1+\lambda E[X])}{b\delta\lambda\left(1+\frac{b}{a}\right)}}\right]. \tag{35}$$

Next, we derive the border value of $E[X]$, E_{border}, which satisfies $3E_{border} = M \cdot T^*|_{E[X]=E_{border}}$ as well as L_{border} in case 1). Note that the a and b given by (32) include $E[X]$. E_{border} is

$$E_{border} = \frac{1}{18\delta\lambda}\Big\{\lambda\left[4CM^2 - 3\delta(M+2)\right]$$
$$+ \sqrt{\lambda^2\left[3\delta(M+2) - 4CM^2\right]^2 + 72\delta\lambda M(2CM - \delta\lambda)}\Big\}. \tag{36}$$

When $E[X] > E_{border}$, we should assume the latter subcase holds, i.e., the recovery point in the interval $[MT^*, 3E[X]]$ must be treated. The expected rollback overhead $R(T)$ is derived using (12), (31), and (13):

$$R(T) = \frac{a\delta}{192}\left(1 + 24M^2 - 32M^3\right)T^3$$
$$+ \frac{aC+\delta\left[16bM^2+4M(a-2b)-3a-2b\right]}{32}T^2$$
$$+ \left[\frac{\delta}{12a}(2M-1)\left(a^2-3b^2\right) - \frac{C}{8}(a+2b)\right]T$$
$$+ C\cdot\frac{b}{2a}(a+b) + \frac{b\delta}{6a^2}(a+b)(b-a). \qquad (37)$$

Therefore, we can obtain the optimal checkpoint interval using (29) and (30), letting A_3 and A_0 be those given by (22) and (24), respectively, and letting A_2 and A_1 be as follows:

$$A_2 \equiv \frac{2}{3a^2\left(32M^3 - 24M^2 - 1\right)}\left[\frac{3aC(a+8b)}{\delta} + 64a^2M^3 - 48a(a+3b)M^2\right.$$
$$\left. + 4\left(-13a^2 + 18ab + 12b^2\right)M + 3\left(11a^2 + 6ab - 8b^2\right)\right], \qquad (38)$$

$$A_1 \equiv \frac{1}{3a^2\left(32M^3 - 24M^2 - 1\right)}\left[\frac{6aC(3a+8b)}{\delta} + 32a^2M^3 - 24a(a+4b)M^2\right.$$
$$\left. + 8\left(-7a^2 + 6ab + 12b^2\right)M + 3\left(11a^2 + 4ab - 16b^2\right)\right], \qquad (39)$$

and P, Q, R, α, β and γ in T^* are the same as those in (25)–(28).

4 Implementation of A Concrete Application: A Parallel Distributed Logic Circuit Simulator

We implemented a parallel distributed simulator for logic circuits using Java. Processes in the system communicate with each other by Remote Method Invocation (RMI) employed in the Java runtime environment. Our distributed system has 16 Linux machines having AMD Athlon XP 1800+, 256Mbytes memory, and J2SE 1.5 Java runtime environments. All of the machines are connected to a single 1000Base-T LAN segment via two hubs.

In order to enable retrial of the out-of-order events in Time Warp simulation, Jefferson proposed that processes have the data structures illustrated in Figure 10. The state queue is a list of copies of the process state. The input and output queues hold incoming/outgoing messages. Generally, a process cannot forecast the timestamp values of incoming messages; therefore, the process should be able to return to arbitrary time points. To realize this, a process saves its state into the state queue before processing every event. The state queue can be regarded as a set of uncoordinated checkpoints in our checkpointing language.

In the previous section, a process usually does not save its entire state but only bi-directional logs in HSS. The process also saves its entire state on periodic uncoordinated checkpointing. Therefore, in our simulator, checkpoints and logs logically live together in the state queue.

Checkpoint and log data are saved in the main memory in each machine. In checkpointing, a process saves the signal values of all input/output ports of all gates hosted by the process. Signal values have three-valued logic, *i.e.*, the value can be 1, 0, or indeterministic value X. Because an output port of a gate is connected to input ports of other gates, saving all signal values of input/output ports is redundant. We introduced such redundancy for avoiding additional query messages in recovery for signal values of gates hosted by other processes. Log data hold both before/after signal values of a port of a gate updated by an event.

Figure 8 illustrates the data flow between processes in the simulator. One machine works as a controller and the other 16 worker machines perform simulation tasks, hosting one process per machine. The controller process performs only setup and termination of the distributed simulation and therefore is not a bottleneck in the simulation. We input a netlist file describing a circuit under simulation, an input vector list file, and a circuit division file to the controller process. The controller process divides the given netlist into 16 portions according to the information in the circuit division file. In addition, the controller process makes 16 subsets of input vectors, each of which includes only the vectors that are necessary for each portion of the divided netlists. Dividing a netlist into 16 pieces is performed offline, as mentioned below. The controller sets up workers, distributes information about the divided netlist and inputs vectors to the workers and waits for their completion. Worker processes directly exchange information during simulation and report simulation results to the controller.

This simulator supports several kinds of logic gates, including AND, OR, NOT, NAND, NOR, XOR, and D-flip flop. Logic gates can have an arbitrary number of inputs. An event expresses a signal change at a port of a gate. When an event which changes the signal value of an input port of a gate, whose output is connected to more than one ports of other gates, we generate an event per connected port, *i.e.*, an event is constructed as shown

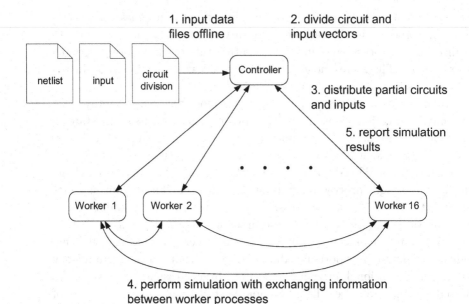

Fig. 8 Data flow between processes in the simulator

in Figure 9. This event structure is held as the **event** field in a message shown in Figure 1.

We divide a circuit into several pieces offline and make a process simulate one of them. The algorithm for dividing a circuit consisting of a set of gates and a set of signal lines is a simple depth-first algorithm starting from primary inputs. Figure 11 shows an example of dividing a small sequential logic benchmark circuit called "s27" proposed in ISCAS'89 [10] into three pieces:

LogicCircuitEvent
gate_id
port_no
new_signal_value

Fig. 9 Event structure

The specifications of machines used in the experiment are shown in Table 1:

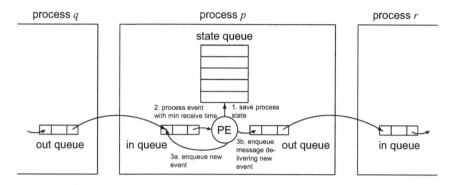

Fig. 10 Data structures of a process in a Time Warp simulation

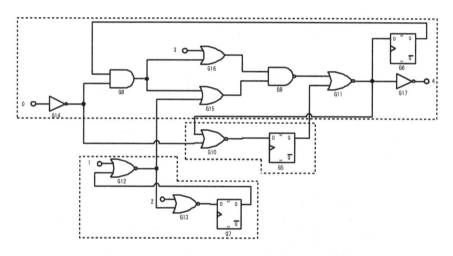

Fig. 11 An 's27' example circuit and its divisions

5 Numerical Examples

We performed simulations for some combinational and sequential benchmark circuits, which were proposed in [9] and [10] mainly for evaluating the effectiveness of testing techniques for logic circuits. In addition, we constructed NAND network circuits illustrated in Figure 12 in order to evaluate the scalability of the Time Warp simulation. This combinational circuit consists of 16 NAND gate series. Each NAND series in the circuit was sometimes connected to other series with probability r, which, in this

Table 1 Specifications of machines used in the experimental system

CPU	AMD Athlon XP 1800+
Memory	256 MBytes
LAN	1000Base-T
OS	RedHat Linux 7.2
Java	J2SE 1.5

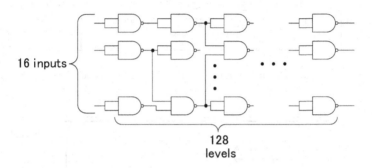

Fig. 12 Construction of the simulated NAND network

section, is called the crossing probability. Obviously, the lower the crossing probability, the greater the parallelism of the circuit. The specifications of the simulated circuits are shown in Table 2:

Table 2 Specifications of benchmark circuits

	NAND network	c6288	s13207
#inputs	16	32	31+1 (CLK)
#outputs	16	32	121
#gates	2080	2479	8773
#D flipflops	0	0	669

We periodically input 256 randomly generated 16-bit vectors for the NAND networks and 50 random vectors having widths of 32 and 31 for c6288 and s13207, respectively. For the sequential circuit s13207, a periodic clock is input and shared between all D flip-flops.

Figure 13 shows the scalability of the Time Warp simulation for the NAND networks. The values on the vertical axis indicate the relative performance compared to that of a single process. When we use only one process for simulation, the process does not need to generate checkpoints

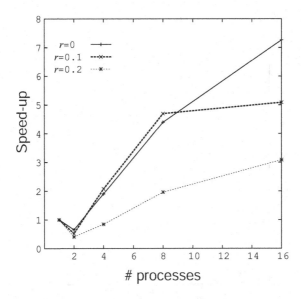

Fig. 13 Scalability of the simulator

and logs because in Time Warp simulation, rollbacks never occur without message passing. Simulations with two or more processes cause message exchanges and rollbacks, and therefore need to generate checkpoints. As shown in the figure, simulations with two processes had the worst performance. This is because the overheads for checkpointing, logging, message-passing and rollbacks negated the benefits of parallel simulation. Using more processes achieved a better performance than that obtained in single-process simulations. In addition, we could obtain greater scalability for less crossing probability r. On the other hand, lower performance was obtained by using two or more processes for s13207. We assume that this was because s13207 has several flip-flops that share a single clock input. Intuitively, shared inputs between processes easily cause rollbacks.

In the previous section, In our stochastic model, we assumed that the rollback distance would obey a truncated geometric distribution. Figure 14 shows an example rollback distance distribution for a process simulating a NAND network with $r = 0.1$. This supports our assumption. In addition, we require that the rollback distance distribution is not affected by checkpoint interval T, that is, $f(x)$ must not be a function of T. We also confirmed this assumption through the simulation. $f(x)$ seemed to depend not on T but on the construction of simulated circuits or how to divide them.

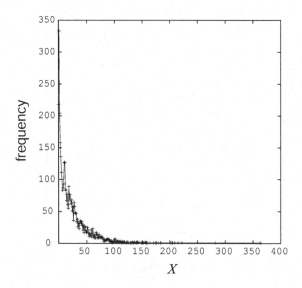

Fig. 14 Rollback distance distribution for the NAND network with $r = 0.1$

Figures 15 and 16 compare the $H(T)$ values obtained from the analyses in the foregoing section with the real overhead value measured in the simulations, for c6288 and s13207. This fits well with the simulation results. We measured the checkpointing overhead C, the logging overhead δ, the recovery frequency λ, the expected rollback distance $E[X]$, and the maximum rollback distance L in the simulations. Table 3 summarizes the averages of such parameters for the circuits:

Table 3 Measured values of parameters C, δ, λ, $E[X]$, and L

circuit	C	δ	λ	$E[X]$	L
c6288	0.67	0.061	0.24	12.0	161
s13207	7.9	0.23	4.4	24.0	476

The analytical model presented in the foregoing section describes the overhead per event imposed on a single process. We did not discuss how changes in the checkpoint interval of a process concerns the behavior of other processes. The rollback distance distribution may be influenced by the

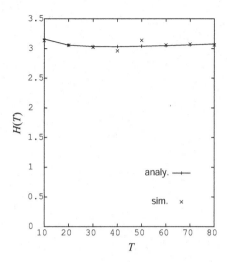

Fig. 15 $H(T)$ measured in the simulation for c6288

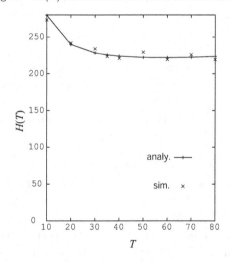

Fig. 16 $H(T)$ measured in the simulation for s13207

checkpoint interval of other processes. This means that we cannot obtain global optimal checkpoint intervals by independently adjusting T in each process. However, intuitively, it is supposed that repeatedly optimizing each process can finally reach asymptotic conversion to quasi global optimal. In order to examine the asymptotic stability, we perform simulation of s13207

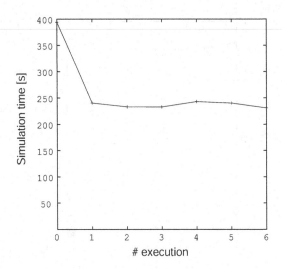

Fig. 17   Reduction of simulation time by repetitive optimization for s13207

in the following steps:

1) Run the simulator with setting $T = 3$ for all processes.
2) Identify parameters C, δ, λ, $E[X]$, and L for each process from the simulation results.
3) Derive T^* for each process from the identified parameters.
4) Run the simulator with the new T^* and repeat steps 2–4.

Figure 17 shows the reduction of the simulation time by the optimization cycles. Zero on the horizontal axis denotes the result obtained in Step 1. The figure shows that we could achieve relatively short simulation time with one-time optimization only.

Figure 18 presents changes in the optimal checkpoint intervals derived in Step 3 for each process. Each line in the figure denotes T of a process, respectively. Most processes reached a constant T with relatively small-time repetition. We expect that we can finally obtain constant T^*'s for all processes with long-time repetition. However, the simulation time might not be decreased significantly. We can obtain a good set of checkpoint intervals, which provides sufficiently short simulation time, by only one-time per-process optimization of T.

Although, we could achieve good optimization by the proposed analytical model, we can still obtain more effective checkpoint intervals by

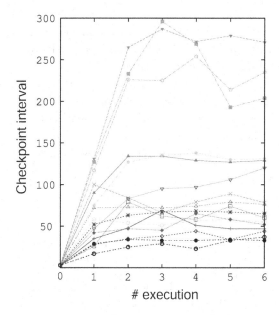

Fig. 18 Changes in T^* of processes by repetitive optimization for the s13207 circuit

constructing analytical models, which can describe the relation between rollback behavior and interaction among processes. In the future, we will attempt to construct analytical models considering such relations and providing the global optimal checkpoint interval.

6 Concluding Remarks

We implemented a parallel distributed logic circuit simulator on a PC cluster consisting of 16 PCs. Using the simulator, we simulated some combinational and sequential circuits and measured scalability, simulation time, and several parameters included in the proposed analysis model.

The simulator had good scalability for combinational circuits, especially for NAND networks with low r. However, this advantage was lessened in simulations of sequential circuits, in which flip-flops share their clock inputs.

From the obtained simulation results, we confirmed that the assumption in the analytical models, that is, the assumption of a geometric distribution of rollback distance, was true in the real application.

We observed that the optimal checkpoint interval T^* obtained by the analytical model could provide shorter simulation time for some benchmark circuits. We also found a similar trend in the measured overhead values compared to those of the analytically derived $H(T)$.

The per-process optimization obtained by the proposed analytical model could provide a relatively short total simulation time. However, we may be able to obtain more effective checkpoint intervals by analytical models describing the relations between process interactions and the state-saving/recovering overheads. Constructing such analytical models is a subject of future study.

References

1. Soliman, H. M. and Elmaghraby, A. S. (1998). *An analytical model for hybrid checkpointing in Time Warp distributed simulation*, IEEE Trans. Parallel Distrib. Syst., **9**, 10, pp. 947–951.
2. Lin, Y. Preiss, B. Loucks, W. and Lazowska, E. (1993). *Selecting the Checkpoint Interval in Time Warp Simulation*, Proc. Workshop on Parallel and Distributed Simulation (PADS) 1993, pp. 3–10.
3. Jefferson, D. (1985). *Virtual time*, ACM Trans. Prog. Lang. Syst., **7**, 3, pp. 404–425.
4. Elnozahy, E. N. Alvisi, L. Wang, Y. and Johnson, D. B. (2002). *A Survey of Rollback-Recovery Protocols in Message-Passing Systems*, ACM Computing Surveys, **34**, 3, pp. 375–408.
5. Fujimoto, R. (1990). *Parallel discrete event simulation*, Commun. ACM, **33**, 10, pp. 30–53.
6. Miller, J. (1992). *Simulation of Database Transaction Management Protocols: Hybrids and Variants of Time Warp*, Proc. 24th Conf. Winter Simulation, pp. 1232–1241.
7. Ohara, M. Suzuki, R. Arai, M. Fukumoto, S. and Iwasaki, K. (2006). *Analytical model on Hybrid State Saving with a limited number of checkpoints and bound rollbacks*, IEICE Trans. Fund., **E89-A**, 9, pp. 2386–2395.
8. Dohi, D., Nomura, K., Kaio, N. and Osaki, S. (2000). *A simulation study to analyze unreliable file systems with checkpointing and rollback recovery*, IEICE Trans. Fund., **E83-A**, 5, pp. 804–811.
9. Brglez, F. and Fujiwara, H. (1985). *A neutral netlist of 10 combinational benchmark circuits and target translator in Fortran*, Proc. Int'l Sympo. Circuits and Systems 1985, pp. 663–698.
10. Brglez, F. Bryan, D. and Kozminski, K. (1989). *Combinational profiles of sequential benchmark circuits*, Proc. Int'l Sympo. Circuits and Systems 1989, pp. 1929–1934.

Chapter 5

Reliability Analysis of a System Connected with Networks

MITSUHIRO IMAIZUMI

College of Business Administration
Aichi Gakusen University
1 Shiotori, Ohike-cho, Toyota 471-8532, Japan
E-mail: imaizumi@gakusen.ac.jp

1 Introduction

We survey the system which is connected with networks. In such a system, system failures sometimes happen by changes in the environment or bad programming code which is downloaded from the network or illegal access such as DoS Attack, and so on. To prevent the system from its failures, we need to have the preventive maintenance policy and have the security management policy. We have considered the reliability of a microprocessor system from the view point of fault tolerance [38]. In this chapter, we pay attention to a system of connected with networks, and treat three stochastic models. Using the theory of Markov renewal processes, we derive the reliability measures such as the mean times to system failure and to completion of the process.

The theory of Markov renewal processes is used to analyze the system: Markov renewal processes were first studied by Lévy [23] and Smith [24]. Pyke [25, 26] gave a careful definition and discussions in detail. Recently, Çinlar [27, 28] surveyed many results and gave diverse applications in an extensive bibliography. In reliability theory, these processes are one of the most powerful mathematical techniques for analyzing complex systems. Nakagawa and Osaki [30] analyzed two-unit systems using a unique modification of the regeneration point techniques of Markov renewal processes.

In Section 2, we formulate a stochastic model for a microprocessor (μP) system with network processing. As a computer network technology has remarkably developed, μPs which form a data terminal equipment (DTE) in a communication network have been used in many practical fields. Recently, a new communication network combining the information processing and communication plays an important role as the infrastructure in the information society. Therefore, the demand for improvement of reliabilities and functions for devices of a communication network have greatly increased [31, 32].

In fact, a μP which is one of vital devices of a communication network often fails through some faults due to noise, changes in the environment and programming bugs. Hence, it is necessary to make the preventive maintenance for occurrences of such errors. Generally, when we consider the reliability of the system on an operational stage, we should regard the cause of error occurrences of a μP as faults of software, such as mistakes of operational control and memory access, rather than faults of hardware. That is, when errors of a μP occur, it would be effective to recover the system by the operation of reset [20].

In Section 2, the expected reset number are derived. Further, an optimal reset number, which minimizes the expected cost until a network processing is successful, is analytically discussed.

In Section 3, we formulate a stochastic model for execution process of an applet. As computer systems have been widely used, the internet has been greatly developed and rapidly spread on all over the world. Recently, an applet has been widely attracted as one of the techniques which realize practical application architecture on WWW (World Wide Web). An applet is a programming code which downloads from WWW server to a client and can execute as security management of a client [1]. Java applet is well-known as a typical one.

Since an applet is loaded from a server, system failures sometimes happen by bad code. In order to cope with this problem, several schemes have been proposed [1–5]. As one of schemes, a security checker monitors the access to system resource. We call the access to system resource *system call*. A security checker is a module which inspects system call based on security management policy which is set up by a user, and can restrict execution of a programming code as the need arises [1]. The security inspection which checks contents of system call is performed by executing an inspection program. System failures caused by bad code are prevented by this security inspection, but, it is the problem that the overhead by this inspection can not be disregarded.

In Section 4, we formulate a stochastic model for a server system with DoS attacks. A server has the function of IDS. The system which is connected with networks has a problem in the illegal access which attacks a server intentionally. In particular, DoS (Denial of Service) attack which sends a huge number of packets has been a serious problem, and the measure is hurried.

In order to cope with this problem, several schemes have been considered [8–10, 16]. As one of schemes to minimize damages by DoS attacks, IDS (Intrusion Detection System) has been widely used. IDS can detect DoS attacks by monitoring packets which flow on the network. IDS judges an abnormal condition by comparing packets which flow on the network to the pattern of DoS attacks registered in advance or by analyzing statistically.

Although the simulation about the policy for the monitoring and detection of bad code or illegal access has already introduced by [6, 7, 11, 13, 15–17], there are few formalized stochastic models.

In Section 3, the expected number of system failures is analytically derived. Further, an optimal policy which minimizes the expected cost until an applet processing is successful is discussed. In Section 4, the mean time and the expected monitoring number until a server system becomes faulty are derived. Further, an optimal policy which minimizes the expected cost is discussed.

2 Optimal Reset Number of a Microprocessor System with Network Processing

We consider the maintenance problem for improving the reliability of a μP system with network processing: After the system has made a stand-alone processing, it executes successively communication procedures of a network processing. When either μP failures or application software errors in the system have occurred, a μP is reset to the beginning of its initial state and restarts again. Most reliability evaluation models of a μP system until now assumed that both errors of a μP and failures of the data transmission occur unlimitedly [33–37]. In this model, we assume that if the reset due to errors has occurred N times intermittently, then a μP interrupts its processing and restarts again from the beginning of its initial state after a constant time. That is, if the reset has occurred frequently, the system has latent faults, and the preventive maintenance to check the operational environment and to eliminate errors.

We derive the reliability quantities such as the mean time and the expected reset number until a network processing is successful. Further, we regard the losses which are the times for the reset and the interruption of processing, and for the maintenance to restart the system as expected costs, and discuss optimal policies which minimize them. A numerical example is finally given.

2.1 Model and Analysis

We pay attention to only a certain DTE which consists of a workstation or a personal computer and connects with some networks, and consider the problem for improving its reliability.

Suppose that errors of a μP system occur according to an exponential distribution $F(t) = 1 - e^{-\lambda t}$ with finite mean $1/\lambda$. If errors of a μP have occurred, a μP is reset to the beginning of its initial state and restarts again. It is assumed that any reset times are neglected:

1) After a μP begins to operate, it executes an initial processing immediately and a stand-alone processing.
2) The times for an initial processing and a stand-alone processing have a general distribution $V(t)$ with finite mean $1/v$ and an exponential distribution $A(t) = 1 - e^{-\alpha t}$ with finite mean $1/\alpha$, respectively.
3) After a μP completes a stand-alone processing, it begins to execute a network connection processing.

   a) A connection processing needs time according to a general distribution $B(t)$ with finite mean $1/\beta$ and fails with probability γ $(0 \leq \gamma < 1)$.
   b) If a connection processing has failed, a μP executes the same processing again after a constant time w, where $W(t) \equiv 0$ for $t < w$ and 1 for $t \geq w$.

4) After a connection processing has been successful, a μP executes a network processing. A network processing needs the time according to a general distribution $U(t)$ with finite mean $1/u$, and is successful with probability 1 if it has not failed.
5) If the Nth reset has occurred since a μP begins to operate, once it interrupts the processing, and restarts again from the beginning after a constant time μ, where $G(t) \equiv 0$ for $t < \mu$ and 1 for $t \geq \mu$.

Under the above assumptions, we define the following states of the system:

State 0: An initial processing begins.
State 1: A stand-alone processing begins.
State 2: A stand-alone processing is completed and a network connection processing begins.
State 3: A network connection processing succeeds and a network processing begins.
State F: A processing is interrupted.
State S: A network processing succeeds.

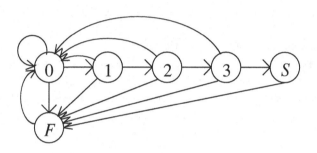

Fig. 1 Transition diagram between system states

The system states defined above form a Markov renewal process [18] where State S is an absorbing state. Transition diagram between system states is shown in Figure 1.

Let $Q_{ij}(t)$ ($i = 0, 1, 2, 3; j = 0, 1, 2, 3, S$) be one-step transition probabilities of a Markov renewal process. Then, by the similar method of [19], we have mass function $Q_{ij}(t)$ from State i at time 0 to State j at time t as follows:

$$Q_{00}(t) = \int_0^t [1 - V(u)] \, dF(u), \tag{1}$$

$$Q_{01}(t) = \int_0^t [1 - F(u)] \, dV(u), \tag{2}$$

$$Q_{10}(t) = \int_0^t [1 - A(u)] \, dF(u), \tag{3}$$

$$Q_{12}(t) = \int_0^t [1 - F(u)]\, dA(u), \tag{4}$$

$$Q_{20}(t) = \sum_{j=1}^{\infty} X^{(j-1)}(t) * \int_0^t \{[1 - B(u)] + \gamma B(u) * [1 - W(u)]\}\, dF(u), \tag{5}$$

$$Q_{23}(t) = \sum_{j=1}^{\infty} X^{(j-1)}(t) * [(1 - \gamma) \int_0^t [1 - F(u)]\, dB(u)], \tag{6}$$

$$Q_{30}(t) = \int_0^t [1 - U(u)]\, dF(u), \tag{7}$$

$$Q_{3S}(t) = \int_0^t [1 - F(u)]\, dU(u), \tag{8}$$

where

$$X(t) \equiv \gamma \int_0^t [1 - F(u)]\, dB(u) * \int_0^t [1 - F(u)]\, dW(u), \tag{9}$$

and the asterisk mark denotes the Stieltjes convolution and $\Phi^{(i)}(t)$ denotes the i-fold Stieltjes convolution of a distribution $\Phi(t)$ with itself, *i.e.*, $\Phi^{(i)}(t) \equiv \Phi^{(i-1)}(t) * \Phi(t), \Phi_1(t) * \Phi_2(t) \equiv \int_0^t \Phi_2(t-u)\, d\Phi_1(u), \Phi^{(0)}(t) \equiv 1$.

We derive the mean time ℓ_S from the beginning of system operation until a network processing is successful. Let $H_{0S}(t)$ be the first-passage time distribution from State 0 to State S. Then,

$$H_{0S}(t) = \sum_{j=1}^{N} D^{(j-1)}(t) * Z(t), \tag{10}$$

where

$$D(t) \equiv Q_{00}(t) + Q_{01}(t) * Q_{10}(t) + Q_{01}(t) * Q_{12}(t) * Q_{20}(t)$$
$$+ Q_{01}(t) * Q_{12}(t) * Q_{23}(t) * Q_{30}(t), \tag{11}$$

$$Z(t) \equiv Q_{01}(t) * Q_{12}(t) * Q_{23}(t) * Q_{3S}(t). \tag{12}$$

It is noted that $D(t)$ is the distribution function which a μP is reset by the occurrence of errors, and $Z(t)$ is the distribution function which the system moves from State 0 to State F directly without being reset. Further, the first-passage time distribution $H_{0F}(t)$ from State 0 to State F by the Nth reset of a μP is

$$H_{0F}(t) \equiv D^{(N)}(t). \tag{13}$$

Therefore, the first-passage time distribution $L_S(t)$ until a network processing is successful is given by the following renewal equation:

$$L_S(t) = H_{0S}(t) + H_{0F}(t) * G(t) * L_S(t). \tag{14}$$

Let $\Phi^*(s)$ be the Laplace-Stieltjes (LS) transform of any function $\Phi(t)$, *i.e.*, $\Phi^*(s) \equiv \int_0^\infty e^{-st} \, d\Phi(t)$. Taking the LS transforms on both sides of (14) and arranging them,

$$L_S^*(s) = \frac{H_{0S}^*(s)}{1 - H_{0F}^*(s)G^*(s)}. \tag{15}$$

Hence, the mean time ℓ_S is

$$\begin{aligned}\ell_S &\equiv \int_0^\infty t \, dL_S(t) = \lim_{s \to 0} -\frac{dL_S^*(s)}{ds} \\ &= -\frac{Z^{*\prime}(0) + D^{*\prime}(0)}{1 - D^*(0)} + \frac{\mu D^*(0)^N}{1 - D^*(0)^N},\end{aligned} \tag{16}$$

where $\Phi'(s)$ is the differential function of $\Phi(s)$, *i.e.*, $\Phi'(s) \equiv d\Phi(s)/ds$. From (16), it is noted that ℓ_S is strictly decreasing in N and is minimized when $N = \infty$.

Next, we derive the mean reset number M_R from the start of system operation or the restart by the reset until a network processing is successful. Let $M_R(t)$ be the expected reset number until a network processing is successful in an interval $(0, t]$. Then,

$$M_R(t) = \sum_{j=1}^{N-1} j D^{(j)}(t) * Z(t). \tag{17}$$

Thus, the mean reset number is given by

$$\begin{aligned}M_R &\equiv \lim_{t \to \infty} M_R(t) = \lim_{s \to 0} \sum_{j=1}^{N-1} j [D^*(s)]^j Z^*(s) \\ &= \frac{D^*(0)}{1 - D^*(0)}[1 - ND^*(0)^{N-1} + (N-1)D^*(0)^N],\end{aligned} \tag{18}$$

where it is noted that $Z^*(0) = 1 - D^*(0)$.

Further, let $M_F(t)$ be the distribution of the expected interruption number of processings from the start of system operation until a network processing is successful. Then, we have the following renewal equation:

$$M_F(t) = H_{0F}(t) * [1 + G(t) * M_F(t)]. \tag{19}$$

Similarly, the expected interruption number M_F until a network processing is successful is

$$M_F = \frac{D^*(0)^N}{1 - D^*(0)^N}. \tag{20}$$

2.2 Optimal Policies

We obtain two objective functions which are the total expected cost $C_1(N)$ and the expected cost rate $C_2(N)$ until a network processing is successful, and discuss the optimal policies which minimize them.

(1) Policy 1

Let c_1 be the cost for the reset and c_2 be the cost for an interruption of processing. Then, we define the total expected cost $C_1(N)$ until a network processing is successful as the following equation:

$$C_1(N) \equiv c_1 M_R + c_2 M_F$$
$$= c_1 \left[\frac{D(1-D^N)}{1-D} - ND^N \right] + \frac{c_2 D^N}{1-D^N} \quad (N=1,2,\cdots), \quad (21)$$

where $D \equiv D^*(0)$ is the probability that a μP is reset.

We seek an optimal number N_1^* which minimizes $C_1(N)$. From the inequality $C_1(N+1) - C_1(N) \geq 0$,

$$N(1-D^{N+1})(1-D^N) \geq \frac{c_2}{c_1}. \quad (22)$$

Denoting the left-hand side of (22) by $L_1(N)$,

$$L_1(1) = (1-D)(1-D^2), \qquad L_1(\infty) = \infty. \quad (23)$$

Hence, $L_1(N)$ is strictly increasing in N from $L_1(1)$ to ∞. Thus, we have the following optimal policy:

(i) If $L_1(1) < c_2/c_1$, then there exists a finite and unique minimum $N_1^* (> 1)$ which satisfies (22).
(ii) If $L_1(1) \geq c_2/c_1$, then $N_1^* = 1$ and the total expected cost is $C_1(1) = (c_2 D)/(1-D)$.

In this model, c_1 is the cost for the increase of system resources such as spaces of memory and times by the reset, and c_2 is the cost for the increase of system resources by the preventive maintenance to eliminate the cause of errors. It could be generally estimated that c_2 is greater than c_1, i.e., $c_2 \geq c_1$. Thus, we have $L_1(1) < c_2/c_1$, and hence, $N_1^* > 1$. Further, it is easily shown that N^* increases with c_2/c_1.

(2) Policy 2

In the Policy 1, we have adopted the total expected cost as an objective function. However, it would be more practical to introduce the measure of the time until a network processing is successful. Next, we consider an optimal policy which minimizes the expected cost rate until a network processing is successful. That is, from (16) and (21), we define the expected cost rate $C_2(N)$ as the following equation:

$$C_2(N) \equiv \frac{C_1(N)}{\ell_S}$$

$$= \frac{c_1 \sum_{j=1}^{N-1} jD^j(1-D) - \frac{A}{\mu}c_2}{A + \frac{\mu D^N}{1-D^N}} + \frac{c_2}{\mu} \quad (N = 1, 2, \cdots), \quad (24)$$

where

$$A \equiv -\frac{Z^*(0) + D^*(0)}{1-D} > 0.$$

We seek an optimal number N_2^* which minimizes $C_2(N)$. From the inequality $C_2(N+1) - C_2(N) \geq 0$,

$$N(1-D^N)(1-D^{N+1}) + \frac{\mu}{A}\left[ND^N(1-D^{N+1}) + (1-D)\sum_{j=1}^{N-1} jD^j\right] \geq \frac{c_2}{c_1}. \quad (25)$$

Denoting the left-hand side of (25) by $L_2(N)$,

$$L_2(1) = (1-D^2)(1-D+\frac{\mu}{A}D), \qquad L_2(\infty) = \infty. \quad (26)$$

Putting the second term on the bracket of the left side of (25) by

$$L_3(N) \equiv ND^N(1-D^{N+1}) + (1-D)\sum_{j=1}^{N-1} jD^j, \quad (27)$$

we have

$$L_3(1) = (1-D^2)D, \quad (28)$$

$$L_3(N+1) - L_3(N) = D^{N+1}[1 - D^{N+2} + ND^N(1-D^2)] > 0. \quad (29)$$

Hence, $L_3(N)$ is strictly increasing in N. Further, since $N(1-D^N)(1-D^{N+1})$ in (25) is also strictly increasing in N, $L_2(N)$ is also strictly increasing in N from $L_2(1)$ to ∞. Thus, we have the following optimal policy:

(i) If $L_2(1) < c_2/c_1$, then there exists a finite and unique minimum N_2^* (> 1) which satisfies (25)

(ii) If $L_2(1) \geq c_2/c_1$, then $N_2^* = 1$, and the resulting cost rate is

$$C_2(1) = \frac{c_2 D}{A(1-D) + \mu D}. \qquad (30)$$

Further, we compare Policy 2 with Policy 1. Since from (22) and (25),

$$L_2(N) - L_1(N) = \frac{\mu}{A}\left[ND^N(1 - D^{N+1}) + (1-D)\sum_{j=1}^{N-1} jD^j\right] > 0$$

$$(N = 1, 2, \cdots), \qquad (31)$$

$N_1^* \geq N_2^*$. This means that when the number N of reset is small, the mean time until a network processing is successful is large, since ℓ_S strictly decreases in N. Thus, it would be better to adopt Policy 2 where N is small when we consider only the cost of the system on the whole. On the other hand, if we want a processing time to be small, we should adopt Policy 1.

2.3 Numerical Example

We compute numerically the optimal number N_2^* which minimizes $C_2(N)$ for Policy 2. Suppose that the mean initial processing time $1/v$ of μP is a unit of time and the mean time to error occurrences is $(1/\lambda)/(1/v) = 30 \sim 60$. Further, the mean stand-alone processing time is $(1/\alpha)/(1/v) = 5 \sim 20$, the mean network connection processing time is $(1/\beta)/(1/v) = 1$, the mean waiting time when a network connection processing fails is $w/(1/v) = 1 \sim 4$, the mean network processing time is $(1/u)/(1/v) = 10$, the mean maintenance time after an interruption of processing is $(1/\mu)/(1/v) = 10$, the probability that a network connection processing fails is $\gamma = 0.2, 0.4, 0.6$, the cost c_1 for the reset is a unit of cost, and the cost rate of an interruption of processing is $c_2/c_1 = 1 \sim 3$.

Table 1 gives the optimal reset number N_2^* which minimizes the expected cost $C_2(N)$. For example, when $(1/\lambda)/(1/v) = 60$, $wv = 2$, $\gamma = 0.2$, $(1/\alpha)/(1/v) = 10$ and $c_2/c_1 = 2$, $N_2^* = 3$. This indicates that N_2^* decreases with $(1/\lambda)/(1/v)$, however, increases with wv, γ, $(1/\alpha)/(1/v)$ and c_2/c_1. This can be interpreted that when the cost for an interruption of processing is large, N_2^* increases with c_2/c_1, and so, the processing should not be excessively interrupted. That is, we should keep on executing the processing as long as possible by the reset. Table 1 also shows that N_2^* depends on

Table 1 Optimal reset number N_2^* to minimize $C_2(N)$

$(1/\lambda)$ /$(1/v)$	wv	γ	$(1/\alpha)/(1/v)$ 5 c_2/c_1			$(1/\alpha)/(1/v)$ 10 c_2/c_1			$(1/\alpha)/(1/v)$ 20 c_2/c_1		
			1	2	3	1	2	3	1	2	3
30	1	0.2	2	3	4	2	3	4	2	3	4
		0.4	2	3	4	2	3	4	2	3	4
		0.6	2	3	4	2	3	4	3	4	4
	2	0.2	2	3	4	2	3	4	2	3	4
		0.4	2	3	4	2	3	4	3	3	4
		0.6	2	3	4	2	3	4	3	4	4
	4	0.2	2	3	4	2	3	4	2	3	4
		0.4	2	3	4	2	3	4	3	4	4
		0.6	2	3	4	2	3	4	3	4	5
60	1	0.2	2	3	4	2	3	4	2	3	4
		0.4	2	3	4	2	3	4	2	3	4
		0.6	2	3	4	2	3	4	2	3	4
	2	0.2	2	3	4	2	3	4	2	3	4
		0.4	2	3	4	2	3	4	2	3	4
		0.6	2	3	4	2	3	4	2	3	4
	4	0.2	2	3	4	2	3	4	2	3	4
		0.4	2	3	4	2	3	4	2	3	4
		0.6	2	3	4	2	3	4	2	3	4

each parameter when $(1/\lambda)/(1/v)$ is small, *i.e.*, when errors of a μP occur frequently, however, N_2^* depends little on wv, γ and $(1/\alpha)/(1/v)$ when $(1/\lambda)/(1/v) \geq 60$, and in this case, N_2^* is almost determined by c_2/c_1.

3 Reliability Analysis for an Applet Execution Process

We formulate a stochastic model for execution process of an applet: When an applet is loaded to a client from a server, JIT (Just-In-Time) compiler is downloaded to a client. Thereby, an applet is changed into an object code and is executed. If a server does not reply because of busy condition, a client waits for loading of an applet.

System call which is executed from an object code is monitored by a security checker. A security checker judges whether its system call needs inspection or not for every system call. The inspection is performed with some probability, and in this case, an inspection program is executed. System call is allowed with some probability by a security checker. In this case, the processing of system call restarts and the control is moved to the operating system. Oppositely, if system call is not allowed, it is returned error code to an applet, and an execution processing of an applet interrupts. After a client eliminates bad code, an execution processing of an applet restarts. If the inspection is not performed by a security checker, system failures occur by bad access to a client system. The expected number of system failures is analytically derived. Further, an optimal policy which minimizes the expected cost until an applet processing is successful is analytically discussed. Finally, a numerical example is given.

3.1 Model and Analysis

Figure 2 draws the outline of an execution system with applet [1]:

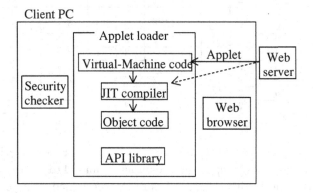

Fig. 2 Outline of an execution system with applet

1) When an applet is loaded to a client from a server, JIT(Just-In-Time) compiler is downloaded to a client. Thereby, an applet is changed into an object code and is executed. The time for loading of an applet has a general distribution $D(t)$ with finite mean d. If a server does not reply because of busy condition, a client waits for loading of an applet for a constant time w, where $W(t) \equiv 0$ for $t < w$ and 1 for $t \geq w$. The time for execution processing of an applet has an exponential distribution $A(t) = (1 - e^{-at})$ $(0 < a < \infty)$, a server becomes a busy condition according to an exponential distribution $(1 - e^{-\alpha t})$ $(0 < \alpha < \infty)$, and becomes a normal condition according to an exponential distribution $(1 - e^{-\beta t})$ $(0 < \beta < \infty)$, where a normal condition is not a busy condition. Then,

we define the following states of a server:

State 0: A server is in a normal condition.
State 1: A server is in a busy condition.

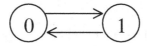

Fig. 3   Transition diagram of a server

The system states defined above form a two-state Markov process [18]. Transition diagram between states of a server is shown in Figure 3. We have the following probabilities under the initial condition that $P_{00}(0) = P_{11}(0) = 1, P_{01}(0) = P_{10}(0) = 0$:

$$P_{00}(t) \equiv \frac{\beta}{\alpha+\beta} + \frac{\alpha}{\alpha+\beta}e^{-(\alpha+\beta)t}, \tag{32}$$

$$P_{11}(t) \equiv \frac{\alpha}{\alpha+\beta} + \frac{\beta}{\alpha+\beta}e^{-(\alpha+\beta)t}, \tag{33}$$

$$P_{01}(t) = 1 - P_{00}(t), \qquad P_{10}(t) = 1 - P_{11}(t). \tag{34}$$

2) System call which is executed from an object code is monitored by a security checker. System call occurs according to an exponential distribution $B(t) = (1 - e^{-bt})$ $(0 < b < \infty)$.

3) A security checker judges whether its system call needs inspection or not for every system call. The inspection is performed with probability p $(0 \leq p \leq 1)$, i.e., a client determines the policy in advance whether a security inspection is performed or not for system call from an applet. In this case, an inspection program is executed. The time for processing of an inspection program has a general distribution $U(t)$ with finite mean u.

4) System call is allowed with probability q $(0 \leq q \leq 1)$ by a security checker. In this case, the processing of system call restarts and the control is moved to the operating system. The time for processing of system call has a general distribution $V(t)$ with finite mean v. Oppositely, if system call is not allowed with probability $1 - q$, it is returned an error code to an applet, and an execution processing of an applet interrupts. After a client eliminates a bad code, an execution processing of an applet restarts. The time to restart has a general distribution $G(t)$ with finite mean μ.

5) If the inspection is not performed by a security checker, system failure occurs by bad access to client system according to an exponential distribution $F(t) = (1 - e^{-\lambda t})$ $(0 < \lambda < \infty)$. In this case, it is maintained and restarts again from the beginning. The time from system failure occurrence to restart has a general distribution $Z(t)$ with finite mean z.

Under the above assumptions, we define the following states of the system:

State 2: Loading of an applet starts.
State 3: Execution processing of an applet starts.
State 4: System call occurs.
State 5: Execution processing of an applet interrupts and elimination of bad code starts.
State F: System failure occurs.
State S: Execution processing of an applet succeeds.
State E: Waiting for loading of an applet starts.

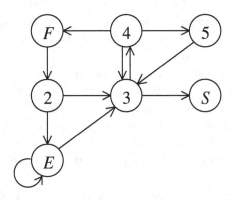

Fig. 4   Transition diagram between system states

The system states defined above form a Markov renewal process, where S is an absorbing state. Transition diagram between system states is shown in Figure 4.

The LS transforms of one-step transition probabilities $Q_{ij}(t)(i = 2, 3, 4, 5, E, F; j = 2, 3, 4, 5, E, F, S)$ of a Markov renewal process are given by the following equations:

$$Q^*_{23}(s) = \int_0^\infty e^{-st} P_{00}(t)\,dD(t), \tag{35}$$

$$Q^*_{2E}(s) = \int_0^\infty e^{-st} P_{01}(t)\,dD(t), \tag{36}$$

$$Q^*_{EE}(s) = \int_0^\infty e^{-st} P_{11}(t)\,dW(t), \tag{37}$$

$$Q^*_{E3}(s) = \int_0^\infty e^{-st} P_{10}(t)\,dW(t), \tag{38}$$

$$Q^*_{34}(s) = \frac{b}{s+a+b}, \tag{39}$$

$$Q^*_{3S}(s) = \frac{a}{s+a+b}, \tag{40}$$

$$Q^*_{43}(s) = pqU^*(s)V^*(s) + (1-p)V^*(s+\lambda), \tag{41}$$

$$Q^*_{45}(s) = p(1-q)U^*(s), \tag{42}$$

$$Q^*_{4F}(s) = (1-p)\frac{\lambda}{s+\lambda}[1 - V^*(s+\lambda)], \tag{43}$$

$$Q^*_{53}(s) = G^*(s), \tag{44}$$

$$Q^*_{F2}(s) = Z^*(s). \tag{45}$$

First, we derive the mean time from the beginning of loading of an applet to an applet processing is successful. Let $H_{3S}(t)$ and $H_{3F}(t)$ be the first-passage time distributions until from the beginning of execution processing of an applet to an applet processing is successful or system failure occurs, respectively. Then,

$$H_{3S}(t) = \sum_{i=1}^\infty \{Q_{34}(t) * [Q_{43}(t) + Q_{45}(t) * Q_{53}(t)]\}^{(i-1)} * Q_{3S}(t), \tag{46}$$

$$H_{3F}(t) = \sum_{i=1}^\infty \{Q_{34}(t) * [Q_{43}(t) + Q_{45}(t) * Q_{53}(t)]\}^{(i-1)} * Q_{34}(t) * Q_{4F}(t). \tag{47}$$

Let $H_{2S}(t)$ be the first-passage time distribution from the beginning of loading of an applet to an applet processing is successful. Then,

$$H_{2S}(t) = \sum_{i=1}^\infty [H_{23}(t) * H_{3F}(t) * H_{F2}(t)]^{(i-1)} * H_{23}(t) * H_{3S}(t), \tag{48}$$

where

$$H_{23}(t) \equiv Q_{23}(t) + Q_{2E}(t) * \sum_{i=1}^\infty Q_{EE}^{(i-1)}(t) * Q_{E3}(t). \tag{49}$$

Hence, forming the LS transform of $H_{2S}(t)$,

$$H^*_{2S}(s) \equiv \int_0^\infty e^{-st}\,dH_{2S}(t)$$

$$= \frac{H^*_{23}(s)Q^*_{3S}(s)}{1-Q^*_{34}(s)Q^*_{4F}(s)Q^*_{F2}(s)H^*_{23}(s)-Q^*_{34}(s)Q^*_{45}(s)Q^*_{53}(s)-Q^*_{34}(s)Q^*_{43}(s)}. \quad (50)$$

Thus, the mean time ℓ_{0S} from the beginning of loading of applet to an applet processing is successful is

$$\ell_{2S} \equiv \lim_{s\to 0} -\frac{dH^*_{0S}(s)}{ds}$$

$$= \frac{w\int_0^\infty P_{01}(t)\,dD(t) + dP_{10}(w)}{1-P_{11}(w)} + \frac{1}{a} + \frac{b}{a}\left\{(1-p)[1-V^*(\lambda)]\left[\frac{1}{\lambda}+z\right.\right.$$

$$\left.\left.+\frac{w\int_0^\infty P_{01}(t)\,dD(t) + dP_{10}(w)}{1-P_{11}(w)}\right] + p[u+qv+(1-q)\mu]\right\}. \quad (51)$$

Next, we derive the expected number of system failures M_F until an applet processing is successful. The LS transform $M^*_F(s)$ of the distribution $M_F(t)$ of system failure number is

$$M^*_F(s) = \left\{Q^*_{23}(s) + Q^*_{2E}(s)\sum_{i=1}^\infty [Q^*_{EE}(s)]^{i-1}Q^*_{E3}(s)\right\}$$

$$\times \sum_{j=1}^\infty \{Q^*_{34}(s)[Q^*_{43}(s) + Q^*_{45}(s)Q^*_{53}(s)]\}^{j-1}$$

$$\times Q^*_{34}(s)Q^*_{4F}(s)[1+Q^*_{F2}(s)M^*_F(s)]. \quad (52)$$

Thus, the expected number of system failures is given by

$$M_F \equiv \lim_{s\to 0} M^*_F(s) = \frac{a}{b}(1-p)[1-V^*(\lambda)]. \quad (53)$$

Moreover, we derive the expected number of interruptions M_5 during the period from the beginning of execution processing of an applet to the time when an applet processing is successful or system failure occurs. The LS transform $M^*_5(s)$ of $M_5(t)$ is given by the following equation:

$$M^*_5(s) = \sum_{i=1}^\infty [Q^*_{34}(s)Q^*_{43}(s)]^{i-1}Q^*_{34}(s)Q^*_{45}(s)[1+Q^*_{53}(s)M^*_5(s)]. \quad (54)$$

Thus, the expected number of interruptions is

$$M_5 \equiv \lim_{s\to 0} M^*_5(s) = \frac{bp(1-q)}{a+b(1-p)[1-V^*(\lambda)]}. \quad (55)$$

3.2 Optimal Policy

We obtain the expected cost and discuss an optimal policy which minimizes it: Let c_1 be the cost for system failure, $c_2 (< c_1)$ be the cost for interruption of processing and $c_3 (< c_2)$ be the cost for system operation. Then, we define the expected cost $C(p)$ until an applet processing is successful as

$$C(p) \equiv c_1 M_F + c_2 M_5 + c_3$$
$$= c_1 \frac{b}{a}(1-p)[1-V^*(\lambda)] + c_2 \frac{bp(1-q)}{a + b(1-p)[1-V^*(\lambda)]} + c_3. \quad (56)$$

We seek an optimal probability p^* of security inspection which minimizes $C(p)$. Differentiating (56) with respect to p and setting it equal to zero,

$$\frac{a(1-q)\{a+b[1-V^*(\lambda)]\}}{[1-V^*(\lambda)]\{a+b(1-p)[1-V^*(\lambda)]\}^2} = \frac{c_1}{c_2}. \quad (57)$$

Denoting the left-hand side of (57) by $L(p)$,

$$L'(p) = \frac{2ab(1-q)\{a+b[1-V^*(\lambda)]\}}{\{a+b(1-p)[1-V^*(\lambda)]\}^3} > 0, \quad (58)$$

$$L(0) = \frac{a(1-q)}{[1-V^*(\lambda)]\{a+b[1-V^*(\lambda)]\}}, \quad (59)$$

$$L(1) = \frac{(1-q)\{a+b[1-V^*(\lambda)]\}}{a[1-V^*(\lambda)]}. \quad (60)$$

Hence, $L(p)$ is strictly increasing in p from $L(0)$ to $L(1)$. Thus, we can characterize the optimal policy:

(i) If $L(0) < c_1/c_2 < L(1)$, i.e.,

$$\frac{a(1-q)\{a+b[1-V^*(\lambda)]\}}{1-V^*(\lambda)} < \frac{c_1}{c_2} < \frac{(1-q)\{a+b[1-V^*(\lambda)]\}}{a[1-V^*(\lambda)]},$$

then there exists a finite and unique optimal probability p^* $(0 < p^* < 1)$ of security inspection which satisfies (57).

(ii) If $L(0) \geq c_1/c_2$, i.e.,

$$\frac{c_1}{c_2} \leq \frac{a(1-q)\{a+b[1-V^*(\lambda)]\}}{1-V^*(\lambda)},$$

then $p^* = 0$. It is optimal to do no inspection of system call.

(iii) If $L(1) \leq c_1/c_2$, i.e.,

$$\frac{c_1}{c_2} \geq \frac{(1-q)\{a+b[1-V^*(\lambda)]\}}{a[1-V^*(\lambda)]},$$

then $p^* = 1$. It is optimal to inspect all system calls.

Table 2 Optimal probability p^* of security inspection to minimize $C(p)$

$(1/\lambda)/v$	q	c_1/c_2				
		100	200	300	400	500
300	0.5	0.51	0.75	0.86	0.92	1
	0.6	0.6	0.82	0.91	0.97	1
	0.7	0.7	0.89	1	1	1
	0.8	0.82	0.97	1	1	1
	0.9	0.97	1	1	1	1
600	0.5	0	0.37	0.61	0.75	0.85
	0.6	0.03	0.51	0.72	0.85	0.94
	0.7	0.25	0.67	0.85	0.96	1
	0.8	0.51	0.85	1	1	1
	0.9	0.85	1	1	1	1
1200	0.5	0	0	0	0.17	0.4
	0.6	0	0	0.1	0.4	0.6
	0.7	0	0	0.4	0.66	0.84
	0.8	0	0.4	0.76	0.97	1
	0.9	0.4	0.97	1	1	1
1800	0.5	0	0	0	0	0
	0.6	0	0	0	0	0.06
	0.7	0	0	0	0.15	0.45
	0.8	0	0	0.32	0.68	0.92
	0.9	0	0.68	1	1	1

3.3 Numerical Example

We compute numerically the probability p^* of security inspection which minimizes the expected cost $C(p)$: The processing time of each system call would be random. Thus, suppose that the processing of system call needs the time according to an exponential distribution $V(t) = (1 - e^{-t/v})$ and its mean processing time v of a system call is a unit time of the system in order to investigate the relative tendency of performance measure. It is assumed that the mean execution processing time of an applet is $(1/a)/v = 1800$, the mean time of system call occurrence is $(1/b)/v = 2$, the mean time to system failure is $(1/\lambda)/v = 300 \sim 1800$, the probability that system call is allowed is $q = 0.5 \sim 0.9$, the cost for interruption is a unit of cost, and the cost rate of system failure for an interruption is $c_1/c_2 = 100 \sim 500$.

Table 2 gives the optimal probability of security inspection p^* which minimizes the expected cost $C(p)$. For example, when $(1/\lambda)/v = 600, q = 0.6$ and $c_1/c_2 = 200$, $p^* = 0.51$. In this case, we should perform the security inspection for system call at one time of a rate to two times.

This indicates that p^* decreases with $(1/\lambda)/v$, however, increases with q and c_1/c_2. This can be interpreted that when the cost for system failure is large, p^* increases with c_1/c_2, so that the system should not become system failure. That is, we should set up the probability of security inspection highly. Table 2 also presents that when $(1/\lambda)/v$ is large and q is small, p^* is nearly 0. In this case, we should not perform the security inspection.

4 Reliability Analysis of a Network Server System with DoS Attacks

We extract the characteristic element from a server system with DoS attacks and formulate a stochastic model. The purpose of this model is to detect DoS Attacks with certain proper time interval and to propose the stochastic model which can continue normal service of a server system.

In this section, we have the following assumptions: DoS attacks taken up here are attacks where the service of a server is not available, and eavesdropping, alteration of data and DDoS (Distributed DoS) are not included. A server has the function of IDS, and DoS Attacks repeat occurrence and disappearance at random. Further, a server becomes faulty owing to DoS attacks according to a certain probability distribution. The monitoring policy for DoS attacks is introduced into a server system, and DoS attacks are detected by monitoring the change of the quantity of a packet, and so on. After a server system starts a service for clients, the monitoring of DoS attacks is performed at a certain interval. When DoS attacks are detected by a server system, a server system interrupts a service for clients and traces its attacks. If its tracing has succeeded, a server system deletes its processing. In this case, a server system returns to a normal condition and restarts a service for clients again. The mean time and the expected monitoring number until a server system becomes faulty are derived. Further, an optimal policy which minimizes the expected cost is discussed. Finally, a numerical example is given.

4.1 Model and Analysis

We pay attention to only a server which is connected with the internet, and propose the stochastic model which can continue normal service of a server system:

1) DoS attacks occur according to an exponential distribution $(1 - e^{-\alpha t})$ $(0 < \alpha < \infty)$. After continuing of DoS attacks according to an exponential distribution $(1 - e^{-\beta t})$ $(0 < \beta < \infty)$, they disappear or a server

becomes faulty. Since DoS attacks are caused at random by many terminals, it is assumed that they occur according to an exponential distribution, their duration times are also random, and their time has an exponential distribution. Let X be the random variable that represents the duration of DoS attacks and Y be the random variable that represents the upper limit time until a server becomes faulty after DoS attacks occurrence. If the duration X of DoS attacks exceeds an upper limit time Y, then a server becomes faulty owing to DoS attacks, and otherwise, DoS attacks disappear and it returns to a normal state. This indicates that if the event $\{X \leq Y\}$ occurs, DoS attacks disappear, and if the event $\{X > Y\}$ occurs, a server becomes faulty. That is, when the time of DoS attacks continues for a long time, the probability that a server becomes faulty is high. Since a server is attacked with various patterns from many terminals, it is assumed that both random variables X and Y are independent, and have exponential distributions, i.e., $\Pr\{X \leq t\} = 1 - e^{-\beta t}$ $(0 < \beta < \infty)$ and $\Pr\{Y \leq t\} = 1 - e^{-\lambda t}$ $(0 < \lambda < \infty)$.

2) After a server system begins to operate, it executes an initial processing and starts a service for clients. The time for the initial processing has a general distribution $A(t)$ with finite mean $1/a$.

3) DoS attacks are detected by monitoring the change of the quantity of a packet and by comparing packets which flow on the network to the pattern of DoS attack. After a server system starts a service for clients, the monitoring of DoS attacks is performed at a certain interval. The interval time has the degenerate distribution $G(t)$ placing unit mass at T. If DoS attacks have occurred, it is certainly detected. Oppositely, if it has not occurred, it is not detected, and after the monitoring, a server system returns to the beginning point of a service. If a server system becomes faulty when it is executing a service for clients, then it stops owing to system down.

4) When DoS attacks are detected by a server system, a server system interrupts a service for clients and traces its attacks. If its tracing has succeeded, a server system deletes its processing. In this case, a server system returns to a normal condition and restarts a service for clients again. This time has a general distribution $B(t)$ with finite mean $1/b$. The probability that the tracing of DoS attacks succeeds is q $(0 < q < 1)$.

5) If a server system fails to trace DoS attacks and deletes its processing, the refreshment processing is performed and restarts again from the beginning. The time for the refreshment processing has a general

distribution $V(t)$ with finite mean $1/v$. If DoS attacks disappear during the refreshment processing, a server system restarts a service for clients. On the other hand, if a server system becomes faulty during the refreshment processing, it stops owing to system down.

First, we define the following states of occurrence or disappearance of DoS attacks:

State 0: DoS attacks disappear and a server is in a normal condition.
State 1: DoS attacks occur.
State 2: A server becomes faulty.

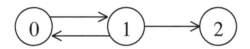

Fig. 5   Transition diagram between states of DoS attacks

The states defined above form a Markov renewal process. Transition diagram between states of occurrence or disappearance of DoS attacks is shown in Figure 5.

Let $Q_{ij}(t)$ $(i = 0, 1; j = 0, 1, 2)$ be one-step transition probabilities of a Markov renewal process. Then,

$$Q_{01}(t) = 1 - e^{-\alpha t}, \tag{61}$$

$$Q_{10}(t) = \int_0^t (1 - \Pr\{Y \le u\}) \, d\Pr\{X \le u\} = \int_0^t \beta e^{-(\beta+\lambda)u} \, du, \tag{62}$$

$$Q_{12}(t) = \int_0^t (1 - \Pr\{X \le u\}) \, d\Pr\{Y \le u\} = \int_0^t \lambda e^{-(\beta+\lambda)u} \, du. \tag{63}$$

From [19], the transition probabilities $P_{0j}(t)$ $(j = 0, 1, 2)$ that the system is in State j at time t when a server is in State 0 at time 0 are:

$$P_{00}(t) = \frac{1}{\gamma_1 - \gamma_2}[(\beta + \lambda - \gamma_2)e^{-\gamma_2 t} - (\beta + \lambda - \gamma_1)e^{-\gamma_1 t}], \tag{64}$$

$$P_{01}(t) = \frac{\alpha}{\gamma_1 - \gamma_2}(e^{-\gamma_2 t} - e^{-\gamma_1 t}), \tag{65}$$

$$P_{02}(t) = 1 - P_{00}(t) - P_{01}(t), \tag{66}$$

where

$$\gamma_1 \equiv \frac{1}{2}[(\alpha + \beta + \lambda) + \sqrt{(\alpha + \beta + \lambda)^2 - 4\alpha\lambda}],$$

$$\gamma_2 \equiv \frac{1}{2}[(\alpha + \beta + \lambda) - \sqrt{(\alpha + \beta + \lambda)^2 - 4\alpha\lambda}].$$

Next, we define the following states of a server system:

State 3: A server system begins to operate.
State 4: A server system starts or restarts a service for clients.
State 5: DoS attacks are detected, and the tracing of DoS attacks starts.
State F: A server system becomes faulty.

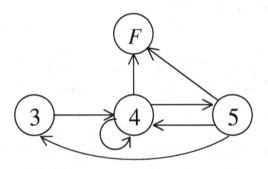

Fig. 6 Transition diagram between states of a server system

The states of a server system defined above form a Markov renewal process. Transition diagram between states of a server system is shown in Figure 6.

The LS transforms of one-step transition probabilities $Q_{ij}(t)(i = 3, 4, 5; j = 3, 4, 5, F)$ of a Markov renewal process are:

$$Q^*_{34}(s) = A^*(s), \tag{67}$$
$$Q^*_{44}(s) = e^{-sT} P_{00}(T), \tag{68}$$
$$Q^*_{45}(s) = e^{-sT} P_{01}(T), \tag{69}$$
$$Q^*_{4F}(s) = \int_0^T e^{-st} d\left\{ \sum_{k=1}^\infty [Q_{01}(t) * Q_{10}(t)]^{(k-1)} * Q_{01}(t) * Q_{12}(t) \right\}, \tag{70}$$
$$Q^*_{53}(s) = (1-q)B^*(s+\beta+\lambda)V^*(s+\beta+\lambda), \tag{71}$$
$$Q^*_{54}(s) = \frac{\beta}{s+\beta+\lambda}[1 - B^*(s+\beta+\lambda)] + qB^*(s+\beta+\lambda)$$
$$+ (1-q)B^*(s+\beta+\lambda)\frac{\beta}{s+\beta+\lambda}[1 - V^*(s+\beta+\lambda)], \tag{72}$$

$$Q^*{}_{5F}(s) = \frac{\lambda}{s+\beta+\lambda}[1 - B^*(s+\beta+\lambda)] + (1-q)B^*(s+\beta+\lambda)$$
$$\times \frac{\lambda}{s+\beta+\lambda}[1 - V^*(s+\beta+\lambda)]. \tag{73}$$

When the state of occurrence or disappearance of DoS attacks transits to State 2, the state of a server system transits to State F as a result.

We derive the mean time $\ell_{3F}(T)$ until a server system becomes faulty. Let $H_{iF}(s)$ ($i = 3, 4, 5$) be the first-passage time distribution from State i to State F. Then,

$$H_{3F}(t) = Q_{34}(t) * H_{4F}(t), \tag{74}$$
$$H_{4F}(t) = Q_{44}(t) * H_{4F}(t) + Q_{45}(t) * H_{5F}(t) + Q_{4F}(t), \tag{75}$$
$$H_{5F}(t) = Q_{54}(t) * H_{4F}(t) + Q_{53}(t) * H_{3F}(t) + Q_{5F}(t). \tag{76}$$

Taking the LS transforms of (74)–(76) and arranging them,

$$H^*{}_{3F}(s) = \frac{Q^*{}_{34}(s)Q^*{}_{4F}(s) + Q^*{}_{34}(s)Q^*{}_{45}(s)Q^*{}_{5F}(s)}{1 - Q^*{}_{44}(s) - Q^*{}_{45}(s)Q^*{}_{54}(s) - Q^*{}_{45}(s)Q^*{}_{53}(s)Q^*{}_{34}(s)}. \tag{77}$$

Hence, the mean time $\ell_{3F}(T)$ until a server system becomes faulty is

$$\ell_{3F}(T) \equiv \lim_{s\to 0} -\frac{dH^*{}_{3F}(s)}{ds} = \frac{J(T)}{1 - I(T)}, \tag{78}$$

where

$$I(T) \equiv P_{00}(T) + P_{01}(T)D = I_1 e^{-\gamma_1 T} + I_2 e^{-\gamma_2 T},$$
$$J(T) \equiv \frac{1}{a} - \frac{1}{a}P_{00}(T) + (E - \frac{1}{a}D)P_{01}(T)$$
$$+ \frac{1}{\gamma_1 - \gamma_2}\int_0^T (\gamma_1 e^{-\gamma_2 t} - \gamma_2 e^{-\gamma_1 t})\, dt$$
$$= \frac{1}{a} + J_1 e^{-\gamma_1 T} + J_2 e^{-\gamma_2 T} + \frac{1}{\gamma_1 - \gamma_2}\int_0^T (\gamma_1 e^{-\gamma_2 t} - \gamma_2 e^{-\gamma_1 t})\, dt,$$
$$I_1 \equiv \frac{\gamma_1 - (\alpha D + \beta + \lambda)}{\gamma_1 - \gamma_2}, \quad I_2 \equiv \frac{(\alpha D + \beta + \lambda) - \gamma_2}{\gamma_1 - \gamma_2},$$
$$J_1 \equiv \frac{(-\frac{1}{a})[\gamma_1 - (\beta + \lambda)] - \alpha(E - \frac{1}{a}D)}{\gamma_1 - \gamma_2},$$

$$J_2 \equiv \frac{(-\frac{1}{a})[(\beta+\lambda)-\gamma_2]+\alpha(E-\frac{1}{a}D)}{\gamma_1-\gamma_2},$$

$$D \equiv Q^*_{53}(0) + Q^*_{54}(0),$$

$$E \equiv \lim_{s\to 0}\left\{\frac{1}{a}Q^*_{53}(s) + P_{01}(T)\{-[Q^{*\prime}_{53}(s) + Q^{*\prime}_{54}(s) + Q^{*\prime}_{5F}(s)]\}\right\}$$

$$= \frac{1}{a}Q^*_{53}(0) - P_{01}(T)[Q^{*\prime}_{53}(0) + Q^{*\prime}_{54}(0) + Q^{*\prime}_{5F}(0)].$$

Next, we derive the expected monitoring number M_4 until a server system becomes faulty. Let $M_{44}(t)$ be the expected monitoring number until a server system becomes faulty in $(0,t]$. Then,

$$M_{44}(t) = [Q_{44}(t) + Q_{45}(t) * Q_{54}(t) + Q_{45}(t) * Q_{53}(t) * Q_{34}(t)] * [1+M_{44}(t)]. \tag{79}$$

Taking the LS transforms of (79),

$$M^*_{44}(s) = \frac{Q^*_{44}(s) + Q^*_{45}(s)Q^*_{54}(s) + Q^*_{45}(s)Q^*_{53}(s)Q^*_{34}(s)}{1 - Q^*_{44}(s) - Q^*_{45}(s)Q^*_{54}(s) - Q^*_{45}(s)Q^*_{53}(s)Q^*_{34}(s)}. \tag{80}$$

Thus, the expected monitoring number is

$$M_4 \equiv \lim_{t\to\infty} M_{44}(t) = \lim_{s\to 0} M^*_{44}(s) = \frac{P_{00}(T) + P_{01}(T)D}{1 - P_{00}(T) - P_{01}(T)D} = \frac{I(T)}{1 - I(T)}. \tag{81}$$

Further, we derive the expected refreshment number M_3 until a server system becomes faulty. Let $M_{33}(t)$ be the expected refreshment number until a server system becomes faulty in $(0,t]$. Then,

$$M_{33}(t) = Q_{34}(t) * M_{43}(t), \tag{82}$$

$$M_{43}(t) = Q_{44}(t) * M_{43}(t) + Q_{45}(t) * M_{53}(t), \tag{83}$$

$$M_{53}(t) = Q_{54}(t) * M_{43}(t) + Q_{53}(t)[1+M_{33}(t)]. \tag{84}$$

Taking the LS transforms of (82)–(84) and arranging them,

$$M^*_{33}(s) = \frac{Q^*_{34}(s)Q^*_{45}(s)Q^*_{53}(s)}{1 - Q^*_{44}(s) - Q^*_{45}(s)Q^*_{54}(s) - Q^*_{45}(s)Q^*_{53}(s)Q^*_{34}(s)}. \tag{85}$$

Thus, the expected refreshment number is

$$M_3 \equiv \lim_{t\to\infty} M_{33}(t) = \lim_{s\to 0} M^*_{33}(s) = \frac{P_{01}(T)G}{1-I(T)}, \tag{86}$$

where $G \equiv Q^*_{53}(0)$.

Moreover, we derive the expected interruption number M_5 until a server system becomes faulty. Let $M_{55}(t)$ be the expected interruption number until a server system becomes faulty in $(0,t]$. Then,

$$M_{45}(t) = Q_{44}(t) * M_{45}(t) + Q_{45}(t) * [1+M_{55}(t)], \tag{87}$$

$$M_{55}(t) = Q_{54}(t) * M_{45}(t) + Q_{53}(t) * Q_{34}(t) * M_{45}(t). \tag{88}$$

Taking the LS transforms of (87) and (88), and arranging them,

$$M^*{}_{55}(s) = \frac{Q^*{}_{45}(s)}{1 - Q^*{}_{44}(s) - Q^*{}_{45}(s)Q^*{}_{54}(s) - Q^*{}_{45}(s)Q^*{}_{53}(s)Q^*{}_{34}(s)}. \quad (89)$$

Thus, the expected interruption number is

$$M_5 \equiv \lim_{t \to \infty} M_{55}(t) = \lim_{s \to 0} M^*{}_{55}(s) = \frac{P_{01}(T)}{1 - I(T)}. \quad (90)$$

4.2 Optimal Policy

We obtain the expected cost and discuss an optimal policy which minimizes it: Let c_4 be the cost for monitoring, c_3 be the cost for a refreshment, c_5 be the cost for an interruption, and c_0 ($c_0 > c_3 > c_5 > c_4$) be the cost for system failure. Then, we define the expected cost $C(T)$ until a server system becomes faulty as

$$\begin{aligned} C(T) &\equiv \frac{c_4 M_4 + c_3 M_3 + c_5 M_5 + c_0}{\ell_{3F}(T)} \\ &= \frac{c_0 - (c_0 - c_4)I(T) + (c_3 G + c_5)P_{01}(T)}{J(T)}. \end{aligned} \quad (91)$$

We seek an optimal time T^* which minimizes $C(T)$. Differentiating (91) with respect to T and setting it equal to zero,

$$\frac{I(T)J'(T) - I'(T)J(T) - \frac{c_3 G + c_5}{c_0 - c_4}[J'(T)P_{01}(T) - J(T)P'_{01}(T)]}{J'(T)} = \frac{c_0}{c_0 - c_4}, \quad (92)$$

where

$$I'(T) = -(\gamma_1 I_1 e^{-\gamma_1 T} + \gamma_2 I_2 e^{-\gamma_2 T}),$$

$$J'(T) = -(\gamma_1 J_1 e^{-\gamma_1 T} + \gamma_2 J_2 e^{-\gamma_2 T}) + \frac{1}{\gamma_1 - \gamma_2}(\gamma_1 e^{-\gamma_2 T} - \gamma_2 e^{-\gamma_1 T}),$$

$$P'_{01}(T) = \frac{\alpha}{\gamma_1 - \gamma_2}(\gamma_1 e^{-\gamma_1 T} - \gamma_2 e^{-\gamma_2 T}).$$

Denoting the left-hand side of equation (92) by $L(T)$,

$$L(0) = 1, \qquad L'(T) = \frac{J(T)\left\{ \begin{array}{l} I'(T)J''(T) - I''(T)J'(T) \\ -\frac{c_3 G + c_5}{c_0 - c_4}[J''(T)P'_{01}(T) - J'(T)P''_{01}(T)] \end{array} \right\}}{J'(T)^2}, \quad (93)$$

where
$$I''(T) = \gamma_1^2 I_1 e^{-\gamma_1 T} + \gamma_2^2 I_2 e^{-\gamma_2 T},$$
$$J''(T) = \gamma_1^2 J_1 e^{-\gamma_1 T} + \gamma_2^2 J_2 e^{-\gamma_2 T} + \frac{\gamma_1 \gamma_2}{\gamma_1 - \gamma_2}(e^{-\gamma_1 T} - e^{-\gamma_2 T}),$$
$$P''_{01}(T) = \frac{\alpha}{\gamma_1 - \gamma_2}(\gamma_2^2 e^{-\gamma_2 T} - \gamma_1^2 e^{-\gamma_1 T}).$$

Hence, when
$$I'(T)J''(T) - I''(T)J'(T) > \frac{c_3 G + c_5}{c_0 - c_4}[J''(T)P'_{01}(T) - J'(T)P''_{01}(T)],$$
$L(T)$ is strictly increasing in T from 1. Thus, we can characterize the optimal policy:

(i) If $L(\infty) > c_0/(c_0 - c_4)$, then there exists a finite and unique T^* (> 0) which satisfies (92).
(ii) If $L(\infty) \leq c_0/(c_0 - c_4)$, then $T^* = \infty$. In this case, it is optimal to do no monitor. The expected cost is
$$C(\infty) = \frac{c_0}{\frac{1}{a} + \frac{\alpha + \beta + \lambda}{\alpha \lambda}}, \tag{94}$$
and the mean time until a server system becomes faulty is
$$\ell_{3F}(\infty) = \frac{\alpha + \beta + \lambda}{\alpha + \lambda}. \tag{95}$$

4.3 Numerical Example

We compute numerically the optimal time T^* from (92): Suppose that $B(t) = 1 - e^{-bt}$, $V(t) = 1 - e^{-vt}$ and the mean time $1/a$ for the initial processing is a unit of time. It is assumed that the mean time of DoS attacks interval is $(1/\alpha)/(1/a) = 60 \sim 600$, the mean duration of DoS attacks occurrence is $(1/\beta)/(1/a) = 10$, the mean upper limit duration of DoS attacks occurrence is $(1/\lambda)/(1/a) = 60 \sim 600$, the mean time for tracing of DoS attacks and deletion of its processing is $(1/b)/(1/a) = 1$, the mean time for the refreshment processing is $(1/v)/(1/a) = 5$, and the probability that the tracing of DoS attacks succeeds is $q = 0.5, 0.9$. Further, the cost c_0 for system failure is a unit of cost, the cost rate of a refreshment is $c_3/c_0 = 10^{-2}$, the cost rate of an interruption is $c_5/c_0 = 10^{-3}$, and the cost rate of monitoring is $c_4/c_0 = 1 \sim 8$ $(\times 10^{-4})$.

Table 3 gives the optimal T^* which minimizes the expected cost $C(T)$. This indicates that T^* decreases with q, however, increases with $(1/\alpha)/(1/a)$, $(1/\lambda)/(1/a)$ and c_4/c_0. When c_4/c_0 and $(1/\lambda)/(1/a)$ are large, $T^* = \infty$. In this case, it is optimal not to monitor.

Table 3 Optimal time T^* to minimize $C(T)$

a/α	a/λ	\multicolumn{8}{c}{q}

a/α	a/λ	0.5				0.9			
		\multicolumn{8}{c}{c_4/c_0 ($\times 10^{-4}$)}
		1	2	4	8	1	2	4	8
60	60	1.0	1.4	2.1	3.1	1.0	1.4	2.1	3.1
	120	1.5	2.2	3.2	4.8	1.5	2.2	3.2	4.8
	180	2.0	2.9	4.3	6.6	1.9	2.8	4.2	6.5
	300	3.0	4.4	6.8	11.4	2.9	4.3	6.7	11.1
	600	9.5	17.6	∞	∞	8.5	15.0	∞	∞
300	60	2.3	3.4	5.2	8.2	2.3	3.4	5.1	8.1
	120	3.5	5.3	8.3	14.4	3.5	5.3	8.2	14.2
	180	4.7	7.2	11.9	24.2	4.6	7.1	11.7	23.6
	300	7.3	11.9	24.0	∞	7.1	11.6	23.0	∞
	600	30.0	∞	∞	∞	24.8	∞	∞	∞
600	60	3.4	5.1	8.1	14.1	3.4	5.1	8.0	13.9
	120	5.3	8.3	14.2	38.1	5.2	8.2	14.0	36.2
	180	7.1	11.7	23.4	∞	7.1	11.5	22.8	∞
	300	11.7	23.1	∞	∞	11.4	22.2	∞	∞
	600	∞	∞	∞	∞	∞	∞	∞	∞

5 Conclusions

We have formulated three stochastic models of a system connected with networks, and have discussed the optimal policy which minimizes the expected cost rate. Moreover, we have given numerical examples of each models and have evaluated them for various standard parameters. If some parameters are estimated from actual data, we could select the best policy.

Finally, we enumerate the following questions for future studies:

1) Is it possible to estimate statistically various parameters in the formulated models?
2) What types of distribution are fit for the observed data?
3) What are appropriate measures which show the reliability of the system?

It would be very important to evaluate and improve the reliability of a system connected with networks. The results derived in this paper would be applied in practical fields by making some suitable modification and extensions. Further studies for such subject would be expected.

References

1. Itabashi, K., Matsubara, K., Moriyama, Y. Soneya, Y., Kato, K., Sekiguchi, T. and Yonezawa, A. (2001). *Implementing an Applet System without Fixing Virtual-Machine Designs (in Japanese)*, Transactions of IEICE Japan, J84-DI, pp. 639–649.
2. Oyama Y. and Kato, K. (2001). *SecurePot: Secure Software Execution System Based on System Call Hooks*, Proceeding of 18th Conference of Japan Society for Software Science and Technology.
3. Kato, K. and Oyama, Y. (2002). *SecurePot: An Encapsulated Transferable File System for Secure Software Circulation*, Technical Report ISE-TR-02-185, Institute of Information Sciences and Electronics, University of Tsukuba.
4. Gosling, J. and McGilton, H., (1995). *The Java Language Environment*, Sun Microsystems.
5. Gong, L. (1999). *Inside Java 2 Platform Security*, Addison Wesley Longman.
6. Shinagawa, T., Kono, K., Takahashi, M. and Masuda, T. (1999). *Kernel Support of Fine-grained Protection Domains for Extension Components*, Transactions of IPS Japan, 40, pp. 2596–2606.
7. Shinagawa, T., Kono, K. and Masuda, T. (2002). *A Secure Execution Environment for Executable Contents (in Japanese)*, Transactions of IPS Japan, 43, pp. 1677–1689.
8. The Institute of Electronics, Information and Communication Engineers (2004). *Information Security Handbook* (in Japanese), Ohmsha, 354.
9. Ohta, K. and Mansfield, G. (2000). *Illegal Access Detection on the Internet -Present status and future directions* (in Japanese), Transactions of IEICE Japan, J83-B, pp. 1209–1216.
10. Miyamoto, T., Kojima, A., Izumi, M. and Fukunaga, K. (2004). *Network Traffic Anomaly Detection Using SVM* (in Japanese), Transactions of IEICE Japan, J87-B, pp. 593–598.
11. Takei, Y., Ohta, K., Kato, N. and Nemoto, Y. (2001). *Detecting and Tracing Illegal Access by using Traffic Pattern Matching Technique* (in Japanese), Transactions of IEICE Japan, J84-B, pp. 1464–1473.
12. Aburakawa, R., Ohta, K., Kato, N. and Nemoto, Y. (2003). *An Early Warning System for Illegal Access based on Distributed Network Monitoring* (in Japanese), Transactions of IEICE Japan, J86-B, pp. 410–418.
13. Sakaguchi, K., Ohta, K., Waizumi, Y., Kato, N. and Nemoto, Y. (2002). *Tracing DDos Attacks by Comparing Traffic Patterns Based on Quadratic Programming Method* (in Japanese), Transactions of IEICE Japan, J85-B, pp. 1295–1303.
14. Miyake, M. (2006). *Attack Detection and Analysis on Internet* (in Japanese), Journal of IEICE Japan, J89, pp. 313–317.
15. Kai, T., Nakatani, H., Shimizu, H. and Suzuki, A. (2006). *Efficient Traceback Method for Detecting DDoS Attacks* (in Japanese), Transactions of IPS Japan, 47, pp. 1266–1275.

16. Kanaoka, A. and Okamoto, E. (2003). *The Problems and Countermeasures of Intrusion Detection Exchange Format in IETF* (in Japanese), Transactions of IPS Japan, 44, pp. 1830-1837.
17. Kuzmanovic, A. and Knightly, E. (2006). *Low-Rate TCP-Targeted Denial of Service Attacks and Counter Strategies*, IEEE Transactions on Networking, 14(4), pp. 683-696.
18. Osaki, S. (1992). *Applied Stochastic System Modeling*, Springer-Verlag, Berlin.
19. Yasui, K., Nakagawa, T. and Sandoh, H. (2002). *Reliability Models in Data Communication Systems*, in Stochastic Models in Reliability and Maintenance, (Edited by. Osaki, S.) 281-301, Springer-Verlag, Berlin.
20. Nanya, T. (1991). *Fault-Tolerant Computer*, Ohmsha, Tokyo.
21. Siewiorek, D. P. and Swarz, R. S. (eds.) (1982). *The Theory and Practice of Reliable System Design*, Digital Press, Bedford, Massachusetts.
22. Mukaidono, M. (ed.) (1988). *Introduction to Highly Reliable Techniques for Computer Systems*, Japanese Standards Association, Tokyo.
23. Lévy, P. (1954). *Processus Semi-Markoviens*, Proc. Int. Congr. Math., 3, pp. 416-426.
24. Smith, W.L. (1955). *Regenerative Stochastic Processes*, Proc. Roy. Soc. London, Ser. A, 232, pp. 6-31.
25. Pyke, R. (1961). *Markov Renewal Processes: Definitions and Preliminary Properties*, Ann. Math. Statist., 32, 4, pp. 1231-1242.
26. Pyke, R. (1961). *Markov Renewal Processes with Finitely Many States*, Ann. Math. Statist., 32, 4, pp. 1243-1259.
27. Çinlar, E. (1975). *Markov Renewal Theory*, Management Science, 21, 7, pp. 727-752.
28. Çinlar, E. (1975). *Introduction to Stochastic Processes*, Prentices-Hall, Englewood Cliffs, N.J.
29. Barlow, R.E. and Proschan, F. (1965). *Mathematical Theory of Reliability*, Wiley, New York.
30. Nakagawa, T. and Osaki, S. (1979). *Bibliography for Availability of Stochastic Systems*, IEEE Transactions on Reliability, R-25, 4, pp. 284-287.
31. Ono, K. (1996). *Computer Communication*, Ohmsha.
32. Akiyama, M. (1997). *The Base of Information Communication Network*, Maruzen.
33. Yasui, K., Nakagawa, T. and Motoori, M. (1991). *Two-State Retry Policies for Fault-Tolerant Systems*, Transactions of IEICE Japan, J74-A, pp. 1053-1058.
34. Yasui, K., Nakagawa, T. and Motoori, M. (1992). *A Two-State Error Control Policy for a Data Transmission System with Intermittent Faults*, Transactions of IEICE Japan, J75-A, pp. 944-948.
35. Sando, H., Nakagawa, T. and Koike, S. (1992). *A Bayesian Approach to an Optimal ARQ Number in Data Transmission*, Transactions of IEICE Japan, J75-A, pp. 1192-1198.

36. Nakagawa, T., Yasui, K. and Sando, H. (1993). *An Optimal ARQ Policy for a Data Transmission System with Intermittent Faults*, Transactions of IEICE Japan, J76-A, pp. 1201–1206.
37. Yasui, K., Nakagawa, T. and Sando, H. (1995). *An ARQ Policy for a Data Transmission with Three Types of Error Probabilities*, Transactions of IEICE Japan, J78-A, pp. 824–830.
38. Imaizumi, M. (2001). *Studies on Reliability Analysis for Microprocessor Systems*, Aichi Institute of Technology.

Chapter 6

Reliability Analysis of Communication Systems

MITSUTAKA KIMURA

Department of International Culture Studies
Gifu City Woman's College
7-1 Hitoichibakita-machi, Gifu 501-0192, Japan
E-mail: kimura@gifu-cwc.ac.jp

1 Introduction

As a computer communication technology has remarkably developed, computing systems have widely spread and have been used in many practical fields. This chapter summarizes reliability analysis of various communication systems. In Section 2, we consider a communication system which consists of several processors. Efficient control mechanisms of a communication system have been actually realized by a number of processors [7]. Hence, the processing of each processor has to be carried out accurately and rapidly. Moreover, the system has to restore a consistent state immediately after transient faults. There are rollback recovery technique to improve the reliability of communications between processors [1–4,6]. That is, a processor saves the state of process in a stable storage, in which it is now executing. This is called *checkpoint* and is set up at periodic times. When faults have occurred and have been detected, the a process is rolled back to the most recent checkpoint and can restore a consistent state.

Generally, there are the following two main approaches to take checkpoint: *Independent* approach and *coordinated* approach [5]. In independent approach, the process takes checkpoint periodically regardless of its state depending on states of the other processes. In coordinated approach, all processes take checkpoint synchronously. Independent approach may be attractive because its checkpointing overhead is lower than that of coordinated one. However, there often arise the following two problems: One is *domino effect* where the processes cannot find out any consistent states even if they roll back to checkpoints and do recursively at many times. In the worst case, some processes may have to roll back to their initial states. The other is that a receiver may accept the same message again [7, 8]. On the other hand, *domino effect* in coordinated approach can be avoid by using the techniques of checkpoint and rollback, in which the process takes into consideration of its state depending on states of the other processes.

In Section 2, all processors take checkpoints and roll back synchronously in order to prevent *domino effect*. Further, a receiver can protect against the acceptance of the same message by adopting the Transmission Control Protocol(TCP). We formulate the stochastic model with the above recovery techniques, using the theory of Markov renewal processes. We obtain the expected cost as an objective function and analytically derive an optimal checkpointing interval T^* which minimizes it [9].

In Section 3, we consider the problem for improving the reliability of a mobile communication system. The problem is very important to realize stable and high-speed network communications in mobile environments [10–12]. However, mobile stations often may become unavailable due to communication errors, which have been generated by disconnections, wireless link failures, and so on. Then, a mobile communication system, which consists of mobile stations, makes the recovery techniques. That is, when a communication error has occurred, the rollback recovery for the mobile station associated with such an event is executed to the most recent checkpoint, and so that, the system can restore a consistent state [13–16]. Recently, some error recovery schemes with the feature of mobile environments have been proposed by several authors [15, 16]. They have dealt mainly with the operation time from the beginning of operation to handoff. We consider a mobile communication system which consists of mobile stations, several base stations and a switching center, and formulate the stochastic model adopting the recovery technique of checkpoint and rollback with the feature of mobile environments. We obtain the expected cost and analytically derive an optimal checkpointing interval which minimizes it [17].

In Section 4, we consider a communication system which consists of several clients and a web server. As the internet has been widely used, its network scheme has been urgently needed for a high reliable communication.

For example, some packets may be discarded at a client due to buffer overflow. The window flow control mechanism to defuse this situation has been implemented [18–21]. That is, the client can throttle a web server by specifying some limit on the amount of data it can transmit. The limit is determined by a window size at the client. The amount of packets, which corresponds to the window size, is successively transmitted to the client by the server.

On the other hand, in the packet delivery process, some packets are sometimes dropped by network congestion (packets loss). Several authors have studied some protocols for dissolving packets loss [22–28]. For example, setting the window to half of the first window size when a sender detects congestion, has been researched [20]. Moreover, some ways of the congestion detection have been already proposed [23–32]. For example, there are congestion detections by dropped packets and ECN (Explicit Congestion Notification) [30]. In the congestion detection by dropped packets, the server detects dropped packets either from the receipt of three duplicate acknowledgements or after time out of a retransmit timer, and the server detects the congestion [20]. In the congestion detection by ECN, Routers set the ECN bit in packet headers when their average queue size exceeds a certain threshold, and the server detects incipient congestion during connection [29–32]. In Section 4, we consider a stochastic model of a communication system using a window flow control scheme considering error of ECN message. We obtain the amount of packets per unit of time until the transmission succeeds (throughput) and analytically derive an optimal policy which maximizes it [33, 34].

2 Communication System with Rollback Recovery

This section considers a communication system which consists of several processors, and studies the problem for improving its reliability by adopting the recovery techniques of checkpoint and rollback : When either processor failure or communication error has occurred, the rollback recovery for processors associated with such an event is executed to the most recent checkpoint, and a consistent state in the whole system is always maintained by coordinated approach. The stochastic model with the above recovery techniques is formulated, using the theory of Markov renewal processes. The mean time to take checkpoint and the expected number of rollback recoveries caused by processor failures and communication errors are derived.

An optimal checkpointing interval which minimizes the expected cost is discussed. Finally, numerical examples are given.

2.1 Reliability Quantities

A communication system consists of several processors and its control mechanisms are realized by communications with each other between processors. We assume that the system is divided into a processor called A and the other processors called B for convenience, and is concerned only about the communication behaviors of A. The system is shown in Figure 1.

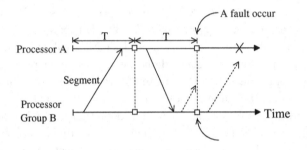

Fig. 1 Outline of the model with rollback recovery

1) The system begins to operate at time 0, and takes checkpoints for all processes that are relevant to the operation of A, at scheduled time T or when the transmissions between A and B have terminated successfully at m times, whichever occurs first. Any transmissions which have not finished until time T are dealt with no transmission with each other.
2) The request time for transmissions between A and B has a general distribution $A(t)$ with finite mean $1/\alpha$.
3) A message is divided into n pieces of segments because it is necessary to ensure the reliability of transmissions, and each segment is sent from a sender to a receiver with acknowledgment by handshake as follows:

   a) Each corresponding answer of ACK(acknowledgment) or NAK(negative ACK) from a receiver to a sender judges whether the transmission of a segment succeeds or does not. The communication of a message terminates when n times of ACK have been accepted from a receiver.

b) When NAK or no answer has been received until a limited time, we retransmit a message of the same segment. If the retransmission does not succeed again, it is judged that communication errors have occurred.

c) The time required for the transmission of a segment has a probability distribution $a(t)$, and the probability of its success is $p(0 < p \leq 1)$.

4) Failures of processor A and processors B occur independently according to distributions $F_A(t)$ and $F_B(t)$, respectively. Then, we define a probability distribution $\overline{F}(t) \equiv \overline{F_A}(t)\overline{F_B}(t)$ with finite mean $1/\lambda$, where $\overline{\Phi}(t) \equiv 1 - \Phi(t)$.

5) When either processor failures or communication errors have occurred, the rollback recovery for processors associated with such events is executed from that time to its most recent checkpoint:

a) Any transmissions which have not finished until that time are dealt with no transmission with each other.

b) The system is regenerated by rollback recovery.

c) The time required for rollback recovery has a general distribution $G(t)$ with finite mean $1/\mu$.

Under the above assumptions, we define the following states of the system:

State S_0: The system begins to operate or restarts.
State S_P: Processor failures occur.
State S_C: Communication errors occur.
State S_T: Checkpoint of the system is carried out when the transmissions of m messages have succeed or at time T, whichever occurs first.

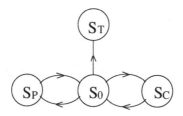

Fig. 2 Transition diagram between system states

The system states defined above form the Markov renewal process, where S_T is an absorbing state and S_0 is a regeneration point. Transition diagram between system states is shown in Figure 2.

Using the mass functions of Markov renewal processes [35], Laplace-Stieltjes (LS) transforms $Q_{ij}^*(s)$ of transition probabilities $Q_{ij}(t)$ from State $i(i = S_0)$ to State $j(j = S_P, S_C, S_T)$ are:

$$Q_{S_0,S_P}^*(s) \equiv \int_0^T e^{-st}[1 - W_n^{(m)}(t) - X_m(t)]\,\mathrm{d}F(t), \qquad (1)$$

$$Q_{S_0,S_C}^*(s) \equiv \int_0^T e^{-st}\overline{F}(t)\,\mathrm{d}X_m(t), \qquad (2)$$

$$Q_{S_0,S_T}^*(s) \equiv e^{-sT}\overline{F}(T)[1 - W_n^{(m)}(T) - X_m(T)] + \int_0^T e^{-st}\overline{F}(t)\,\mathrm{d}W_n^{(m)}(t), \qquad (3)$$

where

$$W_i(t) \equiv A(t) * [pa(t) + (1-p)pa^{(2)}(t)]^{(i)} \quad (i = 1, 2, \cdots, n),$$

$$X_m(t) \equiv \sum_{j=1}^m W_n^{(j-1)}(t) \sum_{i=1}^n W_{i-1}(t) * [(1-p)a(t)]^{(2)},$$

where $\Phi^{(i)}(t)$ is the i-fold convolution of $\Phi(t)$ and $\Phi^{(i)}(t) \equiv \Phi^{(i-1)}(t) * \Phi(t)$, $\Phi_1(t) * \Phi_2(t) \equiv \int_0^t \Phi_2(t-u)\mathrm{d}\Phi_1(u)$, $\Phi^{(0)}(t) \equiv 1$. The asterisk mark denotes the Stieltjes convolution, i.e., $\Phi^*(s) \equiv \int_0^\infty e^{-st}\mathrm{d}\Phi(t)$ for $s > 0$. Then, LS transform $H_{S_0,S_T}^*(s)$ of the first-passage time distribution $H_{S_0,S_T}(t)$ from the beginning of operation to the next checkpoint is

$$H_{S_0,S_T}^*(s) \equiv \frac{Q_{S_0,S_T}^*(s)}{1 - [Q_{S_0,S_P}^*(s) + Q_{S_0,S_C}^*(s)]G^*(s)}. \qquad (4)$$

Therefore, the mean time ℓ_{S_0,S_T} is

$$\ell_{S_0,S_T} \equiv \lim_{s \to 0} \frac{-\mathrm{d}H_{S_0,S_T}^*(s)}{\mathrm{d}s}$$

$$= \frac{\int_0^T \overline{F}(t)[1 - W_n^{(m)}(t) - X_m(t)]\,\mathrm{d}t + \frac{1}{\mu}\{1 - \int_0^T \overline{F}(t)\,\mathrm{d}W_n^{(m)}(t) - \overline{F}(T)[1 - W_n^{(m)}(T) - X_m(T)]\}}{\overline{F}(T)[1 - W_n^{(m)}(T) - X_m(T)] + \int_0^T \overline{F}(t)\,\mathrm{d}W_n^{(m)}(t)}. \qquad (5)$$

Similarly, LS transforms $M_P^*(s)$ and $M_C^*(s)$ of $M_P(t)$ and $M_C(t)$ which are the expected numbers of rollback recoveries caused by processor failures and communication errors are, respectively,

$$M_P^*(s) = \frac{Q_{S_0,S_P}^*(s)G^*(s)}{1 - [Q_{S_0,S_P}^*(s) + Q_{S_0,S_C}^*(s)]G^*(s)}, \tag{6}$$

$$M_C^*(s) = \frac{Q_{S_0,S_C}^*(s)G^*(s)}{1 - [Q_{S_0,S_P}^*(s) + Q_{S_0,S_C}^*(s)]G^*(s)}. \tag{7}$$

Therefore, the respective expected numbers of rollback recoveries M_P and M_C are

$$\begin{aligned} M_P &\equiv \lim_{t \to \infty} M_P(t) = \lim_{s \to 0} M_P^*(s) \\ &= \frac{\int_0^T [1 - W_n^{(m)}(t) - X_m(t)]\,\mathrm{d}F(t)}{\overline{F}(T)[1 - W_n^{(m)}(T) - X_m(T)] + \int_0^T \overline{F}(t)\,\mathrm{d}W_n^{(m)}(t)}, \end{aligned} \tag{8}$$

$$\begin{aligned} M_C &\equiv \lim_{t \to \infty} M_C(t) = \lim_{s \to 0} M_C^*(s) \\ &= \frac{\int_0^T \overline{F}(t)\,\mathrm{d}X_m(t)}{\overline{F}(T)[1 - W_n^{(m)}(T) - X_m(T)] + \int_0^T \overline{F}(t)\,\mathrm{d}W_n^{(m)}(t)}. \end{aligned} \tag{9}$$

2.2 Optimal Policy

Let c_0 be the cost for system operation, c_1 be the cost for rollback recovery of communication errors and c_2 be the cost for rollback recovery of processor failures. We define that the expected cost rate until the next checkpoint is

$$C(T,m) \equiv \frac{c_0 + c_1 M_C + c_2 M_P}{\ell_{S_0,S_T}}. \tag{10}$$

We seek an optimal checkpointing interval T^* which minimizes $C(T,m)$ for $c_2 > c_1 > c_0$ and given m $(m \geq 1)$. We consider the particular case that $A(t)$ is exponential and the transmission time of a segment can be neglected because it is much smaller than the other times, i.e., $A(t) \equiv 1 - e^{-\alpha t}$ and $a(t) \equiv 1$ for $t \geq 0$, and discuss analytically it.

Differentiating $C(T, m)$ in (10) with respect to T and setting it equal to zero,

$$\frac{\left[\lambda(T) + \alpha(1-x)\left(\frac{c_1-c_0}{c_2-c_0}\right)\right]\left[\sum_{k=1}^{m} x^{k-1}\int_0^T \overline{F}(t)H_{k-1}(\alpha t)\,dt + \frac{1}{\mu}\right]}{1 + \frac{1}{\mu}[\lambda(T) + \alpha(1-x)]}$$

$$+ \frac{\left[1 + \frac{\alpha}{\mu}(1-x)\left(\frac{c_2-c_1}{c_2-c_0}\right)\right]\left[\overline{F}(T)\sum_{k=1}^{m}x^{k-1}H_{k-1}(\alpha T)\right.}{1 + \frac{1}{\mu}[\lambda(T)+\alpha(1-x)]}$$

$$\left.+ \alpha x^m \int_0^T \overline{F}(t)H_{m-1}(\alpha t)\,dt\right]$$

$$+ \alpha(1-x)\left(\frac{c_2-c_1}{c_2-c_0}\right)\sum_{k=1}^{m} x^{k-1}\int_0^T \overline{F}(t)H_{k-1}(\alpha t)\,dt = \frac{c_2}{c_2-c_0}, \quad (11)$$

where $x \equiv [p(2-p)]^n$ $(0 < x \le 1)$ and $H_k(\alpha t) \equiv [(\alpha t)^k/k!]e^{-\alpha t}$ ($k = 0, 1, \cdots$). Letting denote the left-hand side of (11) by $Q_m(T)$,

$$Q_m(0) = 1, \quad (12)$$

$$Q_m(\infty) = \frac{\left[\lambda(\infty)+\alpha(1-x)\left(\frac{c_1-c_0}{c_2-c_0}\right)\right]}{1+\frac{1}{\mu}[\lambda(\infty)+\alpha(1-x)]}$$

$$\times \left[\sum_{k=1}^{m}x^{k-1}\int_0^\infty \overline{F}(t)H_{k-1}(\alpha t)\,dt + \frac{1}{\mu}\right]$$

$$+ \frac{\left[1+\frac{\alpha}{\mu}(1-x)\left(\frac{c_2-c_1}{c_2-c_0}\right)\right]\left[\alpha x^m \int_0^\infty \overline{F}(t)H_{m-1}(\alpha t)\,dt\right]}{1+\frac{1}{\mu}[\lambda(\infty)+\alpha(1-x)]}$$

$$+ \alpha(1-x)\left(\frac{c_2-c_1}{c_2-c_0}\right)\sum_{k=1}^{m}x^{k-1}\int_0^\infty \overline{F}(t)H_{k-1}(\alpha t)\,dt, \quad (13)$$

$$Q'_m(T) = \frac{\sum_{k=1}^{m}x^{k-1}\int_0^T \overline{F}(t)H_{k-1}(\alpha t)\,dt}{\{1+\frac{1}{\mu}[\lambda(T)+\alpha(1-x)]\}^2}$$
$$+ \frac{1}{\mu}\left[\begin{array}{c}1-\overline{F}(T)\sum_{k=1}^{m}x^{k-1}H_{k-1}(\alpha T)\\ -\alpha x^m \int_0^T \overline{F}(t)H_{m-1}(\alpha t)\,dt\end{array}\right]$$

$$\times \lambda'(T)\left[\left(\frac{c_1-c_0}{c_2-c_0}\right) + \left(\frac{c_2-c_1}{c_2-c_0}\right)\left[1+\frac{1}{\mu}\alpha(1-x)\right]\right] > 0, (14)$$

where $\Phi'(t)$ represents a density function of any function $\Phi(t)$.

Therefore, we have the following optimal policy:

(i) If $\lambda(t)$ is strictly increasing in t and $Q_m(\infty) > c_2/(c_2-c_0)$, then there exists a finite and unique T^* which satisfies (11), and the expected cost rate is

$$C(T^*, m) = \frac{(c_2-c_0)\lambda(T^*) + (c_1-c_0)\alpha(1-x)}{1+\frac{1}{\mu}[\lambda(T^*)+\alpha(1-x)]}. \quad (15)$$

The right-hand side of (15) is strictly increasing in T^*. Thus, if T^* is increasing, then $C(T^*, m)$ is also increasing.

From the notation of $Q_m(t)$,

$$Q_{m+1}(T) - Q_m(T) = \frac{x^m J(T)}{1 + \frac{1}{\mu}[\lambda(T) + \alpha(1-x)]} \qquad (16)$$

where

$$J(T) \equiv [\lambda(T) + \alpha] \int_0^T \overline{F}(t) H_m(\alpha t)\, dt + \overline{F}(T) H_m(\alpha T)$$

$$-\alpha \int_0^T \overline{F}(t) H_{m-1}(\alpha t)\, dt + \frac{\alpha}{\mu}(1-x)\left(\frac{c_2 - c_1}{c_2 - c_0}\right)\left[\overline{F}(T) H_m(\alpha T)\right.$$

$$\left. + \lambda(T) \int_0^T \overline{F}(t) H_m(\alpha t)\, dt - \alpha \int_0^T \overline{F}(t)[H_{m-1}(\alpha t) - H_m(\alpha t)]\, dt\right].$$

We easily have

$$J(0) = H_m(0)\left[1 + \frac{1}{\mu}\alpha(1-x)\left(\frac{c_2 - c_1}{c_2 - c_0}\right)\right] \geq 0,$$

$$J'(T) = \lambda'(T)\left[1 + \frac{1}{\mu}\alpha(1-x)\left(\frac{c_2 - c_1}{c_2 - c_0}\right)\right]\int_0^T \overline{F}(t) H_m(\alpha t)\, dt > 0.$$

Thus, if $\lambda(t)$ is strictly increasing then $Q_m(T)$ is also strictly increasing in m, and hence, T^* is strictly decreasing in m. Therefore, an optimal T^* reaches a minimum value when $m \to \infty$, and also, $C(T^*, m)$ becomes a minimum.

Therefore, when $m \to \infty$, T^* reaches a minimum value, and then, $C(T^*, m)$ also has its minimum. Consequently, (11) is rewritten as, by setting $m = \infty$,

$$\frac{\left[\lambda(T) + \alpha(1-x)\left(\frac{c_1 - c_0}{c_2 - c_0}\right)\right]\left[\int_0^T \overline{F}(t) e^{-\alpha(1-x)t}\, dt + \frac{1}{\mu}\right]}{1 + \frac{1}{\mu}\left[\lambda(T) + \alpha(1-x)\right]}$$

$$+ \alpha(1-x)\left(\frac{c_2 - c_1}{c_2 - c_0}\right)\int_0^T \overline{F}(t) e^{-\alpha(1-x)t}\, dt = \frac{c_2}{c_2 - c_0}. \qquad (17)$$

Let denote the left-hand side of (17) by $Q_\infty(T)$. Then, from the above discussions, the optimal policy (i) is rewritten as follows:

(i)' If $\lambda(t)$ is strictly increasing in t and $Q_\infty(\infty) > c_2/(c_2 - c_0)$, then there exists a finite and unique T^* which satisfies (17), and the expected cost rate is given in (15).

Example 2.1. We compute numerically an optimal checkpointing interval T^* which minimizes the expected cost $C(T, m)$ in (15). We assume that failures of processor A and processor B are caused by independent random factors. That is, failures occur according to a gamma distribution with order 2, *i.e.*,

$$F(t) \equiv 1 - (1 + 2\lambda t)e^{-2\lambda t}.$$

Then, we compute an optimal T^* which satisfies (11). Suppose that the mean time $1/\mu$ of the rollback recovery is a unit of time, the mean time of fault occurrences is $\mu/\lambda = 1800, 3600$, the mean time requested for communications between A and B is $\mu/\alpha = 20, 30$, the number of segments per a message is $n = 2, 4$, and the number of messages until the next checkpoint is $m = 20 \sim \infty$. When $1/\mu$ is 1 second, $1/\lambda$ is 30 or 60 minutes, and $1/\alpha$ is 20 or 30 seconds.

Suppose that the probability of accepting ACK depends on the length of a message. That is, when the transmission of a message which does not divide into segments fails with a probability q, the probability of accepting ACK for each segment is $p \equiv 1 - q/n$, and we set $q = 0.1, 0.2$. Moreover, suppose that the cost for system operation including checkpoint is $c_0 = 1$, the loss cost for rollback recovery of communication errors is $c_1/c_0 = 10$, and the loss cost for rollback recovery of processor failures is $c_2/c_1 = 2, 4$.

Under the above assumptions, we give numerical values $\mu T^*/60$ in Table 1 which minimize $C(T, m)$. When a unit time is a second, they are scaled to a unit of minute. Table 1 indicates that T^* decrease with c_2/c_1 and n. Further, T^* increase with μ/λ which is the mean interval of failure occurrences for processors and decrease with μ/α which is the mean interval of demand occurrences for communications. When μ/α is given, T^* decrease with m and have a minimum value as $m \to \infty$.

The increase of μ/λ means that the rate of processor failures decrease, *i.e.*, the expected number M_P of rollback recoveries for processor failures decrease with μ/λ. Table 1 indicates that T^* depends on the number m of messages, however, when λ is large, T^* depend little on m. Similarly, the increase of q means that the probability p of accepting ACK for each segment decrease, *i.e.*, the expected number M_C of rollback recoveries for communication errors increase with q. From the above viewpoints, these

Table 1 Optimal intervals $\mu T^*/60$ to minimize $C(T,m)$ when $c_1/c_0 = 10$

q	c_2/c_1	n	m	$\mu/\lambda = 1800$ μ/α 20	$\mu/\lambda = 1800$ μ/α 30	$\mu/\lambda = 3600$ μ/α 20	$\mu/\lambda = 3600$ μ/α 30
0.1	2	2	20	6.3	6.3	16.1	13.3
			50	6.3	6.3	12.8	12.7
			∞	6.3	6.3	12.8	12.7
		4	20	6.3	6.2	15.7	13.1
			50	6.2	6.2	12.6	12.5
			∞	6.2	6.2	12.6	12.5
	4	2	20	4.0	4.0	8.5	8.1
			50	4.0	4.0	8.2	8.1
			∞	4.0	4.0	8.2	8.1
		4	20	4.0	4.0	8.4	8.1
			50	4.0	4.0	8.1	8.0
			∞	4.0	4.0	8.1	8.0
0.2	2	2	20	6.7	6.5	18.5	14.5
			50	6.7	6.5	14.5	13.7
			∞	6.7	6.5	14.5	13.7
		4	20	6.5	6.3	16.8	13.6
			50	6.4	6.3	13.4	13.0
			∞	6.4	6.3	13.3	13.0
	4	2	20	4.2	4.1	8.3	8.5
			50	4.2	4.1	8.0	8.5
			∞	4.2	4.1	8.0	8.5
		4	20	4.1	4.0	8.7	8.3
			50	4.1	4.0	8.4	8.2
			∞	4.1	4.0	8.4	8.2

results indicate that when the expected number of rollback recoveries is large, T^* depends little on m, and when it is small, T^* almost depend on m. Further, from Table 1, when n is large, T^* depend little on m and μ/α, and have a constant value roughly. When the rate of c_2/c_1 is large, T^* also have a constant value roughly, which depends little on n, m and μ/α. □

3 Mobile Communication System with Error Recovery Schemes

This section considers a mobile communication system which consists of mobile stations, several base stations and a switching center. We formulate a stochastic model of a mobile network system which reflects actual behaviors of mobile stations. That is, when a mobile station communicates with a base station, if it moves from one area (cell) to another, in which a base station can connect with it, then it interrupts the operation transiently. Then, the connection of a mobile station is changed from an old base station to a new one, and after that, this process such as handoff has been completed and the operation restarts again.

Under the above stochastic model, we obtain the reliability quantities of a mobile communication system such as the mean time to take checkpoint and the expected numbers of successful transmissions until communication errors occur and of handoff, using the theory of Markov renewal processes. Further, the expected costs are derived and optimal checkpointing intervals which minimize them are analytically discussed. Finally, numerical examples are given.

3.1 Reliability Quantities

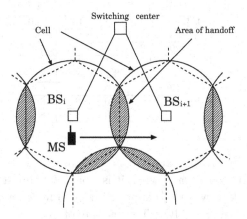

Fig. 3 Outline of a mobile communication system

A mobile communication system consists of mobile stations and several base stations shown in Figure 3: Each base station is connected by wired links through a switching center, and one mobile station communicates with the others by wireless links through a base station $BS_i (i = 0, 1, 2, \cdots)$. An area in which a base station can connect with a mobile station is called *cell*. A mobile station moves one cell to another and its connection changes from BS_i to BS_{i+1}. This process is called *handoff*. The communications between mobile stations can be realized by such control mechanisms.

We concern only about the communication behaviors of the system with mobile station MS and base stations $BS_i (i = 0, 1, 2, \cdots)$, and apply LP(Logging Pessimistic) recovery scheme to the system [15].

In LP recovery scheme, if a mobile station MS communicates with BS_i, the checkpoint and message logs of a mobile station MS are stored in BS_i. When MS moves from one cell of the base station BS_i to another cell of BS_{i+1}, the checkpoint and message logs are sent from BS_i to BS_{i+1}. If communication errors have occurred, a mobile station MS demands the checkpoint and message logs from BS_i. After MS has received the checkpoint and message logs, the rollback recovery for MS associated with such an event is executed from that time to the most recent checkpoint [15].

We assume that:

1) The system begins to operate at time 0 and takes the first checkpoint for BS_0. Next, it takes the checkpoint for BS_i that manages the operation of mobile station MS, when the transmissions between MS and BS_i have terminated successfully at m ($m = 1, 2, \cdots$) times. The state of processes which are executed at MS is sent to BS_i.

2) A mobile station MS begins to move from BS_0. The request time for transmissions between MS and BS_i has a general distribution $A(t)$ with finite mean α and the time required for transmission of one message including the time to save the message log at BS_i has an exponential distribution $(1 - e^{-at})$ $(0 < a < \infty)$.

3) Communication errors of MS occur according to an exponential distribution $(1 - e^{-\lambda t})$ $(0 < \lambda < \infty)$. When communication errors have occurred, the rollback recovery for MS associated with such an event is executed from that time to the most recent checkpoint. The state of processes and message logs are sent from BS_i to MS.

   a) The system is regenerated by rollback recovery.
   b) The time required for rollback recovery has a general distribution $V(t)$ with finite mean v.

4) When the operation of MS moves from BS_i to BS_{i+1}, i.e., when MS goes into the area of handoff, it interrupts its operation transiently.

   a) Handoff occurs according to a general distribution $U(t)$ with finite mean $1/u$.
   b) The time required for handoff including the time to transmit the most recent checkpoint and message logs, has a general distribution $G(t)$ with finite mean $1/\mu$.

Under the above assumptions, we define the following states of the system:

State 0: The system begins to operate or restart.
State 1: Request for transmissions between MS and BS_i occurs.
State 2: Communication errors occur.
State 3: Handoff occurs.
State 4: Transmission of one message succeeds.
State S: Transmissions of m messages have succeeded, and the system takes the checkpoint for BS_i.

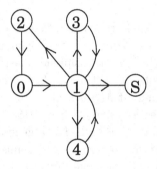

Fig. 4 Transition diagram between system states

The system states defined above form a Markov renewal process [35], where State S is an absorbing state and States 0–4 are regeneration points. Transition diagram between system states is shown in Figure 4.

Using the mass functions of Markov renewal processes, LS transforms $Q_{1j}^*(s)$ of transition probabilities $Q_{1j}(t)$ from State 1 to State j ($j = 2, 3, 4$) are:

$$Q_{1,2}^*(s) = \int_0^\infty \lambda e^{-(s+\lambda+a)t} \overline{U}(t)\,dt = \frac{\lambda}{s+\lambda+a}[1 - U^*(s+\lambda+a)], \quad (18)$$

$$Q_{1,3}^*(s) = \int_0^\infty e^{-(s+\lambda+a)t}\,dU(t) = U^*(s+\lambda+a), \quad (19)$$

$$Q_{1,4}^*(s) = \int_0^\infty a e^{-(s+\lambda+a)t} \overline{U}(t)\,dt = \frac{a}{s+\lambda+a}[1 - U^*(s+\lambda+a)]. \quad (20)$$

Thus, LS transform $H_{0,S}^*(s)$ of the first-passage time distribution $H_{0S}(t)$ from the beginning of operation to the next checkpoint is

$$H_{0,S}^*(s) \equiv \frac{[A^*(s)M^*(s)]^m}{1 - Z^*(s)V^*(s)}, \qquad (21)$$

where

$$M(t) \equiv \sum_{i=1}^{\infty} [Q_{13}(t) * G(t)]^{(i-1)} * Q_{14}(t),$$

$$X(t) \equiv \sum_{i=1}^{\infty} [Q_{13}(t) * G(t)]^{(i-1)} * Q_{12}(t),$$

$$Z(t) \equiv \sum_{j=1}^{m} [A(t) * M(t)]^{(j-1)} * [A(t) * X(t)].$$

Note that $M(t)$ is a probability distribution that one transmission succeeds and $X(t)$ is a probability distribution that communication errors occur. Therefore, the mean time $\ell_{0,S}(m)$ is

$$\begin{aligned}\ell_{0,S}(m) &\equiv \lim_{s \to 0} \frac{-\mathrm{d}H_{0,S}^*(s)}{\mathrm{d}s} \\ &= \frac{1 - M^m}{(1-M)M^m} \left[\alpha + \frac{1}{\mu}D + \frac{1}{\lambda + a} + v(1-M) \right], \qquad (22)\end{aligned}$$

where

$$D \equiv U^*(\lambda + a)/[1 - U^*(\lambda + a)], \qquad M \equiv a/(\lambda + a).$$

We derive the expected numbers of rollback recoveries caused by communication errors and of handoff. LS transforms $M_R^*(s)$ of $M_R(t)$ which is the expected number of rollback recoveries caused by communication errors and LS transform $M_H^*(s)$ of $M_H(t)$ which is the expected number of handoff from the beginning of operation to the next checkpoint are, respectively,

$$M_R^*(s) = \frac{Z^*(s)V^*(s)}{1 - Z^*(s)V^*(s)}, \qquad (23)$$

$$M_H^*(s) = \frac{\sum_{j=1}^{m} [A^*(s)M^*(s)]^{j-1} \frac{Q_{13}^*(s)}{1 - Q_{13}^*(s)G^*(s)}}{1 - \sum_{j=1}^{m} [A^*(s)M^*(s)]^{j-1} A^*(s)X^*(s)V^*(s)}. \qquad (24)$$

Therefore, the expected numbers of rollback recoveries $M_R(m)$ and the expected number $M_H(m)$ are

$$M_R(m) \equiv \lim_{s \to 0} M_R^*(s) = \frac{1 - M^m}{M^m}, \qquad (25)$$

$$M_H(m) \equiv \lim_{s \to 0} M_H^*(s) = \left[\frac{1 - M^m}{(1-M)M^m} \right] D. \qquad (26)$$

3.2 Optimal Policy

Let c_1 be the cost for system operation, c_2 be the cost for handoff and c_3 be the cost for rollback recovery of communication errors. Then, we give the expected cost rate of transmission number until the next checkpoint as

$$C(m) \equiv \frac{c_1 + c_2 M_H(m) + c_3 M_R(m)}{m} \quad (m = 1, 2, \cdots). \qquad (27)$$

We seek an optimal checkpointing interval m^* which minimizes $C(m)$ for $c_3 \geq c_2 > c_1$. From the inequality $C(m+1) - C(m) \geq 0$,

$$\frac{m}{M^{m+1}} - \left(\frac{M}{1-M}\right)\left(\frac{1-M^m}{M^{m+1}}\right) \geq \frac{c_1}{Dc_2 + (1-M)c_3} \quad (m = 1, 2, \cdots). \qquad (28)$$

Denoting the left-hand side of (28) by $L(m)$, Then, $L(1) = (1-M)/M^2$ and $L(\infty) = \infty$, and hence, there exist a finite $m^*(1 \leq m^* < \infty)$ which satisfies (28). Further, we have the following optimal policy:

1) If $L(1) < c_1/[Dc_2 + (1-M)c_3]$, then there exists a finite and unique $m^*(>1)$ which satisfies (28).
2) If $L(1) \geq c_1/[Dc_2 + (1-M)c_3]$, then $m^* = 1$.

Example 3.1. We compute numerically optimal checkpointing intervals m^* which satisfy (28). It is assumed that the handoff is caused by some random factors of a mobile station, and occurs according to an exponential distribution, i.e., $U(t) \equiv 1 - e^{-ut}$. Suppose that the mean time $1/\mu$ of handoff is a unit of time, the request time for transmissions is $\mu/\alpha = 30$, the mean time of communication errors is $\mu/\lambda = 1800, 3600$, the mean time required for transmissions is $\mu/a = 30 \sim 120$ and the mean time of handoff occurrence is $1/u = 30 \sim 1800$. For example, when $1/\mu = 1$ second, $1/\lambda = 30 \sim 60$ minutes. Introduce the following costs: The cost for system operation is $c_1 = 1$, the loss cost for handoff is $c_2/c_1 = 2, 5$ and the loss cost for rollback recovery is $c_3/c_1 = 10, 50$.

Under the above assumptions, we give numerical values m^* in Table 2. Table 2 indicates that m^* decrease with μ/a and increase with μ/λ. This can be interpreted that when the time required for transmissions is large, the checkpoint should be made at a small number m^*. Moreover, m^* increase with μ/u and decrease with c_2/c_1. Similarly, m^* decrease with c_3/c_1. However, when μ/λ are small, and $\mu/a, c_3/c_1$ are large, m^* depend little on μ/u and become constant. □

Table 2 Optimal numbers m^* to minimize $C(m)$

c_2/c_1	c_3/c_1	μ/u	$\mu/\lambda = 1800$ μ/a			$\mu/\lambda = 3600$ μ/a		
			30	60	120	30	60	120
2	10	30	7	3	1	10	5	2
		60	9	4	2	14	7	3
		300	16	8	4	27	13	6
		600	19	9	4	33	16	8
		1800	21	11	5	40	20	10
	50	30	6	3	1	9	4	2
		60	7	3	1	12	6	3
		300	10	5	2	18	9	4
		600	10	5	2	20	10	5
		1800	11	5	2	21	10	5
5	10	30	4	2	1	6	3	1
		60	6	3	1	9	4	2
		300	12	6	3	19	9	4
		600	15	7	4	25	12	6
		1800	19	9	5	34	17	8
	50	30	4	2	1	6	3	1
		60	5	2	1	8	4	2
		300	9	4	2	15	7	3
		600	10	5	2	18	9	4
		1800	10	5	2	20	10	5

4 Communication System with Window Flow Control Scheme

This section consider a communication system which consists of several clients and a web server, and formulate a stochastic model of a communication system using a window flow control scheme considering error of ECN message: If the ECN bit is not set from the absence of congestion, the number of packets, which corresponds to a window size, is successively transmitted to a client by a web server. If it is set from the presence of congestion, the number of packets, which correspond to half of the first window size, are transmitted. The mean time until packet transmissions succeed is derived. An optimal policy which maximizes the amount of packets per unit of time until the transmission succeeds is analytically discussed. Finally, numerical examples are given.

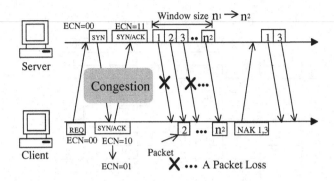

Fig. 5   Outline of Window Flow Control Scheme with ECN

4.1 Reliability Quantities

We consider a communication system which consists of several clients and a web server as shown in Figure 5. The data transmission is implemented by the Selective-Repeat Protocol which is the usual retransmission control between the server and the client [21,36]. Then, we formulate the stochastic model as follows:

1) Congestion happens in the network according to an exponential distribution $(1-e^{-\lambda t})(0 < \lambda < \infty)$ and continues according to an exponential distribution $(1 - e^{-\gamma t})(0 < \gamma < \infty)$. We define the following states of a network system:

   State 0: No congestion occurs and the network system is in a normal condition.
   State 1: Congestion occurs.

The network system states defined above form a two-state Markov process. A transition diagram of a network system states is shown in Figure 6.

Fig. 6   Transition diagram of a network system states

Thus, we have the following probabilities under the initial condition that $P_{00}(0) = P_{11}(0) = 1, P_{01}(0) = P_{10}(0) = 0$:

$$P_{00}(t) \equiv \frac{\gamma}{\lambda+\gamma} + \frac{\lambda}{\lambda+\gamma}e^{-(\lambda+\gamma)t},$$

$$P_{11}(t) \equiv \frac{\lambda}{\lambda+\gamma} + \frac{\gamma}{\lambda+\gamma}e^{-(\lambda+\gamma)t},$$

$$P_{01}(t) = 1 - P_{00}(t), \qquad P_{10}(t) = 1 - P_{11}(t),$$

where $P_{ij}(t)$ are probabilities that the network system is in State i ($i = 0, 1$) at time 0 and State j ($j = 0, 1$) at time t (> 0).

2) A client transmits the request for data packets to the server. The request information is included in a window size, and the request requires a time according to the general distribution $A(t)$ with finite mean a.

3) The server establishes connection when a client requests data packets. Then, the server and the client confirm whether the ECN bit is set. That is, when congestion happens in the network, routers have set the ECN bit in the packet header. Then, we assume the probability that ECN bit is error: If no congestion occurs, the probability that a ECN bit is set from router error, is α ($0 < \alpha < 1$). If congestion occurs, the probability that a ECN bit is not set from router error, is β ($0 < \beta < 1$). The server transmits the notification for connection completion to the client, and the notification requires the time according to an exponential distribution $A_1(t) = 1 - e^{-t/a_1}$ with finite mean a_1. The client transmits the acknowledgement for the notification to the sever, and the acknowledgement requires the time according to an exponential distribution $A_2(t) = 1 - e^{-t/a_2}$ with finite mean a_2.

4) If the ECN bit is not set, the number of packets n_1, which correspond to a window size, is successively transmitted to the client from the server. Then, if no congestion occurs, the probability that a packet loss occurs is p_0 ($0 < p_0 < 1$). If congestion occurs, the probability that a packet loss occurs is p_1 ($> p_0$).

5) If ECN bit is set, n_2 ($< n_1$) packets, which corresponds to half of the first window size, are transmitted. Then, the probability that a packet loss occurs is p_0. If the server has received ACK for the first time, the remaining packets n_2 are transmitted again. The process of editing and transmitting the remaining packets n_2 requires the time according to a general distribution $W(t)$ with finite mean w.

6) The process of editing and transmitting the data requires the time according to a general distribution $B(t)$ with finite mean b.

7) When the client has received n_1 or n_2 packets correctly, it returns ACK. When the server has received NAK, the retransmission for only loss packets is made. The time the server takes to transmit the last packet to receive ACK or NAK has a general distribution $D(t)$ with finite mean d.
8) If the retransmission has failed at k times again, the server interrupts, and the transmission are made again from the beginning of its initial state after a constant time μ, where $G(t) \equiv 0$ for $t < \mu$ and 1 for $t \geq \mu$.
9) If the server has received ACK for all packets n_1, the transmission succeeds

Under the above assumptions, we define the following states of the system:

State 2: System begins to operate.
State 3: Connection establishment from the client begins.
State 4: n_1 packet transmission begins (no congestion occurs and no ECN bit has been set).
State 5: n_1 packet transmission begins (congestion occurs and no ECN bit has been set).
State 6: n_2 packet transmission begins (congestion occurs and ECN bit has been set).
State 7: n_2 packet transmission begins (no congestion occurs and ECN bit has been set).
State F: Retransmission fails k times and interrupted.
State S_2: n_2 packet transmission of first time succeeds and n_2 packet transmission of second time begins.
State S_1: n_1 packet transmission succeeds.

The system states defined above form a Markov renewal process, where S_1 is an absorbing state. A transition diagram between system states is shown in Figure 7.

We derive transition probabilities from State i to State j. Let $Q_{i,j}(t)(i = 3, 4, 5, 6, 7, S_2; j = 4, 5, 6, 7, S_1, S_2, F)$ be one-step transition probabilities of a Markov renewal process. For convenience, we define $B_i(n, m, p)$ as a probability distribution that m packet losses occur when n packets are successively transmitted to the client, and a packet loss occurs with probability p $(0 < p < 1)$:

Reliability Analysis of Communication Systems

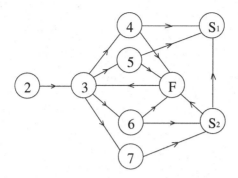

Fig. 7 Transition diagram between system states

$$B_i(n,m,p) \equiv \binom{n}{m} p^m (1-p)^{n-m} B(t)^{(n)} * D(t).$$

Moreover, we define $Q_S(t|n,p,k)$ as a probability distribution that the transmission of n packets succeeds at k times, and define $Q_F(t|n,p,k)$ as the probability distribution that the transmission of n packets fails at k times:

$Q_S(t|n,p,k) \equiv$

$(1-p)^n B^{(n)}(t) * D(t) + \sum_{m_1=1}^{n} B_i(n,m_1,p) * (1-p)^{m_1} B^{(m_1)}(t) * D(t)$

$+ \sum_{m_1=1}^{n} B_i(n,m_1,p) * \sum_{m_2=1}^{m_1} B_i(m_1,m_2,p) * (1-p)^{m_2} B^{(m_2)}(t) * D(t) + \cdots$

$+ \sum_{m_1=1}^{n} B_i(n,m_1,p) * \sum_{m_2=1}^{m_1} B_i(m_1,m_2,p) * \cdots * \sum_{m_{k-1}=1}^{m_{k-2}} B_i(m_{k-2},m_{k-1},p)$

$* (1-p)^{m_{k-1}} B^{(m_{k-1})}(t) * D(t) \quad (k=3,4,\cdots),$ \hfill (29)

$Q_F(t|n,p,k) \equiv$

$\sum_{m_1=1}^{n} B_i(n,m_1,p) * \sum_{m_2=1}^{m_1} B_i(m_1,m_2,p) \cdots * \sum_{m_{k-1}=1}^{m_{k-2}} B_i(m_{k-2},m_{k-1},p)$

$* [1-(1-p)^{m_{k-1}}] B(t)^{(m_{k-1})} * D(t), \quad (k=3,4,\cdots).$ \hfill (30)

LS transforms $Q_{i,j}^*(s)$ ($i=3,4,5,6,7,S_2; j=4,5,6,7,S_1,S_2,F$) of transition probabilities $Q_{i,j}(t)$ ($i=3,4,5,6,7,S_2; j=4,5,6,7,S_1,S_2,F$) are:

$$Q_{3,4}^*(s) = a_1 a_2 (1-\alpha)^2 \widetilde{P}_{0,0}(s+a_1)\widetilde{P}_{0,0}(s+a_2), \tag{31}$$

$$Q_{3,5}^*(s) = a_1 a_2 [(1-\alpha)\beta$$
$$\times \left\{ \widetilde{P}_{0,1}(s+a_1)\widetilde{P}_{1,0}(s+a_2) + \widetilde{P}_{0,0}(s+a_1)\widetilde{P}_{0,1}(s+a_2) \right\}$$
$$+ \beta^2 \widetilde{P}_{0,1}(s+a_1)\widetilde{P}_{1,1}(s+a_2), \tag{32}$$

$$Q_{3,6}^*(s) = a_1 a_2 [[1-(1-\alpha)\beta]$$
$$\times \left\{ \widetilde{P}_{0,0}(s+a_1)\widetilde{P}_{0,1}(s+a_2) + \widetilde{P}_{0,1}(s+a_1)\widetilde{P}_{1,0}(s+a_2) \right\}$$
$$+ (1-\beta^2)\widetilde{P}_{0,1}(s+a_1)\widetilde{P}_{1,1}(s+a_2), \tag{33}$$

$$Q_{3,7}^*(s) = a_1 a_2 \alpha (2-\alpha) \widetilde{P}_{0,0}(s+a_1)\widetilde{P}_{0,0}(s+a_2), \tag{34}$$

$$Q_{4,S_1}^*(s) \equiv Q_S^*(s|n_1,p_0,k), \quad Q_{5,S_1}^*(s) \equiv Q_S^*(s|n_1,p_1,k), \tag{35}$$

$$Q_{6,S_2}^*(s) = Q_{S_2,S_1}^*(s) = Q_{7,S_2}^*(s) \equiv Q_S^*(s|n_2,p_0,k), \tag{36}$$

$$Q_{4,F}^*(s) \equiv Q_F^*(s|n_1,p_0,k), \quad Q_{5,F}^*(s) \equiv Q_F^*(s|n_1,p_1,k), \tag{37}$$

$$Q_{6,F}^*(s) = Q_{S_2,F}^*(s) = Q_{7,F}^*(s) \equiv Q_F^*(s|n_2,p_0,k), \tag{38}$$

where $\widetilde{P}_{i,j}(s) \equiv \int_0^\infty e^{-st} P_{i,j}(t) dt$ ($i,j=0,1$). LS transforms $Q_S^*(s|n,p,k)$ and $Q_F^*(s|n,p,k)$ of (29) and (30) are, respectively,

$$Q_S^*(s|n,p,k) = [(1-p)B^*(s)]^n \sum_{j=1}^k [D^*(s)]^j$$
$$\times \left[\left\{ \sum_{i=0}^{j-1} [pB^*(s)]^i \right\}^n - \left\{ \sum_{i=0}^{j-2} [pB^*(s)]^i \right\}^n \right] \quad (k=1,2,\cdots), \tag{39}$$

$$Q_F^*(s|n,p,k) = [B^*(s)]^n [D^*(s)]^k$$
$$\times \left[\left\{ (1-p)\sum_{i=0}^{k-2} [pB^*(s)]^i + [pB^*(s)]^{k-1} \right\}^n - \left\{ (1-p)\sum_{i=0}^{k-1} [pB^*(s)]^i \right\}^n \right]$$
$$(k=1,2,\cdots), \tag{40}$$

where $\sum_{i=0}^{-1} \equiv 0$ and $Q_S^*(0|n,p,k) + Q_F^*(0|n,p,k) = 1$.

First, we derive the mean time ℓ_{2,S_1} until transmission of n_1 packets succeeds. Let LS transform $H_{2,S_1}^*(s)$ be the first-passage time distribution

$H_{2,S_1}(t)$ from State 2 to State S_1. Then,
$H^*_{2,S_1}(s) =$

$$= \frac{A^*(s) \left\{ \begin{array}{c} Q^*_{3,4}(s)Q^*_S(s|n_1,p_0,k) + Q^*_{3,5}(s)Q^*_S(s|n_1,p_1,k) \\ +Q^*_{3,6}(s)W^*(s)\left[Q^*_S(s|n_2,p_0,k)\right]^2 \\ +Q^*_{3,7}(s)W^*(s)\left[Q^*_S(s|n_2,p_0,k)\right]^2 \end{array} \right\}}{1 - \left\{ \begin{array}{c} Q^*_{3,4}(s)Q^*_F(s|n_1,p_0,k) + Q^*_{3,5}(s)Q^*_F(s|n_1,p_1,k) \\ +Q^*_{3,6}(s)Q^*_F(s|n_2,p_0,k)\left[1 + W^*(s)Q^*_S(s|n_2,p_0,k)\right] \\ +Q^*_{3,7}(s)Q^*_F(s|n_2,p_0,k)\left[1 + W^*(s)Q^*_S(s|n_2,p_0,k)\right] \end{array} \right\} G^*(s)}.$$
(41)

Hence, the mean time ℓ_{2,S_1} is

$$\ell_{2,S_1} \equiv \lim_{s \to 0} \frac{-\mathrm{d}H^*_{2,S_1}(s)}{\mathrm{d}s}$$
$$= (a - \mu)$$

$$+ \frac{\mu + \frac{1}{a_1} + \frac{1}{a_2}}{1 - Q^*_{3,4}(0)(Z^k_0)^{n_1} - Q^*_{3,5}(0)(Z^k_1)^{n_1} - (Q^*_{3,6}(0) + Q^*_{3,7}(0))(Z^k_0)^{2n_2}}$$

$$+ b \left[\begin{array}{c} \left\{ n_1 Q^*_{3,4}(0) + n_2[Q^*_{3,6}(0) + Q^*_{3,7}(0)][2 - (Z^k_0)^{n_2}] \right\} \sum_{i=0}^{k-1} p_0{}^i \\ + n_1 Q^*_{3,5}(0) \sum_{i=0}^{k-1} p_1{}^i \end{array} \right]$$

$$+ d \left[\begin{array}{c} Q^*_{3,4}(0)(X^k_0)^{n_1} + Q^*_{3,5}(0)(X^k_1)^{n_1} \\ + [Q^*_{3,6}(0) + Q^*_{3,7}(0)][2 - (Z^k_0)^{n_2}](X^k_0)^{n_2} \end{array} \right]$$

$$+ w[Q^*_{3,6}(0) + Q^*_{3,7}(0)](1 - p_0{}^k)^{n_2}$$
(42)

where $(Z^{j_2}_{j_1})^{j_3} \equiv 1 - [1 - (p_{j_1})^{j_2}]^{j_3}$ and $(X^{j_2}_{j_1})^{j_3} \equiv j_2 - \sum_{i=0}^{j_2-1}(1-p_{j_1}{}^i)^{j_3}$,

$$Q^*_{3,4}(0) \equiv \frac{(1-\alpha)^2(a_1+\gamma)(a_2+\gamma)}{(a_1+\lambda+\gamma)(a_2+\lambda+\gamma)},$$

$$Q^*_{3,5}(0) \equiv \frac{\beta\lambda[(1-\alpha)(a_1+2\gamma) + \beta(a_2+\lambda)]}{(a_1+\lambda+\gamma)(a_2+\lambda+\gamma)},$$

$$Q^*_{3,6}(0) \equiv \frac{\lambda \left[\begin{array}{c} (1 - (1-\alpha)\beta))(a_1+2\gamma) \\ +(1-\beta^2)(a_2+\lambda) \end{array} \right]}{(a_1+\lambda+\gamma)(a_2+\lambda+\gamma)},$$

$$Q_{3,7}^*(0) \equiv \frac{\alpha(2-\alpha)(a_1+\gamma)(a_2+\gamma)}{(a_1+\lambda+\gamma)(a_2+\lambda+\gamma)}.$$

4.2 Optimal Policy

We discuss an optimal policy which maximizes the transmission of all packets per unit of time until it succeeds. We consider the particular case that $n_1 \equiv 2n_2$ to simplify the optimal policy. We define the throughput $E(n_2)$, which represents the rate of n_1 packets to their mean transmission times, as the following equation:

$$E(n_2) \equiv \frac{2n_2}{\ell_{2,S_1}(n_2)} = \frac{2n_2}{Y(n_2)+(a-\mu)}, \tag{43}$$

where

$$Y(n_2) \equiv \frac{\mu + \frac{1}{a_1} + \frac{1}{a_2}}{1 - Q_{3,5}^*(0)(Z_1^k)^{2n_2} - (Q_{3,4}^*(0)+Q_{3,6}^*(0)+Q_{3,7}^*(0))(Z_0^k)^{2n_2}}.$$

(The numerator also contains:
$$+bn_2\left[\begin{array}{c}\{2Q_{3,4}^*(0)+[Q_{3,6}^*(0)+Q_{3,7}^*(0)][2-(Z_0^k)^{n_2}]\}\sum_{i=0}^{k-1}p_0^i\\ +2Q_{3,5}^*(0)\sum_{i=0}^{k-1}p_1^i\end{array}\right]$$
$$+d\left[\begin{array}{c}Q_{3,4}^*(0)(X_0^k)^{2n_2}+Q_{3,5}^*(0)(X_1^k)^{2n_2}\\ +[Q_{3,6}^*(0)+Q_{3,7}^*(0)][2-(Z_0^k)^{n_2}](X_0^k)^{n_2}\end{array}\right]$$
$$+w[Q_{3,6}^*(0)+Q_{3,7}^*(0)](1-p_0^k)^{n_2}$$
)

We seek an optimal window size n_2^* which maximizes $E(n_2)$ in (43). We put formally that $A_1(n_2) \equiv 1/E(n_2)$ and seek an optimal n_2^* which minimizes $A_1(n_2)$. From the inequality $A_1(n_2+1) - A_1(n_2) \geq 0$,

$$n_2 Y(n_2+1) - (n_2+1)Y(n_2) \geq a - \mu. \tag{44}$$

Denoting the left side of (44) by $L(n_2)$,

$$L(n_2+1) - L(n_2) = (n_2+1)D(n_2), \tag{45}$$

where

$$D(n_2) \equiv [Y(n_2+2) - Y(n_2+1)] - [Y(n_2+1) - Y(n_2)].$$

From (45), when $Y(n_2)$ is a convex function and $D(1) > 0$, $L(n_2)$ is strictly increasing in n_2 from $L(1)$ to $L(\infty)$. Therefore, we have the following optimal policy:

1) If $L(1) < a - \mu$ and $L(\infty) > a - \mu$, then there exists a finite and unique $n_2{}^*(> 1)$ which satisfies (44).
2) If $L(1) \geq a - \mu$, then $n_2{}^* = 1$.
3) If $L(\infty) \leq a - \mu$, then $n_2{}^* = \infty$.

Example 4.1. We give a numerical problem to maximize the throughput $E(n_2)$. We assume the case of $k = 2$ and suppose that the mean time b until editing the data and transmitting one packet is a unit time. It is assumed that the mean time required for data packets is $a/b = 10$, the mean generation interval of network congestion is $(1/\lambda)/b = 60, 600$, the mean time until the congestion clears up is $(1/\gamma)/b = 10, 100$, the mean time required for the notification of connection completion is $(1/a_1)/b = 5$, the mean time required for the acknowledgement of connection completion is $(1/a_2)/b = 5$, the probability that the ECN bit is set from router error when no congestion happens is $\alpha = 0 \sim 0.5$, the probability that the ECN bit is not set from router error when congestion happens is $\beta = 0 \sim 0.5$, the mean time for the server to transmit all packets to receive ACK or NAK is $d/b = 2 \sim 32$, the mean time from editing the data to n_2 transmit again is $w/b = 2 \sim 10$, the mean time for the server to interrupt n_2 retransmission to restart again is $\mu/b = 30$ and the probability that loss packets occur is $p_0 = 0.04, 0.05$ and $p_1 = 0.1, 0.2$.

Table 3 gives the optimal window size n_2^* which maximizes the throughput $E(n_2{}^*)$, the mean time $\ell_{2,S_1}(n_2{}^*)$ and the throughput $E(n_2{}^*)$. For example, when $p_1 = 0.1, p_2 = 0.05, d/b = 8, (1/\lambda)/b = 60$ and $\gamma/b = 10$, $n_2{}^* = 54$, $\ell_{2,S_1}(n_2^*) = 207.9$ and $E(n_2^*) = 0.520$. This indicates that n_2^* increase with d/b and decrease with p_0. Under the same value p_0, n_2^* decrease with p_1. Under the same value p_0, p_1, $\ell_{2,S_1}(n_2^*)$ decrease with $1/\lambda$ and $E(n_2^*)$ increase with $1/\lambda$.

Further, Figure 8 gives $E(n_2^*)$ for $w/b = 2, \alpha$ and β, and Figure 9 gives $E(n_2^*)$ for $w/b = 10, \alpha$ and β when $p_0 = 0.05, p_1 = 0.2, d/b = 2, (1/\lambda)/b = 60, (1/\gamma)/b = 10$. $E(n_2^*)$ decrease with β. When $w/b = 2$, $E(n_2^*)$ increase with α. But, when $w/b = 10$, $E(n_2^*)$ decrease with α. The increase of α means that the probability of transmitting half of the first window size increase, i.e., the probability that loss packets occur decreases with α and $E(n_2^*)$ increases with α. On the other hand, when the mean time w/b from

Table 3 Optimal window size $n_2{}^*$ to maximize $E(n_2)$

			$(1/\lambda)/b = 60, (1/\gamma)/b = 10$			$(1/\lambda)/b = 600, (1/\gamma)/b = 100$		
p_0	p_1	d/b	$n_2{}^*$	$\ell_{2S_1}(n_2{}^*)$	$E(n_2{}^*)$	$n_2{}^*$	$\ell_{2S_1}(n_2{}^*)$	$E(n_2{}^*)$
0.04	0.1	2	58	181.3	0.6397	58	180.1	0.6423
		8	71	241.7	0.5875	69	232.7	0.5930
		16	84	313.2	0.5364	82	302.2	0.5426
	0.2	2	59	185.0	0.6378	58	180.1	0.6440
		8	71	242.3	0.5860	70	236.2	0.5928
		16	84	313.8	0.5353	82	302.3	0.5425
0.05	0.1	2	45	156.4	0.5753	44	151.9	0.5793
		8	54	207.9	0.5195	53	202.1	0.5244
		16	64	274.6	0.4661	63	267.1	0.4717
	0.2	2	45	157.0	0.5731	44	152.0	0.5790
		8	55	212.5	0.5177	53	202.2	0.5242
		16	64	275.4	0.4647	63	267.2	0.4715

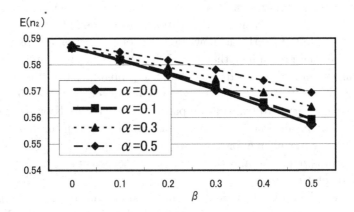

Fig. 8 Throughput $E(n_2^*)$ for $w/b = 2$

editing the data to n_2 transmit again is large, the mean time $\ell_{2,S_1}(n_2^*)$ increases and $E(n_2^*)$ decreases with α. □

5 Conclusions

This chapter has studied analytically the three stochastic models of communication systems such as a communication system by applying the recovery techniques of checkpoint and rollback, a mobile communication system with recovery scheme and a communication system using a window flow control

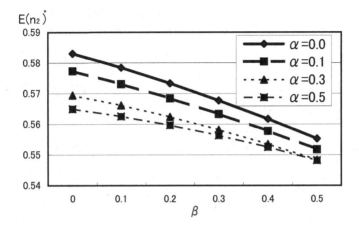

Fig. 9   Throughput $E(n_2^*)$ for $w/b = 10$

scheme. Further, we have derived the reliability measures by using the theory of Markov renewal processes, and have discussed the optimal policy which minimizes the expected cost. Finally, we have given the numerical examples of each model in order to understand the results easily, and have evaluated them under some standard parameters. If these parameters would be statistically estimated from actual systems, we could determine the best policy.

References

1. Strom, R. and Yemini, S. (1985). *Optimistic recovery in distributed systems*, ACM Transactions on Computer Systems, **3**, 3, pp. 204–226.
2. David B. J. and Willy, Z. (1990). *Recovery in distributed systems using optimistic message logging and check-pointing*, Journal of Algorithms, **11**, 3, pp. 462–491.
3. Elmootazbellah N.E. and Willy, Z. (1992). *Manetho:Transparent Rollback-Recovery with Low Overhead, Limited Rollback and Fast Output Commit*, IEEE Transactions on Computers Special Issue on Fault-Tolerant Computing, **41**, 5, pp. 526–531.
4. Wang, Y. M. and Fuchs, W. K.(1993). *Lazy checkpoint coordination for bounding rollback propagation*, Reliable Distributed Systems, 1993. Proceedings., 12th Symposium on, pp. 78–85.
5. Kim, J. L. and Park, T. (1993). *An efficient protocol forcheckpointing recovery in distributed systems*, IEEE Transactions on Parallel and Distributed systems, **4**, 8, pp. 955–960.

6. Chandy, K. M. and Lamport (1985). *Distributed snapshots : Determining global states of distributed systems*, ACM Transactions on Computer Systems, **3**, 1, pp. 63–75.
7. Yoneda, K., Matsubara, T. and Koga, Y. (1997). *Investigation of multiprocessor system with rollback function*, The Transactions of the Institute of Electronics, Information and Communication Engineers of Japan, **FTS97-20**, pp. 27–33.
8. Randell, B. (1975). *System structure for software fault tolerance*, IEEE Transactions on Software Engineering, **SE-1**, 2, pp. 220–232.
9. Kimura, M., Yasui, K. and Nakagawa, T. (1999). *Rollback-Recovery for a Communication System with Two Failure States*, The Transactions of the Institute of Electronics, Information and Communication Engineers of Japan, **J82-A** 8, pp. 1383–1390.
10. Rangarajan, S., Ratnam, K. and Dahbura, A. T. (1995). *A fault-tolerant protocol for location directory maintenance in mobile networks*, Proceedings of FTCS-25, pp. 164–173.
11. Biaz, S. and Vaidya, N. H. (1998). *Tolerating visitor location register failure in mobile environments*, Reliable Distributed Systems-SRDS'98, pp. 109–117.
12. Fang, Y., Chlamtac, I. and Fei, H. (2000). *Analytical results for optimal choice of location update interval for mobility database failure restoration in PCS Networks*, IEEE Transactions on Parallel and Distributed Systems, **11**, 6, pp. 615–624.
13. Prakash, R. and Singhal, M. (1996). *Low-cost checkpointing and failure recovery in mobile computing systems*, IEEE Transactions on Parallel and Distributed Systems, **7**, 10, pp.1035–1048.
14. Acharya, A. and Badrinath, B. R. (1994). *Checkpointing distributed applications on mobile computers*, IEEE Proceedings of 3rd International Conference on PDIS, pp. 73–80.
15. Pradhan, D. K. and Krishna, P. (1996). *Recoverable mobile environment, Design and trade-off analysis*, Proceedings of FTCS-26, pp. 16–25.
16. Yagi, M., Kaneko, K. and Ito, H.(2000). *Error recovery schemes in mobile environments*, The Transactions of the Institute of Electronics, Information and Communication Engineers of Japan, **J83-D-I**, 3, pp. 348–359.
17. Kimura, M., Yasui, K., Nakagawa, T. and Ishii, N. (2001). *Reliability Considerations of a Mobile Network System with Error Recovery Schemes*, The Transactions of Information Processing Society of Japan, **42**, 7, pp. 1811–1818.
18. Jacobson, V. (1988). *Congestion Avoidance and Control*, Computer Communication Review, **18**, 4, pp. 314–329.
19. Stevens, W. R. (1994). *TCP/IP illustrated, volume 1:The protocols*, Addison-Wesley, Reading, Massachusetts.
20. Stevens, W. (1997). *TCP Slow Start, Congestion Avoidance, Fast Retransmit and Fast Recovery Algorithms*, RFC2001.
21. Mathis, M., Mahdavi, J., Floyd, S. and Romanow, A. (1996). *TCP Selective Acknowledgements Options*, RFC2018.

22. Wang, Z. and Crowcroft, J.(1992). *Eliminating periodic packet losses in 4.3-Tahoe BSD TCP congestion control*, ACM Computer Communication Review, **22**, 2, pp. 9–16.
23. Brakmo, L. S., Malley, S. O. and Paterson, L. L. (1994). *TCP Vegas: New Techniques for congestion detection and avoidance*, Proceedings of ACM SIGCOMM'94, 4, pp. 24–35.
24. Brakmo, L. S. and Paterson, L. L. (1995). *TCP Vegas: End to end congestion avoidance on a global Internet*, IEEE Journal on Selected Areas in Communications, **13**, 8, pp. 1465–1480.
25. Anantharam, J. V. and Walrand, J. (1999). *Analysis and comparison of TCP reno and vegas*, Proceedings of INFOCOM'99, **3**, pp. 1556–1563.
26. Hasegawa, G., Murata, M. and Miyahara, H. (1999). *Fairness and Stability of congestion control mechanisms of TCP*, The Transactions of the Institute of Electronics, Information and Communication Engineers of Japan, **J82-B-I**, 1, pp. 1–9.
27. Ohsaki, H., Murata, M., Suzuki, H., Ikeda, C., and Miyahara, H. (1995). *Rate-based congestion control for ATM networks*, ACM SIGCOMM Computer Communication Review, **25**, pp. 60–72.
28. Floyd, S. and Fall, K. (1999). *Promoting the Use of End-to-End Congestion Control in the Internet*, IEEE/ACM Transaction on Networking, **7**, 4, pp. 458–472.
29. Floyd, S. and Jacobson, V.(1994). *Random Early Detection gateways for Congestion Avoidance*, IEEE/ACM Transactions on Networking, **1**, 4, pp. 397–413.
30. Floyd, S. (1994). *Tcp and explicit congestion notification*, ACM Computer. Communication. Review, **24**, 5, pp. 10–23.
31. Ramakrishnan, K. and Floyd, S. (1999). *A Proposal to add Explicit Congestion Notification (ECN) to IP*, RFC2481.
32. Ogawa, A., Sugiura, K., Nakamura, O. and Murai, J. (2003). *Design of Packet Lossless TCP Friendly DV over IP*, The Transactions of the Institute of Electronics, Information and Communication Engineers of Japan, **J86-B**, 8, pp. 1561–1569.
33. Kimura, M., Imaizumi, M. and Yasui, K. (2005). *Reliability Consideration of Window Flow Control Scheme for a Communication System with Explicit Congestion Notification*, Proceedings of International Workshop on Recent Advances in Stochastic Operations Research, pp.110–117.
34. Kimura, M., Imaizumi, M. and Yasui, K. (2006). *Optimal Policy of a Window Flow Control Scheme with Explicit Congestion Notification*, Proceedings of 12th International Conference on Reliability and Quality in Design, pp. 154–158.
35. Osaki, S., (1992). *Applied Stochastic System Modeling*, Springer-Verlag, Berlin.
36. Yasui, K., Nakagawa,T. and Sandoh, H., (2002). *Reliability models in data communication systems*, Stochastic Models in Reliability and Maintenance (edited by S.Osaki), pp. 281–301, Springer-Verlag, Berlin.

Chapter 7

Backup Policies for a Database System

CUN-HUA QIAN

College of Management Science and Engineering
Nanjing University of Technology
Jiamgsu, Nanjing 210009, China
E-mail: qch64317@njut.edu.cn

1 Introduction

In recent years, databases in computer systems have become very important in the high information-oriented society. In particular, a reliable database is the most indispensable instrument in on-line transaction processing systems such as real-time systems used for bank accounts. The data in a computer system are frequently updated by adding or deleting them, and are stored in secondary media. Even high reliable computers might sometimes break down eventually by several errors due to noises, human errors and hardware faults. It would be possible to replace hardware and software when they fail, but it would be impossible to do a database. In this case, we have to reconstruct the same files from the beginning.

The most simple and dependable method to ensure the safety of data would be always to shut down a database, and to make the backup copies of all data, log and control files in other places, and to remove them immediately when some data in the original secondary media are corrupted. This is called *total backup*. But, this method has to be made while a database is off-line and unavailable to its users, and would be time-consuming and costly when files become large. To overcome these disadvantages according to the attributes of archives while backing up only modified files, we might dump only modified files since a prior backup. The resources required for such backup are proportional to the transactional activities which have taken place in a database, and not to its size because only a small percentage changes in most applications between successive backups. This can shorten backup times and can decrease the required resources, and would be more useful for larger databases [20]. This is called *export backup*.

Total backup is a physical backup scheme which copies all files. On the other hand, export backup is a logical backup scheme which copies the data and the definition of the database where they are stored in the operating system in binary notation. This approach is generally classified into three schemes: *incremental backup, cumulative backup*, and *full backup* or *complete backup* [22].

Full backup exports all files, and by this backup, the attributes of archives are updated, that is, a database system returns to an initial state. When full backup copies are made frequently, all images of a database can be secured, but its operation cost is greatly increased. Thus, the scheme of incremental or cumulative backup is usually adopted and is suitably executed between the operations of full backup.

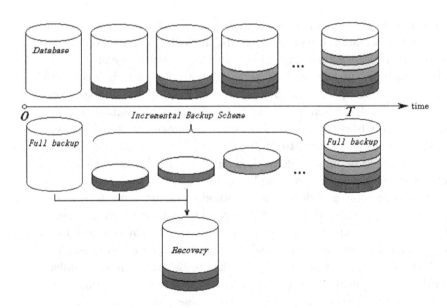

Fig. 1   Incremental backup scheme

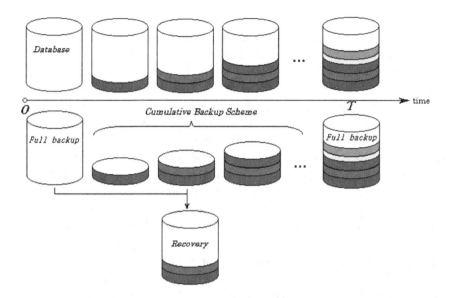

Fig. 2 Cumulative backup scheme

Incremental backup exports only files which have changed or are new since the last incremental or full backup, and updates the attributes of archives (Figure 1). On the other hand, cumulative backup exports only modified files since the last full backup, however, does not update the attributes of archives (Figure 2). When some errors have occurred in storage media, we can recover a database by importing files of all incremental backups and the full backup for the incremental backup scheme, and by importing files of the last cumulative backups and the full backup for the cumulative backup scheme. Cumulative backup exports more files than the incremental one, but it imports fewer files than the incremental one when we recover a database. It would be important to compare two schemes of incremental and cumulative backups.

On the other hand, an important problem in actual backup schemes is when to create the full backup. The full backup with large overhead is done at long intervals and the incremental or cumulative backup with small overhead is done at short intervals. We want to lessen the number of full backups with large overhead. However, the overhead of cumulative backup is increasing with the number of newly updated trucks, because the list of modified files is growing up each day until the next full backup which will clear all attributes of archives. For the incremental backup, the amount of data transfer is constant approximately. However, the overhead of recovery which imports files of all incremental and the full backups is remarkably increased with the number of incremental backups, when some errors have occurred in storage media. That is, both overheads of cumulative backup and recovery of incremental backup increase adaptively with the amount of newly updated trucks. From this point of view, we should decide the full backup interval by comparing two overheads of backup and recovery.

Chandy, et al. [2] and Reuter [17] considered some recovery techniques for database failures. The optimal checkpoint intervals of such models which minimize the total overhead were studied by many authors [4, 7, 8]. Further, Sandoh, et al. [18] discussed optimal backup policies for hard disks.

On the other hand, cumulative damage models in reliability theory, where a system suffers damage due to shocks and fails when the total amount of damage exceeds a failure level K, generate a cumulative reliability viewpoints were discussed [5] and [13]. It is of great interest that a system is replaced before failure as preventive maintenance. The replacement policies where a system is replaced before failure at time T [21], at shock N [9], or at damage Z [6, 10] were considered. Nakagawa and Kijima [11] applied the periodic replacement with minimal repair [1] at failure to a cumulative damage model and obtained optimal values T^*, N^* and Z^* which minimize the expected cost. Satow, et al. [19] applied the cumulative damage model to garbage collection policies for a database system.

In this chapter, we apply the cumulative damage model to the backup of files for database media failures, by putting *shock* by *update* and *damage* by *dumped files* [13–16]. We have to pay only attention to the matter what are essential laws of governing systems, and take a grasp of updated processes and try to formulate it simply, avoiding small points. In other words, it would be necessary to form mathematical models of backup schemes which outline the observational and theoretical features of complex phenomena.

2 Cumulative Damage Model

Consider a cumulative damage model in the context of [3]: A system is subjected to shocks and suffers damage due to shocks. As damage, we can consider wear, stress, fatigue, corrosion, erosion, or garbage. Let random variables X_i ($i = 1, 2, \cdots$) denote a sequence of inter-arrival times between successive shocks, and random variables W_i $(i = 1, 2, \cdots)$ denote the damage produced by the ith shock. It is assumed that the $\{W_i\}$ are non-negative, independent, and identically distributed, and that W_i is independent of X_j $(j \neq i)$.

Let $N(t)$ denote the random variable which is the total number of shocks up to time t. Then, define a random variable

$$Z(t) \equiv \begin{cases} \sum_{i=1}^{N(t)} W_i & (N(t) = 1, 2, \cdots), \\ 0 & (N(t) = 0), \end{cases} \tag{1}$$

where $Z(t)$ represents the total damage at time t. It is assumed that a system fails when the total damage has exceeded a prespecified failure level K (> 0) for the first time. Of interest is a random variable $Y \equiv \min\{t; Z(t) > K\}$, i.e., $\Pr\{Y \leq t\}$ represents the distribution of time to system failure.

In this section, we consider a standard cumulative damage model and a modified cumulative damage model.

2.1 *Standard Cumulative Damage Model*

Consider a standard cumulative damage model: Successive shocks occur at time interval X_i and each shock causes damage to a system in the amount W_i. The total damage to a system is additive.

Assume that $1/\lambda \equiv E\{X_i\} < \infty$, $1/\mu \equiv E\{W_i\} < \infty$, and $H(t) \equiv \Pr\{X_i \leq t\}$, $G(x) \equiv \Pr\{W_i \leq x\}$ $(i = 1, 2, \cdots)$. Then, the total damage $Z_j \equiv \sum_{i=1}^{j} W_i$ to the jth shock, where $Z_0 \equiv 0$, has a probability distribution

$$\Pr\{Z_j \leq x\} \equiv G^{(j)}(x) \quad (j = 1, 2, \cdots), \tag{2}$$

which is the j-fold Stieltjes convolution of the distribution $G(x)$ of itself, i.e., $G^{(j)}(x) \equiv \int_0^x G^{(j-1)}(x-u) dG(u)$ with $G^{(0)}(x) = 1(x)$ (a step function) and $G^{(1)}(x) = G(x)$.

Thus, it is evident from a renewal theory [12] that

$$\Pr\{N(t) = j\} = H^{(j)}(t) - H^{(j+1)}(t) \quad (j = 0, 1, \cdots), \tag{3}$$

and

$$\Pr\{Z_j \leq x, N(t) = j\} = G^{(j)}(x)[H^{(j)}(t) - H^{(j+1)}(t)] \quad (j = 0, 1, \cdots). \tag{4}$$

Then, the distribution of $Z(t)$ defined in (1) is

$$\Pr\{Z(t) \le x\} = \sum_{j=0}^{\infty} \Pr\{Z_j \le x, N(t) = j\}$$

$$= \sum_{j=0}^{\infty} G^{(j)}(x)[H^{(j)}(t) - H^{(j+1)}(t)], \quad (5)$$

and the survival probability is

$$\Pr\{Z(t) > x\} = \sum_{j=0}^{\infty} [G^{(j)}(x) - G^{(j+1)}(x)] H^{(j+1)}(t). \quad (6)$$

The total expected damage at time t is

$$E\{Z(t)\} = \int_0^{\infty} x \, d\Pr\{Z(t) \le x\} = \frac{1}{\mu} \sum_{j=1}^{\infty} H^{(j)}(t). \quad (7)$$

The first-passage time distribution to system failure, when a failure level is a constant K (> 0), is

$$\Phi(t) = \Pr\{Y \le t\} = \Pr\{Z(t) > K\}, \quad (8)$$

which is derived from (6). Further, the Laplace-Stieltjes (LS) transform is

$$\Phi^*(s) \equiv \int_0^{\infty} e^{-st} d\Phi(t) = \sum_{j=0}^{\infty} [G^{(j)}(K) - G^{(j+1)}(K)][H^*(s)]^{j+1}, \quad (9)$$

where $H^*(s)$ represents the LS transform of $H(t)$, i.e., $H^*(s) \equiv \int_0^{\infty} e^{-st} dH(t)$ for $\mathrm{Re}(s) > 0$. Thus, the mean time to system failure is

$$E\{Y\} \equiv \int_0^{\infty} t \, d\Phi(t) = -\left.\frac{d\Phi^*(s)}{ds}\right|_{s=0} = \frac{1}{\lambda} \sum_{j=0}^{\infty} G^{(j)}(K). \quad (10)$$

Next, suppose that shocks occur at a nonhomogeneous Poisson process with an intensity function $\lambda(t)$ and a mean-value function $R(t)$, i.e., $R(t) \equiv \int_0^t \lambda(u) du$. Then, the probability that the jth shock occurs exactly during any interval $(u, t]$ is [5]

$$H_j(t, u) \equiv \Pr\{N(t) - N(u) = j\}$$

$$= \frac{[R(t) - R(u)]^j}{j!} e^{-[R(t) - R(u)]} \quad (j = 0, 1, 2, \cdots), \quad (11)$$

where $R(0) \equiv 0$. In particular, the probability of occurrences of j shocks during $(0, t]$ is

$$H_j(t) \equiv \Pr\{N(t) = j\} = \frac{[R(t)]^j}{j!} e^{-R(t)} \quad (j = 0, 1, 2, \cdots). \quad (12)$$

Thus, by putting that $H^{(j)}(t) \equiv \sum_{i=j}^{\infty} H_i(t)$ formally, we can rewrite all reliability quantities:

$$\Pr\{Z(t) \leq x\} = \sum_{j=0}^{\infty} G^{(j)}(x) H_j(t), \tag{13}$$

$$E\{Z(t)\} = \frac{1}{\mu} R(t), \tag{14}$$

$$E\{Y\} = \sum_{j=0}^{\infty} G^{(j)}(K) \int_0^{\infty} H_j(t) dt. \tag{15}$$

2.2 Modified Cumulative Damage Model

Consider the case where the total damage due to shocks is additive if it has not exceeded a threshold level k, where $k \leq K$, and conversely, when it has exceeded a threshold level k, it is not additive by minimal maintenances at each shock and its level is constant. The same assumptions of cumulative damage model are made except that the total damage is additive.

In this case, the distribution of $Z(t)$ defined in (1) is

$$\Pr\{Z(t) \leq x\} = \begin{cases} \sum_{j=0}^{\infty} G^{(j)}(x)[H^{(j)}(t) - H^{(j+1)}(t)] & (x \leq k), \\ 1 & (x > k), \end{cases} \tag{16}$$

and the survival probability is

$$\Pr\{Z(t) > x\} = \begin{cases} \sum_{j=0}^{\infty} [G^{(j)}(x) - G^{(j+1)}(x)] H^{(j+1)}(t) & (x \leq k), \\ 0 & (x > k). \end{cases} \tag{17}$$

The total expected damage at time t is given by

$$E\{Z(t)\} = \sum_{j=1}^{\infty} [H^{(j)}(t) - H^{(j+1)}(t)] \int_0^k x \, dG^{(j)}(x). \tag{18}$$

If shocks occur at a nonhomogeneous Poisson process with a mean-value function $R(t)$, then

$$\Pr\{Z(t) \leq k\} = \sum_{j=0}^{\infty} G^{(j)}(k) H_j(t), \tag{19}$$

and the total expected damage at time t is

$$E\{Z(t)\} = \sum_{j=1}^{\infty} H_j(t) \int_0^k x \, dG^{(j)}(x). \tag{20}$$

3 Comparison of Backup Schemes and Policies

It is an important problem to determine which of backup schemes should be adopted as the backup policy. In this section, we compare two schemes of incremental and cumulative backups, and compare two schemes of total and cumulative backups, using the results of the modified cumulative damage model in Section 2.2. Further, we discuss optimal full backup times for the incremental and cumulative backups, and compare them numerically.

Taking the above considerations into account, we formulate the following stochastic model of the backup policy for a database system: Suppose that a database in a computer system is updated at a nonhomogeneous Poisson process with an intensity function $\lambda(t)$ and a mean-value function $R(t)$, i.e., $R(t) \equiv \int_0^t \lambda(u) du$. Then, the probability that the jth update occurs exactly during $(u, v]$ is

$$H_j(u,v) \equiv \frac{[R(v) - R(u)]^j}{j!} e^{-[R(v)-R(u)]} \quad (j = 0, 1, 2, \cdots), \quad (21)$$

where $R(0) \equiv 0$ and $R(\infty) \equiv \infty$. In particular, $H_j(t) \equiv H_j(0, t)$.

Further, an amount W_j of files, which have changed or are new, arises from the jth update and is dumped. It is assumed that each W_j has an identical probability distribution $G(x) \equiv \Pr\{W_j \le x\}$ $(j = 1, 2, \cdots)$ with finite mean $1/\mu$. Then, the total amount of files $Z_j \equiv \sum_{i=1}^j W_i$ to the jth update has a probability distribution

$$\Pr\{Z_j \le x\} = G^{(j)}(x) \quad (j = 1, 2, \cdots), \quad (22)$$

with mean j/μ. Let $Z(t)$ be the total amount of updated files at time t. Then, the distribution of $Z(t)$ is

$$\Pr\{Z(t) \le x\} = \sum_{j=0}^{\infty} H_j(t) G^{(j)}(x). \quad (23)$$

3.1 Incremental and Cumulative Backups

The incremental backup exports only files which have changed or are new since the last incremental backup or full backup. On the other hand, the cumulative backup exports only files which have changed or are new since the last full backup. When some errors have occurred in storage media, we can make the recovery of a database by importing files of all incremental backups and the full backup for the incremental backup scheme, and by importing files of the last cumulative and full backups for the cumulative backup scheme. That is, the cumulative backup exports more files than the incremental one at each update, however, imports less files than the incremental one when we make the recovery of a database.

First, we compare two schemes of incremental and cumulative backups. It is assumed that a database in secondary media fails according to a general distribution $F(t)$ with finite mean $1/\gamma$. Suppose that the full backup is performed at time T $(0 < T \geq \infty)$ or when a database fails, whichever occurs first. The increment backup or cumulative one is done at each update between the full backups.

Let introduce the following costs: Cost c_1 is suffered for the full backup, cost $c_2 + c_0 x$ is suffered for the incremental backup when an amount of export files at the backup time is x, and for the cumulative backup when the total amount of export files at the backup time is x. Recovery cost is $c_3 + c_0 x$ for the cumulative backup if the database fails when the total amount of import files at the recovery time is x, and is $c_3 + c_0 x + j c_4$ for the incremental backup when the number of backups is j.

Thus, the expected costs $A_c(T)$ for the cumulative backup and $A_i(T)$ for the incremental backup to full backup are, respectively,

$$A_c(T) = c_1 + \overline{F}(T) \sum_{j=0}^{\infty} H_j(T) \sum_{i=1}^{j} M_i + \sum_{j=0}^{\infty} \int_0^T H_j(t) \mathrm{d}F(t) \left(\sum_{i=1}^{j} M_i + N_j \right)$$

$$= \frac{c_0}{\mu} \int_0^T \overline{F}(t)[\lambda(t) + R(t)r(t)]\mathrm{d}t + c_1 + c_2 \int_0^T \overline{F}(t)\lambda(t)\mathrm{d}t + c_3 F(T)$$

$$+ \frac{c_0}{\mu} \int_0^T \overline{F}(t) R(t) \lambda(t) \mathrm{d}t, \tag{24}$$

$$A_i(T) = c_1 + \overline{F}(T) \sum_{j=0}^{\infty} H_j(T) j M_1 + \sum_{j=0}^{\infty} \int_0^T H_j(t) \mathrm{d}F(t)(jM_1 + N_j + jc_4)$$

$$= \frac{c_0}{\mu} \int_0^T \overline{F}(t)[\lambda(t) + R(t)r(t)]\mathrm{d}t + c_1 + c_2 \int_0^T \overline{F}(t)\lambda(t)\mathrm{d}t + c_3 F(T)$$

$$+ c_4 \int_0^T R(t) \mathrm{d}F(t), \tag{25}$$

where $\sum_{i=1}^{0} \equiv 0$, $\overline{F}(t) \equiv 1 - F(t)$, $f(t)$ is a density of $F(t)$, the failure rate is $r(t) \equiv f(t)/\overline{F}(t)$, and

$$M_j \equiv \int_0^{\infty} (c_2 + c_0 x) \mathrm{d}G^{(j)}(x) = c_2 + \frac{jc_0}{\mu},$$

$$N_j \equiv \int_0^{\infty} (c_3 + c_0 x) \mathrm{d}G^{(j)}(x) = c_3 + \frac{jc_0}{\mu}.$$

To compare the two expected costs, we find the difference between them as follows:

$$A_c(T) - A_i(T) = \frac{c_0}{\mu}\int_0^T \overline{F}(t)R(t)\lambda(t)dt - c_4\int_0^T \overline{F}(t)R(t)r(t)dt. \quad (26)$$

Hence, if $(c_0/\mu)\lambda(t) > c_4 r(t)$ for all t, then the incremental backup is better than the cumulative backup.

3.2 Incremental Backup Policy

Consider an optimal policy for the incremental backup. The mean time to full backup is

$$T\overline{F}(T) + \int_0^T t\,dF(t) = \int_0^T \overline{F}(t)dt, \quad (27)$$

and the expected cost rate is

$$C_i(T) = \frac{A_i(T)}{\int_0^T \overline{F}(t)dt}, \quad (28)$$

where $A_i T)$ is given in (25).

A necessary condition that a finite T minimizes $C_i(T)$ is given by differentiating $C_i(T)$ with respect to T and setting it equal to zero. Hence, from (28),

$$\left(c_2 + \frac{c_0}{\mu}\right)\int_0^T \overline{F}(t)[\lambda(T) - \lambda(t)]dt + c_3\left[r(T)\int_0^T \overline{F}(t)dt - F(T)\right]$$

$$+ \left(c_4 + \frac{c_0}{\mu}\right)\int_0^T \overline{F}(t)[R(T)r(T) - R(t)r(t)]dt = c_1. \quad (29)$$

Letting $U(T)$ be the left-hand side of (29),

$$U(0) \equiv \lim_{T\to 0} U(T) = 0,$$

$$U'(T) = \left[\left(c_2 + \frac{c_0}{\mu}\right)\lambda'(T) + c_3 r'(T)\right]\int_0^T \overline{F}(t)dt$$

$$+ \left(c_4 + \frac{c_0}{\mu}\right)[\lambda(T)r(T) + R(T)r'(T)]\int_0^T \overline{F}(t)dt. \quad (30)$$

Thus, if there exists T_1^* which minimizes $C_i(T)$, then it is unique when both $\lambda(t)$ and $r(t)$ have the property of IFR.

Suppose that a database is updated at a Poisson process with an intensity function λ, i.e., $\lambda(t) \equiv \lambda$ and $R(t) \equiv \lambda t$. In this case,

$$U(\infty) \equiv \lim_{T\to\infty} U(T) \geq c_3\left[\frac{1}{\gamma}r(\infty) - 1\right].$$

It is evident that $U(T)$ is strictly increasing if $r'(t) > 0$. Thus, If $\lambda(t) = \lambda$ and $r'(t) > 0$, then there exists a finite and unique T_1^* which minimizes $C_i(T)$ when $r(\infty) > [(c_1/c_3) + 1]\gamma$, and the resulting cost rate is

$$\frac{C_i(T_1^*)}{\lambda} = c_2 + \frac{c_0}{\mu} + \left[c_3 + \left(c_4 + \frac{c_0}{\mu}\right)\lambda T_1^*\right]\frac{r(T_1^*)}{\lambda}. \tag{31}$$

3.3 Total and Cumulative Backups

The overhead of cumulative backup, which exports only files that have changed since the last full backup, increases with the number of newly updated trucks. For example, if all updated trucks are included in the previous updated ones, then the amount of data transfer is the same as the previous one. However, if the updated trucks include some ones that differ from the previous ones, then the amount of data transfer is increased by their differences. It is well-known that when the amount of updated trucks exceeds a threshold level k, the overhead of cumulative backup is larger than that of total backup. The value of k/M is about 60% in a usual on-line system, where M is the total trucks in a database. Thus, if the amount of updated trucks exceeds a level k, we should make a total backup instead of a cumulative backup.

Next, we formulate the following stochastic model of the backup policy for a database system: When the total updated files do not exceed a threshold level k, we perform a cumulative backup at each update. On the other hand, when the total updated files exceeds a threshold level k, we perform a total backup at each update where both time and size of backups are constant. It is assumed that the ith total backup time is T_i ($i = 1, 2, \cdots, N-1$) and that T_N is the first update time after the $(N-1)$th total backup. Conversely, we should make the full backup at periodic time T or at the first update time T_N after the $(N-1)$th total backup, whichever occurs first.

We discuss the optimal total backup number and full backup interval for cumulative backup: Suppose that a database is updated at a Poisson process with an intensity function λ, i.e., $\lambda(t) \equiv \lambda$ and $H_j(t) \equiv [(\lambda t)^j/j!]e^{-\lambda t}$. Then, the probability P_T that a database undergoes the full backup at time T is

$$P_T = \sum_{j=0}^{\infty} H_j(T)G^{(j)}(k) + \sum_{j=0}^{\infty}[G^{(j)}(k) - G^{(j+1)}(k)]\sum_{i=j+1}^{j+N-1} H_i(T)$$

$$= \sum_{j=0}^{\infty} [G^{(j)}(k) - G^{(j+1)}(k)] \sum_{i=0}^{j+N-1} H_i(T), \tag{32}$$

and the probability P_N that a database undergoes the full backup at time T_N is

$$P_N = \sum_{j=0}^{\infty} [G^{(j)}(k) - G^{(j+1)}(k)] \sum_{i=j+N}^{\infty} H_i(T)$$

$$= \sum_{j=0}^{\infty} [G^{(j)}(k) - G^{(j+1)}(k)] \int_0^T \lambda H_{j+N-1}(t) dt. \tag{33}$$

It is evident that $P_T + P_N = 1$.

Let $M_1(T)$ and $M_2(T, N)$ denote the expected numbers of cumulative backups and total backups, respectively. Then, from (32) and (33),

$$M_1(T) = \sum_{j=0}^{\infty} j H_j(T) G^{(j)}(k) + \sum_{j=0}^{\infty} j [G^{(j)}(k) - G^{(j+1)}(k)] \sum_{i=j+1}^{\infty} H_i(T)$$

$$= \sum_{j=1}^{\infty} H_j(T) \sum_{i=1}^{j} G^{(i)}(k), \tag{34}$$

$$M_2(T, N) = \sum_{j=0}^{\infty} [G^{(j)}(k) - G^{(j+1)}(k)] \sum_{i=j+1}^{j+N-1} (i-j) H_i(T)$$

$$+ \sum_{j=0}^{\infty} [G^{(j)}(k) - G^{(j+1)}(k)] \sum_{i=j+N}^{\infty} (N-1) H_i(T)$$

$$= \sum_{j=1}^{\infty} H_j(T) \sum_{i=j-N+2}^{j} [1 - G^{(i)}(k)], \tag{35}$$

where $G^{(-j)}(x) \equiv 1$ $(j = 1, 2, \cdots)$. Thus, the total expected backup cost $E(C)$ to full backup is

$$E(C) = \sum_{j=1}^{\infty} H_j(T) \sum_{i=1}^{j} \int_0^k (c_2 + c_0 x) dG^{(i)}(x) + c_5 M_2(T, N) + c_1, \tag{36}$$

when the total backup cost is c_5, where $c_5 < c_1$ and $c_5 = c_2 + c_0 k$.

Let $E(T)$ denote the mean time to full backup. Then, from (32) and (33),

$$E(T) = \sum_{j=0}^{\infty} [G^{(j)}(k) - G^{(j+1)}(k)] \int_0^T t \lambda H_{j+N-1}(t) dt + T P_T$$

$$= \sum_{j=0}^{\infty} G^{(j-N+1)}(k) \int_0^T H_j(t) dt. \tag{37}$$

Therefore, from (36) and (37), by using the theory of the renewal reward process, the expected cost rate is

$$\frac{C(T,N)}{\lambda} \equiv \frac{E(C)}{\lambda E(T)} = c_5 + \frac{c_1 - A(T,N)}{B(T,N)}, \qquad (38)$$

where

$$A(T,N) \equiv c_0 \sum_{j=1}^{\infty} H_j(T) \sum_{i=1}^{j} \int_0^k G^{(i)}(x)\,\mathrm{d}x$$

$$+ c_5 \sum_{j=1}^{\infty} H_j(T) \sum_{i=0}^{j-N} [G^{(i)}(k) - G^{(i+1)}(k)], \qquad (39)$$

$$B(T,N) \equiv \sum_{j=1}^{\infty} H_j(T) \sum_{i=1}^{j} G^{(i-N)}(k). \qquad (40)$$

If $M(k) \equiv \sum_{j=1}^{\infty} G^{(j)}(k) < \infty$, then $C(0,N) \equiv \lim_{T \to 0} C(T,N) = \infty$ for all N and $C(\infty,\infty) \equiv \lim_{T \to \infty, N \to \infty} C(T,N) = \lambda c_5$. Thus, there exists a positive pair (T^*, N^*) $(0 < T^*, N^* \le \infty)$ which minimizes $C(T,N)$.

In general, let an optimal pair (T^*, N^*) denote a positive solution which minimizes $C(T,N)$. It is evident that $\mathrm{d}A(T,N)/\mathrm{d}T > 0$, $\mathrm{d}B(T,N)/\mathrm{d}T > 0$ and

$$A(T, N+1) - A(T,N) = -c_5 \sum_{j=0}^{\infty} H_{j+N}(T)[G^{(j)}(k) - G^{(j+1)}(k)] < 0,$$

$$B(T, N+1) - B(T,N) = \sum_{j=N+1}^{\infty} H_j(T)[1 - G^{(j-N)}(k)] > 0.$$

Thus, we have the following property for (T^*, N^*): If

$$c_1 > A(\infty, 1) = c_5 + c_0 \int_0^k M(x)\,\mathrm{d}x, \qquad (41)$$

then $(T^*, N^*) = (\infty, \infty)$, that is, we should make only the total backup, instead of the cumulative and full backups. Conversely, if

$$c_1 \le c_5 + c_0 \int_0^k M(x)\,\mathrm{d}x, \qquad (42)$$

then $N^* = 1$, that is, the full backup is performed at periodic time T^* or when the amount of updated trucks is k, whichever occurs first.

From this result, we have derived the sufficient condition to determine whether a system should adopt the export backup or the total backup.

3.4 Cumulative Backup Policy

We derive an optimal full backup interval for the cumulative backup. We make the same assumptions as those of Section 3.1, and $K \equiv (c_1 - c_2)/c_0$, i.e., the cumulative backup cost is greater than or equal to the full backup cost when the total files have exceeded K. In this case, we should make the full backup, instead of the cumulative backup. This corresponds to the case of $N^* = 1$ in the Section 3.3. We perform the full backup at periodic time T, when the total files have exceeded K or when the recovery is completed on failure of a database, whichever occurs first.

The probability that a database undergoes the full backup at time T is

$$\overline{F}(T) \sum_{j=0}^{\infty} H_j(T) G^{(j)}(K), \qquad (43)$$

the probability that it undergoes the full backup when the total files have exceeded K is

$$\sum_{j=0}^{\infty} [G^{(j)}(K) - G^{(j+1)}(K)] \int_0^T \overline{F}(t) H_j(t) \lambda(t) dt, \qquad (44)$$

the probability that it undergoes the full backup, when the recovery is completed if it fails, is

$$\sum_{j=0}^{\infty} G^{(j)}(K) \int_0^T H_j(t) dF(t). \qquad (45)$$

It is evident that (43) + (44) + (45) = 1.

Thus, the mean time $E(T)$ to full backup is

$$E(T) = \sum_{j=0}^{\infty} [G^{(j)}(K) - G^{(j+1)}(K)] \int_0^T t\overline{F}(t) H_j(t) \lambda(t) dt$$

$$+ \overline{F}(T) T \sum_{j=0}^{\infty} H_j(T) G^{(j)}(K) + \sum_{j=0}^{\infty} G^{(j)}(K) \int_0^T t H_j(t) dF(t)$$

$$= \sum_{j=0}^{\infty} G^{(j)}(K) \int_0^T H_j(t) \overline{F}(t) dt, \qquad (46)$$

and the expected number of cumulative backups is

$$\sum_{j=0}^{\infty}(j+1)[G^{(j)}(K)-G^{(j+1)}(K)]\int_0^T \overline{F}(t)H_j(t)\lambda(t)\mathrm{d}t$$

$$+\overline{F}(T)\sum_{j=0}^{\infty}jH_j(T)G^{(j)}(K)+\sum_{j=0}^{\infty}jG^{(j)}(K)\int_0^T H_j(t)\mathrm{d}F(t)$$

$$=\sum_{j=0}^{\infty}G^{(j)}(K)\int_0^T H_j(t)\lambda(t)\overline{F}(t)\mathrm{d}t. \qquad (47)$$

Thus, the total expected cost $A(T)$ to full backup is

$$A(T)=\sum_{j=0}^{\infty}\int_0^K (c_2+c_0 x)\mathrm{d}G^{(j)}(x)\int_0^T H_j(t)\lambda(t)\overline{F}(t)\mathrm{d}t$$

$$+\sum_{j=0}^{\infty}\int_0^K (c_3+c_0 x)\mathrm{d}G^{(j)}(x)\int_0^T H_j(t)\mathrm{d}F(t)+c_1. \qquad (48)$$

Therefore, dividing the total expected cost in (48) by the mean time in (46), we have the expected cost rate

$$C_c(T)\equiv\frac{A(T)}{E(T)}. \qquad (49)$$

It is very difficult to discuss an optimal T^* analytically. In particular, it is assumed that a database is updated at a Poisson process with rate $p\lambda$, i.e., $\lambda(t)=p\lambda$, $R(t)=p\lambda t$, $H_j(t)=[(p\lambda t)^j/j!]\mathrm{e}^{-p\lambda t}$ ($j=0,1,2,\cdots$) and $F(t)=1-\mathrm{e}^{-q\lambda t}$, where $0<p\leq 1$, $q\equiv 1-p$. In this case, from (49),

$$\frac{C_c(T)}{\lambda}=\frac{c_0\sum_{j=0}^{\infty}\int_0^K x\mathrm{d}G^{(j)}(x)I_j(T)+c_1}{\sum_{j=0}^{\infty}G^{(j)}(K)I_j(T)}+pc_2+qc_3, \qquad (50)$$

where

$$I_j(T)\equiv\int_0^T \lambda H_j(t)\mathrm{e}^{-q\lambda t}\mathrm{d}t.$$

From (50), $C_c(0)=\infty$, and hence, there exists a positive T_2^* ($0<T_2^*\leq\infty$) which minimizes $C_c(T)$ in (50). Differentiating $C_c(T)$ with respect to T and setting it equal to zero,

$$V(T)\sum_{j=0}^{\infty}G^{(j)}(K)I_j(T)-\sum_{j=0}^{\infty}\int_0^K x\mathrm{d}G^{(j)}(x)I_j(T)=\frac{c_1}{c_0}, \qquad (51)$$

where

$$V(T)\equiv\frac{\sum_{j=0}^{\infty}H_j(T)\int_0^K x\mathrm{d}G^{(j)}(x)}{\sum_{j=0}^{\infty}H_j(T)G^{(j)}(K)}. \qquad (52)$$

Denote the left-hand side of (51) by $Q(T)$. Then,

$$Q(0) \equiv \lim_{T \to 0} Q(T) = 0,$$

$$Q(\infty) \equiv \lim_{T \to \infty} Q(T) = V(\infty)N(K) - \int_0^K x \mathrm{d}N(x),$$

$$Q'(T) = V'(T) \sum_{j=0}^{\infty} G^{(j)}(K) I_j(T), \qquad (53)$$

where $N(x) \equiv \sum_{j=0}^{\infty} p^j G^{(j)}(x)$, $V(\infty) \equiv \lim_{T \to \infty} V(T)$. Thus, if $V'(T) > 0$ and $Q(\infty) > c_1/c_0$, then there exists a finite and unique T_2^* which satisfies (51).

Using the above results, we compute optimal policies numerically for the increment and cumulative backups.

Example 3.1. Suppose that a database is updated in a Poisson process with rate λ, the backup is done with probability p $(0 < p \le 1)$, and it fails with probability $q \equiv 1 - p$ at each update time, i.e., $\lambda(t) = p\lambda$, $R(t) = p\lambda t$, $H_j(t) = [(p\lambda t)^j/j!]\mathrm{e}^{-p\lambda t}$ $(j = 0, 1, 2, \cdots)$ and $F(t) = 1 - \mathrm{e}^{-q\lambda t}$. In this case, (26) is rewritten as

$$A_\mathrm{c}(T) - A_\mathrm{i}(T) = \left(\frac{c_0}{\mu}p - c_4 q\right) \int_0^T p\lambda^2 t \mathrm{e}^{-q\lambda t} \mathrm{d}t. \qquad (54)$$

Thus, if $c_4/(c_0/\mu) > p/q$, then the cumulative backup is better than the incremental one. Conversely, if $c_4/(c_0/\mu) < p/q$, then the incremental backup is better than the cumulative one.

First, when the incremental backup is adopted, (29) is

$$p\lambda T - \frac{p}{q}(1 - \mathrm{e}^{-q\lambda T}) = \frac{c_1}{c_4 + c_0/\mu}. \qquad (55)$$

Then, the left-hand side of (55) is strictly increasing from 0 to ∞. Thus, there exists a finite and unique T_1^* which satisfies (55), and the resulting expected cost rate is

$$\frac{C_\mathrm{i}(T_1^*)}{\lambda} = pc_2 + qc_3 + pq\lambda T_1^* c_4 + p(1 + q\lambda T_1^*)\frac{c_0}{\mu}. \qquad (56)$$

It is known from (55) that the optimal full backup times λT_1^* do not depend on both costs c_2 and c_3. Table 1 gives λT_1^* and $C_\mathrm{i}(T_1^*)/(\lambda c_0/\mu)$ of incremental backup for $c_4/(c_0/\mu) = 20, 30, 40, 50$ when $c_1 = 64$, $c_2 = 40$, $c_3 = 100$ and $p = 0.98$. Note that all costs are relative to cost c_0/μ and all times are relative to $1/\lambda$. For example, when $c_4/(c_0/\mu) = 30$, λT_1^* is about

Table 1 Optimal time λT_1^* and resulting cost $C_i(T_1^*)/(\lambda c_0/\mu)$ for $c_4/(c_0/\mu)$ when $c_1 = 64$, $c_2 = 40$, $c_3 = 100$ and $p = 0.98$

$c_4/(c_0/\mu)$	20	30	40	50
λT_1^*	18.74	15.25	13.17	11.76
$C_i(T_1^*)/(\lambda c_0/\mu)$	49.89	51.45	52.76	53.94

15.25. That is, when the mean time of update is $1/\lambda = 1$ day, T_1^* is about 15 days.

These indicate that λT_1^* is decreasing when $c_4/(c_0/\mu)$ is increasing, and $C_i(T_1^*)/(\lambda c_0/\mu)$ is increasing with $c_4/(c_0/\mu)$. When $c_4/(c_0/\mu)$ is increasing to $c_4/(c_0/\mu) = 50 > p/q = 49$, $C_i(T_1^*)/(\lambda c_0/\mu)$ is about 53.94, and from Table 2, it is greater than $C_c(T_2^*)/(\lambda c_0/\mu) = 52.78$ of the cumulative backup. That is, when $c_4/(c_0/\mu) > p/q = 49$, the cumulative backup is better than the incremental one.

Second, when the cumulative backup is adopted, it is assumed in Section 3.3 that $G(x) = 1 - e^{-\mu x}$, i.e., $G^{(j)}(x) = 1 - \sum_{i=0}^{j-1}[(\mu x)^i/i!]e^{-\mu x}$ ($j = 1, 2, \cdots$), $M(K) \equiv \sum_{j=1}^{\infty} G^{(j)}(K) = \mu K$, and

$$N(x) = \begin{cases} \frac{1}{q} - \frac{p}{q}e^{-q\mu x} & (p < 1), \\ 1 + \mu x & (p = 1). \end{cases} \quad (57)$$

In this case, $V(T)$ is strictly increasing from 0 to K, and hence, $Q(T)$ is also strictly increasing from 0 to $\int_0^K N(x)\mathrm{d}x$, i.e.,

$$Q(\infty) = \int_0^K N(x)\mathrm{d}x, \quad Q'(T) = V'(T)\sum_{j=0}^{\infty} G^{(j)}(K)I_j(T) > 0. \quad (58)$$

Thus, if $p < 1$, then

$$Q(\infty) = \frac{K}{q} - \frac{p}{q^2\mu}(1 - e^{-q\mu K}), \quad (59)$$

and we have the following optimal policy:

(i) If $Q(\infty) > c_1/c_0$, then there exists a finite and unique T_2^* which satisfies (51), and the resulting cost rate is

$$\frac{C_c(T_2^*)}{\lambda} = pc_2 + qc_3 + c_0 V(T_2^*). \quad (60)$$

(ii) If $Q(\infty) \leq c_1/c_0$, then $T_2^* = \infty$, and the resulting cost rate is

$$\frac{C_c(\infty)}{\lambda} = pc_2 + qc_3 + \frac{c_0 \int_0^K x\mathrm{d}N(x) + c_1}{N(K)}. \quad (61)$$

Table 2 Optimal time λT_2^* and resulting cost $C_c(T_2^*)/(\lambda c_0/\mu)$ when $p = 0.98$ and $\mu K = 24$

$c_2/(c_0/\mu)$	$c_3/(c_0/\mu)$			
	100		500	
	λT_2^*	$C_c(T_2^*)/(\lambda c_0/\mu)$	λT_2^*	$C_c(T_2^*)/(\lambda c_0/\mu)$
40	11.02	52.78	11.02	60.78
70	14.15	84.72	14.15	92.72
100	17.45	166.26	17.45	174.26
160	27.10	178.60	27.10	186.60
250	∞	271.41	∞	279.41

Note that $c_1/(c_0/\mu) = c_2/(c_0/\mu) + \mu K$.

It is easily seen that $Q(T)$ is strictly increasing in p since $V(T)$ is also strictly increasing in p. Hence, the optimal time T_2^* in case (i) is a decreasing function of p.

In particular, $Q(\infty) = K + \mu K^2/2$ when $p = 1$. Noting that $c_1 = c_2 + c_0 K$, the optimal policy is rewritten as follows:

(iii) If $\mu K(c_0 K/2) > c_2$, then there exists a finite and unique T_2^* which satisfies (51), and the resulting cost rate is

$$\frac{C_c(T_2^*)}{\lambda} = c_2 + c_0 V(T_2^*). \tag{62}$$

(iv) If $\mu K(c_0 K/2) \leq c_2$, then $T_2^* = \infty$, and the resulting cost rate is

$$\frac{C_c(\infty)}{\lambda} = c_2 + \frac{c_1 + c_0 \mu K^2/2}{1 + \mu K}. \tag{63}$$

Note that all costs are relative to cost c_0/μ and all times are relative to $1/\lambda$. Table 2 gives the optimal full backup time λT^* and the resulting cost $C_c(T_2^*)/(\lambda c_0/\mu)$ of the cumulative backup for $c_2/(c_0/\mu) = 40, 70, 100, 160, 250$ and $c_3/(c_0/\mu) = 100, 500$ when $p = 0.98$ and $\mu K = 24$. Similarly, Table 3 gives λT_2^* and $C_c(T_2^*)/(\lambda c_0/\mu)$ of the cumulative backup for $p = 0.90, 0.95, 1.00$ when $c_2 = 30$ $c_3 = 90$ and $\mu K = 16$. These indicate that both λT^* and $C_c(T_2^*)/(\lambda c_0/\mu)$ are increasing with $c_2/(c_0/\mu)$, and conversely, are decreasing when p is increasing.

In Table 2, from the optimal policy (i), a finite T_2^* exists uniquely if

$$\frac{c_2}{c_0/\mu} < \frac{p\mu K}{q} - \frac{p(1 - e^{-q\mu K})}{q^2} \approx 242.02.$$

Taking another point of view, we can note that $N(K) - 1 = \sum_{j=1}^{\infty} p^j G^{(j)}(K)$ represents the expected number of cumulative backups until a database

Table 3 Optimal time λT_2^* and resulting cost $C_c(T_2^*)/(\lambda c_0/\mu)$ when $c_2 = 30$, $c_3 = 90$ and $\mu K = 16$

p	1.00	0.95	0.90
λT_2^*	9.48	11.07	13.36
$C_c(T_2^*)/(\lambda c_0/\mu)$	39.39	42.98	46.74

fails or the total files have exceeded a threshold level K, whichever occurs first. When $c_2/(c_0/\mu) = 40, 70, 100$, λT_2^* is less than $N(K) - 1 \approx 18.68$. That is, we should make the full backup more frequently. Further, when $c_2/(c_0/\mu) = 250 > 242.02$, $T_2^* = \infty$, i.e., we should make the full backup when the total files have exceeded K, or a database fails, whichever occurs first.

In Table 3, when $p = 1$, λT_2^* is about 9.48, and $C_c(T_2^*)/(\lambda c_0/\mu)$ is about 39.39. However, from (41), if $\mu k < \sqrt{2\mu K + 1} - 1 = \sqrt{33} - 1 \approx 4.74$, i.e. $c_5/(c_0/\mu) = c_2/(c_0/\mu) + \mu k < 34.74$, then $(T^*, N^*) = (\infty, \infty)$, and $C(\infty, \infty)/(\lambda c_0/\mu) = c_5$ is about 34.74, that is, the total backup is better than the cumulative backup. □

4 Periodic Backup Policies

Backup frequencies of a database would usually depend on the factors such as its size and availability, and sometimes frequency in use and criticality of data. To make the backup efficiently, incremental or cumulative backups are usually performed at periodic times iT ($i = 1, 2, \cdots, N - 1$), e.g., daily or weekly, and log backups are made at each update between the incremental or cumulative backups in most database systems. For most failures, a database can recover from these points by log files and restore a consistent state by the full backup and all incremental backups for the periodic incremental backup scheme, and by the last cumulative backups and the full backup for the periodic cumulative backup scheme. In this section, we apply the cumulative damage model to the backup policy for a database system with periodic incremental backup or periodic cumulative backup, and derive an optimal full backup interval.

4.1 Periodic Incremental Backup Policy

First, consider a periodic incremental backup policy: Suppose that a database in secondary media fails according to a general distribution $F(t)$, incremental backups are performed at periodic times iT $(i = 1, 2, \cdots, N - 1)$, and full backup is performed at time NT $(N = 1, 2, \ldots)$ or when the database fails, whichever occurs first.

Taking the above considerations into account, we formulate the following stochastic model of the backup policy for a database system: Suppose that a database is updated at a nonhomogeneous Poisson process with an intensity function $\lambda(t)$ and a mean-value function $R(t)$.

Let us introduce the following costs: Cost c_1 is suffered for the full backup, cost $c_2 + c_0(x)$ is suffered for the incremental backup when the amount of export files at the backup time is x, and $c_0(x)$ is increasing with x, cost $c_3 + jc_4 + c_0(x)$ is suffered for recovery if the database fails when the total amount of import files at the recovery time is x, where jc_4 denotes recovery cost of incremental backups when the number of incremental backups is j, and c_3 denotes recovery cost of the last full backup.

Then, the expected cost of incremental backup, when it is performed at time iT $(i = 1, 2, \cdots, N - 1)$, is

$$C_{\text{EI}}(i) \equiv \sum_{j=0}^{\infty} H_j((i-1)T, iT) \int_0^{\infty} [c_2 + c_0(x)] \mathrm{d}G^{(j)}(x). \tag{64}$$

Further, if the database fails during $[iT, (i+1)T]$, then the excepted recovery cost until time iT is

$$C_{\text{ER}}(i) \equiv \sum_{j=0}^{\infty} H_j(iT) \int_0^{\infty} [c_3 + ic_4 + c_0(x)] \mathrm{d}G^{(j)}(x). \tag{65}$$

The mean time to full backup is

$$E(L) \equiv NT\overline{F}(NT) + \int_0^{NT} t\,\mathrm{d}F(t) = \int_0^{NT} \overline{F}(t)\mathrm{d}t. \tag{66}$$

From (64), the total expected backup cost to full backup is

$$C_{\text{TI}}(N) \equiv c_1 + \overline{F}((N-1)T) \sum_{m=1}^{N-1} C_{\text{EI}}(m)$$

$$+ \sum_{i=2}^{N-1} [F(iT) - F((i-1)T)] \sum_{m=1}^{i-1} C_{\text{EI}}(m)$$

$$= c_1 + \sum_{i=1}^{N-1} C_{\text{EI}}(i)\overline{F}(iT), \tag{67}$$

and from (65), the total expected recovery cost to full backup is
$$C_{\mathrm{TR}}(N) \equiv \sum_{i=1}^{N}[F(iT) - F((i-1)T)]C_{\mathrm{ER}}(i-1). \tag{68}$$
Therefore, the expected cost rate is
$$C(N) \equiv \frac{C_{\mathrm{TI}}(N) + C_{\mathrm{TR}}(N)}{E(L)} \quad (N = 1, 2, \ldots). \tag{69}$$
We discuss an optimal value N^* which minimize the expected cost rate $C(N)$. Suppose that the database is updated at a homogeneous Poisson process with an intensity function $\lambda(t) = p\lambda$, and it fails according to distribution $F(t) = 1 - e^{-q\lambda t}$, where $p + q = 1$. Further, suppose that $E\{W_j\} = 1/\mu$ and $c_0(x) = c_0 x$. Then, the expected cost rate $C(N)$ in (69) is
$$\frac{C(N)}{q\lambda} \equiv \left(c_3 - c_4 - \frac{c_0 p\lambda T}{\mu}\right) + \frac{c_2 + c_4 + \frac{2c_0 p\lambda T}{\mu}}{1 - e^{-q\lambda T}} + \widetilde{C}(N), \tag{70}$$
where
$$\widetilde{C}(N) \equiv \frac{\left(c_1 - c_2 - \frac{c_0 p\lambda T}{\mu}\right) - \left(c_4 + \frac{c_0 p\lambda T}{\mu}\right) N e^{-q\lambda NT}}{1 - e^{-q\lambda NT}} \quad (N = 1, 2, \ldots). \tag{71}$$
From (70), we know that an optimal N^* which minimizes $C(N)$ is equality to an optimal N^* which minimizes $\widetilde{C}(N)$.

From the inequality $\widetilde{C}(N+1) - \widetilde{C}(N) \geq 0$,
$$Q(N+1) \geq \frac{c_1 - c_2 - \frac{c_0 p\lambda T}{\mu}}{c_4 + \frac{c_0 p\lambda T}{\mu}}, \tag{72}$$
where
$$Q(N) \equiv N - \frac{1 - e^{-q\lambda NT}}{1 - e^{-q\lambda T}}. \tag{73}$$
Clearly, $Q(1) = 0$, $\lim_{N \to \infty} Q(N) = \infty$ and $Q(N+1) - Q(N) = 1 - e^{-q\lambda NT} > 0$, i.e., $Q(N)$ is strictly increasing with N.

Therefore, we have the following optimal policy:

(i) If $c_1 - c_2 > (c_4 + \frac{c_0 p\lambda T}{\mu})(1 - e^{-q\lambda T}) + \frac{c_0 p\lambda T}{\mu}$, then there exists a finite and unique minimum N^* ($1 < N^* < \infty$) satisfying (72), and
$$\left(c_4 + \frac{c_0 p\lambda T}{\mu}\right) \frac{Q(N^*) - N^* e^{-q\lambda NT}}{1 - e^{-q\lambda NT}} < \widetilde{C}(N^*)$$
$$\leq \left(c_4 + \frac{c_0 p\lambda T}{\mu}\right) \frac{Q(N^*+1) - N^* e^{-q\lambda NT}}{1 - e^{-q\lambda NT}}. \tag{74}$$

(ii) If $c_1 - c_2 \leq (c_4 + \frac{c_0 p\lambda T}{\mu})(1 - e^{-q\lambda T}) + \frac{c_0 p\lambda T}{\mu}$, then $N^* = 1$, i.e., only full backup needs to be done, and the resulting cost is

$$\frac{C(1)}{q\lambda} = c_3 + \frac{c_1}{1 - e^{-q\lambda T}}. \qquad (75)$$

Example 4.1. Suppose that $c_1/(c_0/\mu) = 2000$, $c_2/(c_0/\mu) = 40$, $c_3/(c_0/\mu) = 2400$ and $c_4/(c_0/\mu) = 50$. Table 4 gives the optimal number N^* and the resulting cost rate $C(N^*)/(\lambda c_0/\mu)$ for $q = 10^{-2}, 10^{-3}, 10^{-4}$ and $\lambda T = 200, 400, 800, 1000$. Note that all costs are relative to cost c_0/μ and all times are relative to $1/\lambda$. In this case, it is evident that $(50 + p\lambda T)(1 - e^{-q\lambda T}) + p\lambda T < 1960$ except for $q = 10^{-2}$ and $\lambda T = 1000$. This indicates that N^* decreases with both λT and q, and $C(N^*)/(\lambda c_0/\mu)$ decreases with λT, and conversely, increases with q. □

Table 4 Optimal number N^* and resulting cost rate $C(N^*)/(\lambda c_0/\mu)$ when $c_1/(c_0/\mu) = 2000$, $c_2/(c_0/\mu) = 40$, $c_3/(c_0/\mu) = 2400$ and $c_4/(c_0/\mu) = 50$

λT	q					
	10^{-2}		10^{-3}		10^{-4}	
	N^*	$C(N^*)/(\lambda c_0/\mu)$	N^*	$C(N^*)/(\lambda c_0/\mu)$	N^*	$C(N^*)/(\lambda c_0/\mu)$
200	8	44.7607	12	6.4879	29	2.1670
400	3	44.1646	6	6.0945	14	1.9892
800	2	44.0056	3	5.6392	6	1.8295
1000	1	44.0009	2	5.4372	5	1.7658

4.2 Periodic Cumulative Backup Policies

Consider a periodic cumulative backup policy: Suppose that a database should be operating for an infinite time span. Cumulative backups are performed at periodic time iT ($i = 1, 2, \cdots$), and make the copies of only updated files which have changed or are new since the last full backup. But, because the time and resources required for cumulative backups are growing up every time, full backup is performed at iT, when the total updated files have exactly exceeded a managerial level K during the interval $((i-1)T, iT]$, or NT ($i = 1, 2, \cdots, N-1; N = 1, 2, \cdots$), whichever occurs first, and makes the copies of all files. A database returns to an initial state by such full backups.

Taking the above considerations into account, we formulate the following stochastic model of the backup policy for a database system: Suppose that a database is updated at a nonhomogeneous Poisson process with an intensity function $\lambda(t)$ and a mean-value function $R(t)$. Further, let W_j denote an amount of files, which changes or is new at the jth update. It is assumed that each W_j has an identical probability distribution $G(x) \equiv \Pr\{W_j \leq x\}$ $(j = 1, 2, \cdots)$. Then, the total amount of updated files $Z_j \equiv \sum_{i=1}^{j} W_i$ up to the jth update, where $Z_0 \equiv 0$, has a distribution

$$\Pr\{Z_j \leq x\} \equiv G^{(j)}(x) \quad (j = 0, 1, 2, \cdots). \tag{76}$$

Then, the probability that the total amount of updated files exceeds exactly a managerial level K at jth update is $G^{(j-1)}(K) - G^{(j)}(K)$. Let $Z(t)$ be the total amount of updated files at time t. Then, the distribution of $Z(t)$ is

$$\Pr\{Z(t) \leq x\} = \sum_{j=0}^{\infty} H_j(t) G^{(j)}(x). \tag{77}$$

Since the probability that the total amount of updated files does not exceed a managerial level K at time iT is, from (77),

$$F_i(K) \equiv \sum_{j=0}^{\infty} H_j(iT) G^{(j)}(K) \quad (i = 1, 2, \cdots), \tag{78}$$

where $F_0(K) \equiv 1$, the probability that its total amount exceeds exactly a level K during $((i-1)T, iT]$ is $F_{i-1}(K) - F_i(K)$.

Suppose that full backup cost is c_1, and cumulative backup cost is $c_2 + c_0(x)$ when the total amount of updated files is x $(0 \leq x < K)$. It is assumed that the function $c_0(x)$ is continuous and strictly increasing with $c_0(0) \equiv 0$ and $c_2 < c_1 \leq c_2 + c_0(\infty)$. Then, from (78), the expected cost of cumulative backup, when it is performed at time iT $(i = 1, 2, \cdots)$, is

$$C_{\text{EC}}(i, K) \equiv \frac{\sum_{j=0}^{\infty} H_j(iT) \int_0^K [c_2 + c_0(x)] \mathrm{d}G^{(j)}(x)}{F_i(K)}, \tag{79}$$

where $C_{\text{EC}}(0, K) \equiv 0$. Further, the mean time to full backup is

$$\sum_{i=1}^{N-1} (iT)[F_{i-1}(K) - F_i(K)] + (NT)F_{N-1}(K) = T \sum_{i=0}^{N-1} F_i(K), \tag{80}$$

and the total expected cost to full backup is

$$c_1 + \sum_{i=1}^{N-1} \sum_{j=0}^{i-1} C_{\text{EC}}(j, K)[F_{i-1}(K) - F_i(K)] + \sum_{j=0}^{N-1} C_{\text{EC}}(j, K) F_{N-1}(K)$$

$$= c_1 + \sum_{i=1}^{N-1} C_{\text{EC}}(i, K) F_i(K). \tag{81}$$

Therefore, the expected cost rate is

$$C(K,N) \equiv \frac{c_1 + \sum_{i=1}^{N-1} C_{\mathrm{EC}}(i,K) F_i(K)}{T \sum_{i=0}^{N-1} F_i(K)}. \tag{82}$$

Discuss optimal values K^* and N^* which minimize the expected cost rate $C(K,N)$. Differentiating $C(K,N)$ with respect to K and setting it equal to zero,

$$\sum_{i=0}^{N-1} [c_2 + c_0(K) - C_{\mathrm{EC}}(i,K)] F_i(K) = c_1 - c_2. \tag{83}$$

Forming the inequalities $C(K, N+1) \geq C(K,N)$ and $C(K,N) < C(K,N-1)$,

$$L(N,K) \geq c_1 - c_2 \text{ and } L(N-1,K) < c_1 - c_2, \tag{84}$$

where

$$L(N,K) \equiv \sum_{i=0}^{N-1} [C_{\mathrm{EC}}(N,K) - C_{\mathrm{EC}}(i,K)] F_i(K). \tag{85}$$

Noting that $C_{\mathrm{EC}}(N,K) < c_2 + c_0(K)$ from (79) when $c_0(x)$ is continuous and strictly increasing, there does not exist a positive pair (K^*, N^*) $(0 < K, N < \infty)$ which satisfies (83) and (84), simultaneously. Thus, if such a positive pair (K^*, N^*) exists which minimizes $C(K,N)$ in (82), then $K^* = \infty$ or $N^* = \infty$.

Consider an optimal level for full backup, i.e., a database undergoes full backup at time iT $(i = 1, 2, \cdots)$ only when the total updated files have exceeded exactly a level K during $((i-1)T, iT]$. Putting $N = \infty$ in (82), the expected cost rate is

$$C_1(K) \equiv \lim_{N \to \infty} C(K,N)$$

$$= \frac{c_1 + \sum_{i=1}^{\infty} \sum_{j=0}^{\infty} H_j(iT) \int_0^K [c_2 + c_0(x)] \mathrm{d}G^{(j)}(x)}{T \sum_{i=0}^{\infty} F_i(K)}. \tag{86}$$

A necessary condition that an optimal K^* minimizes $C_1(K)$ is

$$\sum_{i=0}^{\infty} \sum_{j=0}^{\infty} H_j(iT) \int_0^K G^{(j)}(x) \mathrm{d}c_0(x) = c_1 - c_2. \tag{87}$$

In particular, suppose that $c_0(x) = a(1 - e^{-sx})$ $(a, s > 0)$. Letting $Q(K)$ be the left-hand side of (87),

$$Q(K) = a \sum_{i=0}^{\infty} \left[\sum_{j=0}^{\infty} H_j(iT) \int_0^K e^{-sx} \mathrm{d}G^{(j)}(x) - e^{-sK} F_i(K) \right]. \tag{88}$$

It can be easily seen that $Q(K)$ is strictly increasing from 0 to $Q(\infty) = a\sum_{i=0}^{\infty} e^{-R(iT)[1-G^*(s)]} > a$, where $G^*(s) \equiv \int_0^{\infty} e^{-sx} dG(x)$ denotes the Laplace-Stieltjes transform of $G(x)$.

Noting that $c_1 \leq c_2 + a$, there exists a finite and unique K^* ($0 < K^* < \infty$) which minimizes $C_1(K)$ and satisfies (87). In this case, the resulting cost rate is

$$C_1(K^*) = \frac{1}{T}[c_2 + a(1 - e^{-sK^*})]. \tag{89}$$

Next, consider an optimal number N^* when a database undergoes full backup only at time NT ($N = 1, 2, \cdots$). Putting that $K = \infty$ in (82), the expected cost rate is

$$C_2(N) \equiv \lim_{K \to \infty} C(K, N)$$
$$= \frac{c_1 + \sum_{i=1}^{N-1}\sum_{j=0}^{\infty} H_j(iT) \int_0^{\infty} [c_2 + c_0(x)] dG^{(j)}(x)}{NT}. \tag{90}$$

From the inequality $C_2(N+1) - C_2(N) \geq 0$,

$$\sum_{i=0}^{N-1}\sum_{j=0}^{\infty} [H_j(NT) - H_j(iT)] \int_0^{\infty} c_0(x) dG^{(j)}(x) \geq c_1 - c_2. \tag{91}$$

In particular, suppose that $c_0(x) = a(1 - e^{-sx})$. Letting $L(N)$ be the left-hand side of (91),

$$L(N) = a \sum_{i=0}^{N-1} \left\{ e^{-R(iT)[1-G^*(s)]} - e^{-R(NT)[1-G^*(s)]} \right\}, \tag{92}$$

which is strictly increasing to $L(\infty) = a\sum_{i=0}^{\infty} e^{-R(iT)[1-G^*(s)]}$. Noting that $L(\infty) = Q(\infty)$, there exists a finite and unique minimum N^* ($1 \leq N^* < \infty$) which satisfies (91).

Example 4.2. Suppose that a database system is updated according to a Poisson process with rate λ, i.e., $\lambda(t) = \lambda$. Further, it is assumed that $G(x) = 1 - e^{-\mu x}$, i.e., $G^{(j)}(x) = 1 - \sum_{i=0}^{j-1}[(\mu x)^i/i!]e^{-\mu x}$ and $M(K) = \mu K$. Then, (87) is

$$\sum_{i=0}^{\infty}\sum_{j=0}^{\infty} \frac{(i\lambda T)^j}{j!} e^{-i\lambda T} a \sum_{m=j}^{\infty} \frac{[\mu^j(s+\mu)^{m-j} - \mu^m]K^m}{m!} e^{-(s+\mu)K} = c_1 - c_2. \tag{93}$$

The left-hand side of (93) is a strictly increasing function of K from 0 to $a/(1 - e^{-\frac{s\lambda T}{s+\mu}}) > a > c_1 - c_2$. Thus, there exists a finite and unique K^* ($0 < K^* < \infty$) which satisfies (93).

Table 5 Optimal level K^*, number N^* and resulting cost $C_1(K^*)T$, $C_2(N^*)T$ when $c_1 = 3$, $c_2 = 1$, $a = 2$, $\mu = 1$ and $\lambda T = 100$

s	K^*	$C_1(K^*)T$	N^*	$C_2(N^*)T$
2×10^{-2}	133	2.859	2	2.859
2×10^{-3}	337	1.979	4	1.980
2×10^{-4}	1064	1.375	11	1.375
2×10^{-5}	3180	1.123	32	1.123

Similarly, an N^* is given by a finite and unique minimum such that

$$a\left(\frac{1 - e^{-\frac{Ns\lambda T}{s+\mu}}}{1 - e^{-\frac{s\lambda T}{s+\mu}}} - Ne^{-\frac{Ns\lambda T}{s+\mu}}\right) \geq c_1 - c_2. \qquad (94)$$

Table 5 gives the optimal level K^*, number N^* and the resulting cost rates $C_1(K^*)T$, $C_2(N^*)T$ for $s = 2 \times 10^{-2}, 2 \times 10^{-3}, 2 \times 10^{-4}$ and 2×10^{-5} when $c_1 = 3$, $c_2 = 1$, $a = 2$, $\mu = 1$ and $\lambda T = 100$. Compared with $C_1(K^*)$ and $C_2(N^*)$, this indicates that $C_1(K^*) \leq C_2(N^*)$. Thus, if two costs of backups are the same, we should adopt the level policy as full backup scheme. However, it would be generally easier to count the number of backup than to check the amount of updated files. From this point of view, the number policy would be better than the level policy. Therefore, how to select among two policies would depend on actual mechanism of a database system. □

5 Conclusions

We have considered two backup models for a database system, and have analytically discussed optimal backup policies which minimize the expected cost, using theory of cumulative processes. These results would be applied to the backup of a database, by estimating the backup costs and the amount of updated files from actual data. However, backup schemes become very important and much complicated, as database systems have been largely used in most computer systems and information technologies have been greatly developed. These formulations and techniques used in this chapter would be useful and helpful for analyzing such backup policies.

Acknowledgements

This work is supported by the Hori Information Science Promotion Foundation of Japan, National Nature Science Foundation(70471017, 70801036), Humanities and Social Sciences Research Foundation of MOE of China(05JA630027).

References

1. Barlow, R. E. and Proschan, F. (1965). *Mathematical Theory of Reliability*, New York, John Wiley & Sons.
2. Chandy, K. M. Browne, J. C. Dissly, C. W. and Uhrig, W. R. (1975). *Analytic models for rollback and recovery strategies in data base systems*, IEEE Trans. Software Engineering, **SE-1**, pp. 100–110.
3. Cox, D. R. (1962). *Renewal Theory*, Methuen, London.
4. Dohi, T. Aoki, T. Kaio, N. and Osaki, S. (1997). *Computational aspects of optimal checkpoint strategy in fault-tolerant database management*, IEICE Trans. Fundamentals of Electronics, Communications and Computer Sciences, **E80-A**, pp. 2006–2015.
5. Esary, J. D. Marshall, A. W. and Proschan, F. (1973). *Shock model and wear processes*, Ann. Probab., **1**, pp. 627–649.
6. Feldman, R. M. (1976). *Optimal replacement with semi-Markov shock models*, J. Appl. Probab., **13**, 1, pp. 108–117.
7. Fukumoto, S. Kaio, N. and Osaki, S. (1992). *A study of checkpoint generations for a database recovery mechanism*, Comput. Math. Appl., **24**, 1/2, pp. 63–70.
8. Gelenbe, E. (1979). *On the optimum checkpoint interval*, J. Association on Computing Machinery, **26**, pp. 259–270.
9. Nakagawa, T. (1976). *On a replacement problem of a cumulative damage model*, Oper. Res. Quarterly, **27**, 4, pp. 895–900.
10. Nakagawa, T. (1984). *A summary of discrete replacement policies*, Eur. J. Oper. Res., **17**, 3, pp. 382–392.
11. Nakagawa, T. and Kijima, M. (1989). *Replacement policies for a cumulative damage model with minimal repair at failure*, IEEE Trans. Reliability, **38**, 4, pp. 581–584.
12. Nakagawa, T. (2005). *Maintenance Theory of Reliability*, Springer-Verlag, London.
13. Nakagawa, T. (2007). *Shock and Damage Models in Reliability Theory*. Springer-Verlag, London.
14. Qian, C. H. Nakamura, S. and Nakagawa, T. (1999). *Cumulative Damage Model with Two Kinds of Shocks and Its Application to the Backup Policy*, J. Oper. Res. Soc. Japan, **42**, 4, pp. 501–511.
15. Qian, C. H. Nakamura, S. and Nakagawa, T.(2002). *Optimal Backup Policies for a Database System with Incremental Backup*, Electronics and Communications in Japan, Part 3, **85**, 4, pp. 1–9.

16. Qian, C. H. Pan, Y. and Nakagawa, T.(2002). *Optimal policies for a database system with two backup schemes*, RAIRO Oper. Res., **36**, pp. 227–235.
17. Reuter, A. (1984). *Performance analysis of recovery techniques*, ACM Trans. Data. Sys., **4**, pp. 526–559.
18. Sandoh, H. Kaio, N. and Kawai, H.(1992). *On backup policies for computer disks*, Reliability Engineering & System Safety, **37**, pp. 29–32.
19. Satow, T. Yasui, K. and Nakagawa, T. (1996)., *Optimal garbage collection policies for a database in a computer system*, RAIRO: Oper. Res., **30**, 4, pp. 359–372.
20. Suzuki, K. and Nakajima, K. (1995), *Storage management software*, Fujitsu, **46**, pp.389–397.
21. Taylor, H. M. (1975). *Optimal replacement under additive damage and other failure models*, Naval Res. Logist Quarterly, **22**, 1, pp. 1–18.
22. Velpuri, R. and Adkoli, A. (1998). *Oracle8 backup and recovery handbook*, McGraw-Hill.
23. Young, J. W. (1974). *A first order approximation to the optimum checkpoint interval*, Communications of ACM, **17**, pp. 530–531.

Chapter 8

Optimal Checkpoint Intervals for Computer Systems

KENICHIRO NARUSE

Information Center
Nagoya Sangyo University
3255-5 Arai-cho, Owariasahi 488-8711, Japan
E-mail: naruse@nagoya-su.ac.jp

SAYORI MAEJI

Institute of Consumer Science and Human Life
Kinjo Gakuin University
1723 Oomori 2, Moriyama, Nagoya 463-8521, Japan

1 Introduction

In recent years, computers have been used not only to live our daily life, but also to make and sell good products in industries. Most things have computers within them and are moved by computers. Computers play more important role in a highly civilized society. Especially, computer systems have been required to operate normally and effectively as communication and information systems have been developed rapidly and complicated remarkably. However, some errors due to noises, human errors, hardware faults, computer viruses, and so on, occur certainty in systems. Lastly, those errors might have become faults and incur system failures. Such failures have sometimes caused a heavy damage to a human society and have fallen into general disorder. To prevent such faults, various kinds of fault tolerant techniques such as the redundancy of processors and memories and the configuration of systems have been provided [4, 10, 11, 20]. The high reliability and effective performance of real systems can be achieved by fault tolerant techniques.

Partial data loss and operational errors in computer systems are generally called *error* and *fault* caused by errors. Failure indicates that faults are recognized on the exterior systems. Three different techniques of decreasing the possibility of fault occurrences can be used [1]: Fault avoidance is to prevent fault occurrences by improving qualities of structure parts and placing well surroundings. Fault masking is to prevent faults by error correction codes and majority voting. Fault tolerance is that systems continue to function correctly in the presence of hardware failures and software errors. There techniques above are called simply fault tolerance into one word.

Some faults due to operational errors may be detected after some time has passed and a system consistency may be lost by them. Then, we should restore a consistent state just before fault occurrences by some recovery techniques. The operation that takes copies of the normal state of the system is called checkpoint. When faults have been occurred, the process goes back to the nearest checkpoint time by rollback operation [2, 5, 16], and its retry is made, using the copy of a consistent state stored in the checkpoint time.

It is supposed that we have to complete the process of one task with a finite execution time. A module is an element such as a logical circuit or a processor that executes certain lumped parts of the task. Then, we consider the checkpoint models of error detection and masking by redundancy, and propose their modified models. Using reliability theory, we analyze these models and discuss analytically optimal checkpoint intervals.

Section 2 considers two-level recovery schemes of soft and hard checkpoints and derives an optimal interval of soft checkpoint between hard checkpoints. Section 3 adopts multiple modular redundant systems as the recovery techniques of error detection and error masking, and derives optimal checkpoint intervals. Section 4 considers the modified checkpoint model in Section 3 where checkpoints are placed at sequential times and error rates increase with the number of checkpoints and with an original execution time. Finally, Section 5 supposes that tasks with random processing times are executed successively, and two types of checkpoints are placed at the end of tasks. Three schemes are considered and are compared numerically.

2 Two-level Recovery Schemes

Checkpoint is the most effective recovery mechanism which stores a consistent state in the secondary storage at suitable times. Even if failures occur, the process goes back to checkpoint and can resume its normal operation [2, 4, 16]. Ling et al. [5] made a good survey of such checkpoint problems.

Vaidya [19, 20] considered two-level recovery schemes in which N-checkpoint can recover from several number of failures, and 1-checkpoint is taken between N-checkpoint and can recover from only a single failure. He presented an analytical approach for evaluating performance of two-level schemes, using a Markov chain. Further, Ssu et al. [17] described an adaptive protocol that manages storage for base stations in mobile environments where *soft checkpoint* is saved in a mobile host, *e.g.*, in a local disk or flash memory, and *hard checkpoint* is saved in a base station. Soft checkpoints will be lost if a mobile host fails, however, hard checkpoints can survive but have higher overheads since they must be transmitted through the wireless channels.

This section considers two-level recovery schemes based on the proposed scheme of [19]: Soft checkpoint (SC) and hard checkpoint (HC) which are useful to recover from only one failure and several failures, respectively. SCs are set up at periodic intervals between HCs, and are less reliable and less overhead than those of HCs. We discuss a checkpointing interval of SCs when HCs are placed on the beginning and end of the process. The total expected overhead of one cycle from HC to HC is obtained, using Markov renewal processes [14], and an optimal interval which minimizes it numerically computed. It is shown in a numerical example that two-level schemes reduce the total overhead of the process.

Suppose that S is an original execution time of one process or task which does not include the overheads of retries and checkpoint generations. Then, to tolerate some failures, we consider two different types of checkpoints:

- *Soft checkpoint* (SC) can recover from some kinds of failures and its overhead is small.
- *Hard checkpoint* (HC) can recover from any kinds of failures and its overhead is large.

We propose the two-level recovery scheme with the following assumptions:

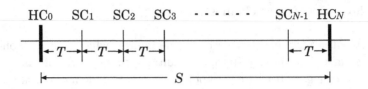

Fig. 1 Soft checkpoints between hard checkpoints.

1) The original execution time of one process is S ($0 < S < \infty$). We divide S equally into N time intervals where $T \equiv S/N$, and take $(N-1)$SC every at times kT ($k = 1, 2, \cdots, N-1$), and two HC at time 0 and time NT, i.e., $SC_1, SC_2, \cdots, SC_{N-1}$ are set up between HC_0 and HC_N (Figure 1).
2) Failures of the process occur at constant rate λ ($\lambda > 0$), i.e., the process has a failure distribution $F(t) = 1 - e^{-\lambda t}$ and $\overline{F}(t) \equiv 1 - F(t) = e^{-\lambda t}$.
3) If failures occur between HC_0 and SC_1, then the process is rolled back to HC_0 and begins its re-execution. If failures occur between SC_j and SC_{j+1} ($j = 1, 2, \cdots, N-1$), then the process is rolled back to SC_j where $SC_N \equiv HC_N$:

   a) The process can recover from their failures with probability q ($0 \leq q \leq 1$) and begins its re-execution from SC_j.

   b) The process cannot recover with probability $1 - q$, and further, is rolled back to HC_0 and begins its re-execution.

4) If there is no failure between SC_j and SC_{j+1} ($j = 0, 1, \cdots, N-1$) where $SC_0 \equiv HC_0$, the process goes forward and begins its execution from SC_{j+1}.
5) The process ends when it attains to HC_N.

2.1 Performance Analysis

We define the following states of the process:

State 0: The process begins to execute its processing from HC_0.
State j: The process begins to execute its processing from SC_j ($j = 1, 2, \cdots, N-1$).
State N: The process attains to HC_N and ends.

The process states defined above form a Markov renewal process [14] in which State N is an absorbing state. All states are regeneration points and the transition diagram between states is shown in Figure 2.

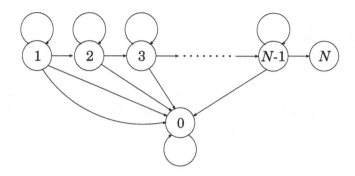

Fig. 2 Transition diagram between states.

Let $Q_{ij}(t)$ ($i, j = 0, 1, 2, \cdots, N$) be one-step transition probabilities of a Markov renewal process. Then, by the similar method [21], mass functions $Q_{ij}(t)$ from State i at time 0 to State j at time t are

$$Q_{00}(t) = \int_0^t F(u)\,\mathrm{d}D(u), \tag{1}$$

$$Q_{jj+1}(t) = \int_0^t \overline{F}(u)\,\mathrm{d}D(u) \qquad (j = 0, 1, \cdots, N-1), \tag{2}$$

$$Q_{jj}(t) = q \int_0^t F(u)\,\mathrm{d}D(u) \qquad (j = 1, 2, \cdots, N-1), \tag{3}$$

$$Q_{j0}(t) = (1-q) \int_0^t F(u)\,\mathrm{d}D(u) \qquad (j = 1, 2, \cdots, N-1), \tag{4}$$

where $D(t)$ is a degenerate distribution placing unit mass at T, i.e., $D(t) \equiv 1$ for $t \geq T$, and 0 for $t < T$.

Further, let $\Phi^*(s)$ be the Laplace-Stieltjes transform of any function $\Phi(t)$, i.e., $\Phi^*(s) \equiv \int_0^\infty e^{-st}\,\mathrm{d}\Phi(t)$ for $s \geq 0$. Then, the LS transforms of $Q_{ij}(t)$ are, from (1)–(4),

$$Q_{00}^*(s) = e^{-sT} F(T), \tag{5}$$

$$Q_{jj+1}^*(s) = e^{-sT} \overline{F}(T) \qquad (j = 0, 1, \cdots, N-1), \tag{6}$$

$$Q_{jj}^*(s) = e^{-sT} q F(T) \qquad (j = 1, 2, \cdots, N-1), \tag{7}$$

$$Q_{j0}^*(s) = e^{-sT} (1-q) F(T) \qquad (j = 1, 2, \cdots, N-1). \tag{8}$$

Denoting $H_{0N}(t)$ by the first-passage time distribution from State 0 to State N, its LS transform is

$$H^*{}_{0N}(s) = Q_{01}^*(s) \frac{Q_{12}^*(s)}{1 - Q_{11}^*(s)} \times \cdots \times \frac{Q_{N-1N}^*(s)}{1 - Q_{N-1N-1}^*(s)}$$

$$+\left\{Q_{00}^*(s)+\sum_{j=1}^{N-1}\left[Q_{01}^*(s)\frac{Q_{12}^*(s)}{1-Q_{11}^*(s)}\times\cdots\times\frac{Q_{j0}^*(s)}{1-Q_{jj}^*(s)}\right]\right\}H^*_{0N}(s). \tag{9}$$

To simplify equations, we put that $Q_{j0}^*(s) \equiv A_0(s)$, $Q_{jj}^*(s) \equiv A_1(s)$ and $Q_{jj+1}^*(s) \equiv A_2(s)$. Then, $Q_{00}^*(s) = A_0(s) + A_1(s)$ and $A_0(0) + A_1(0) + A_2(0) = 1$. Using these notations and solving (9) for $H^*_{0N}(s)$,

$$H^*_{0N}(s) = \frac{A_2(s)\left[\frac{A_2(s)}{1-A_1(s)}\right]^{N-1}}{1-A_0(s)-A_1(s)-\left[\frac{A_0(s)A_2(s)}{1-A_1(s)-A_2(s)}\right]\left\{1-\left[\frac{A_2(s)}{1-A_1(s)}\right]^{N-1}\right\}}. \tag{10}$$

It is evident that $H^*_{0N}(0) = 1$. Thus, the mean first-passage time from State 0 to State N is

$$l_{0N} \equiv \lim_{s\to 0}\frac{1-H^*_{0N}(s)}{s}$$
$$= \frac{T}{\overline{F}(T)}\sum_{j=0}^{N-1}\left(\frac{\overline{F}(T)}{1-qF(T)}\right)^{j-(N-1)} \quad (N=1,2,\cdots). \tag{11}$$

Moreover, the LS transform of the expected number of returning to State 0 is given by a renewal equation

$$M^*_H(s) = \left[Q_{00}^*(s)+\sum_{j=1}^{N-1}Q_{01}^*(s)\frac{Q_{12}^*(s)}{1-Q_{11}^*(s)}\times\cdots\times\frac{Q_{j0}^*(s)}{1-Q_{jj}^*(s)}\right][1+M^*_H(s)]. \tag{12}$$

Solving this equation and arranging it,

$$M^*_H(s) = \frac{A_0(s)+A_1(s)+A_0(s)\sum_{j=1}^{N-1}\left[\frac{A_2(s)}{1-A_1(s)}\right]^j}{1-A_0(s)-A_1(s)-A_0(s)\sum_{j=1}^{N-1}\left[\frac{A_2(s)}{1-A_1(s)}\right]^j}. \tag{13}$$

Thus, the expected number of returning to State 0 is

$$M_H \equiv \lim_{s\to 0}M^*_H(s)$$
$$= \frac{1-\overline{F}(T)\left[\frac{\overline{F}(T)}{1-qF(T)}\right]^{N-1}}{\overline{F}(T)\left[\frac{\overline{F}(T)}{1-qF(T)}\right]^{N-1}} \quad (N=1,2,\cdots). \tag{14}$$

Note that M_H represents the total expected number of rollbacks to HC until the process ends.

Next, we compute the expected number of rollbacks to SC. The expected numbers of returning to State j when the process transits from State j to State $j+1$ and State 0 are, respectively,

$$\sum_{i=1}^{\infty} i\, [Q_{jj}^*(s)]^i Q_{jj+1}^*(s) = \frac{Q_{jj}^*(s) Q_{jj+1}^*(s)}{[1-Q_{jj}^*(s)]^2},$$

$$\sum_{i=1}^{\infty} i\, [Q_{jj}^*(s)]^i Q_{j0}^*(s) = \frac{Q_{jj}^*(s) Q_{j0}^*(s)}{[1-Q_{jj}^*(s)]^2}.$$

Thus, the LS transform of the expected number of returning to State j $(j = 1, 2, \cdots, N-1)$ is

$$M^*{}_S(s) = \sum_{j=1}^{N-1} Q_{01}^*(s) \frac{Q_{12}^*(s)}{1-Q_{11}^*(s)} \times \cdots \times \frac{Q_{j-1\,j}^*(s)}{1-Q_{j-1\,j-1}^*(s)} \frac{Q_{jj}^*(s)[Q_{j j+1}^*(s)+Q_{j0}^*(s)]}{[1-Q_{jj}^*(s)]^2}$$

$$+ \left[Q_{00}^*(s) + \sum_{j=1}^{N-1} Q_{01}^*(s) \frac{Q_{12}^*(s)}{1-Q_{11}^*(s)} \times \cdots \times \frac{Q_{j-1\,j}^*(s)}{1-Q_{j-1\,j-1}^*(s)} \frac{Q_{j0}^*(s)}{1-Q_{jj}^*(s)} \right] M^*{}_S(s). \tag{15}$$

Solving this equation,

$$M^*{}_S(s) = \frac{A_1(s) A_2(s) [A_0(s)+A_2(s)] \dfrac{1 - \left[\dfrac{A_2(s)}{1-A_1(s)}\right]^{N-1}}{1-A_1(s)-A_2(s)}}{1 - A_0(s) - A_1(s) - A_0(s) A_2(s) \dfrac{1 - \left[\dfrac{A_2(s)}{1-A_1(s)}\right]^{N-1}}{1-A_1(s)-A_2(s)}}. \tag{16}$$

Therefore, the total expected number of returning to State j is, for $0 \le q < 1$,

$$M_S \equiv \lim_{s \to 0} M^*{}_S(s) = \frac{q}{1-q} \frac{1 - \left[\dfrac{\overline{F}(T)}{1-qF(T)}\right]^{N-1}}{\left[\dfrac{\overline{F}(T)}{1-qF(T)}\right]^{N-1}} \quad (N=1,2,\cdots), \tag{17}$$

and for $q = 1$,

$$M_S \equiv \frac{(N-1)F(T)}{\overline{F}(T)} \quad (N=1,2,\cdots). \tag{18}$$

Note also that M_S represents the total expected number of rollbacks to SC until the process ends.

2.2 Expected Overhead

Assume that the overheads for rollbacks to HC and SC are C_H and C_S ($C_S < C_H$), respectively, and C_T for setting up one SC. The other overheads except C_H, C_S and C_T would be neglected because they are small. Then, the total expected overhead is, from (11), (14), and (17),

$$C_1(N) \equiv l_{0N} + C_H M_H + C_S M_S + (N-1) C_T - S$$

$$= \frac{T + C_H + [T + qF(T)C_S] \sum_{j=1}^{N-1} \left[\frac{\overline{F}(T)}{1-qF(T)}\right]^j}{\overline{F}(T) \left[\frac{\overline{F}(T)}{1-qF(T)}\right]^{N-1}} - C_H + (N-1)C_T - S$$

$$(N = 1, 2, \cdots), \quad (19)$$

where $\sum_{j=1}^{0} \equiv 0$. In particular, when $N = 1$, *i.e.*, SC is not set up,

$$C_1(1) = \frac{S + C_H F(S)}{\overline{F}(S)} - S. \quad (20)$$

When $q = 1$, *i.e.*, SC can recover from all failures,

$$C_1(N) = \frac{S + F(T)[C_H + (N-1)C_S]}{\overline{F}(T)} + (N-1)C_T - S, \quad (21)$$

and when $q = 0$, *i.e.*, SC cannot recover from any failures,

$$C_1(N) = \frac{1 - [\overline{F}(T)]^N}{[\overline{F}(T)]^N} \left[\frac{T}{\overline{F}(T)} + C_H\right] + (N-1)C_T - S. \quad (22)$$

Example 2.1. We compute an optimal number N^* of SC which minimizes the total expected overhead $C_1(N)$. Since $F(t) = 1 - e^{-\lambda t}$ and $T = S/N$, (19) becomes

$$\lambda C_1(N) = \frac{\frac{\lambda S}{N} + \lambda C_H + \left[\frac{\lambda S}{N} + q(1 - e^{-\lambda S/N})\lambda C_S\right] \sum_{j=1}^{N-1} \left[\frac{e^{-\lambda S/N}}{1-q(1-e^{-\lambda S/N})}\right]^j}{e^{-\lambda S/N} \left[\frac{e^{-\lambda S/N}}{1-q(1-e^{-\lambda S/N})}\right]^{N-1}}$$

$$-\lambda C_H + (N-1)\lambda C_T - \lambda S \quad (N = 1, 2, \cdots). \quad (23)$$

Table 1 gives the optimal number N^* for $q = 0.0, 0.2, 0.4, 0.6, 0.8, 1.0$ and $\lambda C_T = 0.0001, 0.0005, 0.001, 0.005$ when $\lambda S = 0.1$, $\lambda C_H = 0.001$ and $\lambda C_S = 0.0002$. For example, when $q = 0.8$ and $\lambda C_T = 5 \times 10^{-4}$, $N^* = 4$ and $\lambda C_1(4) = 4.866 \times 10^{-3}$, *i.e.*, we should take 3SCs between HCs. It is evident that N^* decrease and $C_1(N^*)$ increase as C_T increase. This also indicates that N^* increase as q increase, because SC becomes useful to recover from failures. Further, the overhead of two-level schemes is smaller than that of one-level scheme in the case of $N = 1$. From this example, two-level recovery schemes would achieve better performances, compared with one-level scheme. □

Table 1 Optimal number N^* and total expected overhead $C(N^*)$ when $\lambda S = 0.1$, $\lambda C_H = 0.001$, $\lambda C_S = 0.0002$

q	$\lambda C_T = 1 \times 10^{-4}$		$\lambda C_T = 5 \times 10^{-4}$		$\lambda C_T = 1 \times 10^{-3}$		$\lambda C_T = 5 \times 10^{-3}$	
	N^*	$\lambda C(N^*) \times 10^3$	N^*	$\lambda C(N^*) \times 10^3$	N^*	$\lambda C(N^*) \times 10^3$	N^*	$\lambda C(N^*) \times 10^3$
0.0	7	6.610	3	8.028	2	8.927	1	10.622
0.2	7	5.656	4	7.280	3	8.301	1	10.622
0.4	9	4.735	4	6.470	3	7.577	1	10.622
0.6	9	3.792	4	5.665	3	6.857	1	10.622
0.8	12	2.871	4	4.866	3	6.140	1	10.622
1.0	12	1.903	5	4.056	3	5.426	2	10.189

3 Error Detection by Multiple Modular Redundancies

This section considers the redundant techniques of error detection and error masking on a finite process execution. First, an error detection of the process can be made by two independent modules where they compare two results at suitable checkpoint times. If their results do not match with each other, we go back to the newest checkpoint and make a retrial of the processes. Secondly, a majority decision system with multiple modules is adopted as the technique of an error masking and the result is decided by its majority of modules. In this case, we determine numerically what a majority system is optimal.

In such situations, if we compare results frequently, then the time required for rollback could decrease, however, the total overhead for comparisons at checkpoints would increase. Thus, this is one kind of trade-off problems how to decide an optimal checkpoint interval.

Several studies of deciding a checkpoint frequency have been discussed for the hardware redundancy above. Pradhan and Vaidya [15] evaluated the performance and reliability of a duplex system with a spare processor. Ziv and Bruck [22,23] considered the checkpoint schemes with task duplication and evaluated the performance of schemes. Kim and Shin [3] derived the optimal instruction-retry period which minimizes the probability of the dynamic failure on the triple modular redundant controller. Evaluation models with finite checkpoints and bounded rollback were discussed [13].

This section considers a double modular redundancy as redundant techniques of error detection and summarizes the results [6,7]. Next, we consider a redundant system of a majority decision with $(2n+1)$ modules as an error masking system, and compute the mean time to completion of the process and decide numerically what a majority system is optimal.

3.1 Multiple Modular System

3.2 Performance Analysis

Suppose that S is a native execution time of the process which does not include the overheads of retries and checkpoint generations. Then, we divide S equally into N parts and create a checkpoint at planned times $kT(k = 1, 2, \cdots, N-1)$ where $S = NT$ (Figure 3).

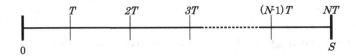

Fig. 3 Checkpoint intervals.

To detect errors, we firstly provide two independent modules where they compare two results at periodic checkpoint times. If two results agree with each other, two processes are correct and go forward. However, if two results do not agree, it is judged that some errors have occurred. Then, we make a rollback operation to the newest checkpoint and a retry of the processes. The process completes when two processes are succeeded in all intervals above.

Let us introduce a constant overhead C_1 for the comparison of two results. We neglect any failures of the system caused by common mode faults to make clear an error detection of the processes. Further, it is assumed that some errors of one process occur at constant rate λ, i.e., the probability that any errors do not occur during $(0, t]$ is given by $e^{-\lambda t}$. Thus, the probability that two processes have no error during $(0, t]$ is $\overline{F}_1(T) = e^{-2\lambda T}$ [14].

The mean time $L_1(N)$ to completion of the process is the summation of the processing times and the overhead C_1 of comparison of two processes. From the assumption that two processes are rolled back to the previous checkpoint when an error has been detected at a checkpoint, the mean execution time of the process for one checkpoint interval $(0, T]$ is given by a renewal equation:

$$L_1(1) \equiv (T + C_1) e^{-2\lambda T} + [T + C_1 + L_1(1)] \left(1 - e^{-2\lambda T}\right), \quad (24)$$

and solving it,
$$L_1(1) = (T + C_1)\,e^{2\lambda T}. \tag{25}$$

Thus, the mean time to completion of the process is
$$L_1(N) \equiv NL_1(1) = N(T + C_1)\,e^{2\lambda T} = (S + NC_1)\,e^{2\lambda S/N}. \tag{26}$$

We seek an optimal number N_1^* which minimizes $L_1(N)$ for a specified S. Evidently, $L_1(\infty) = \infty$ and
$$L_1(1) = (S + C_1)\,e^{2\lambda S}. \tag{27}$$

Thus, there exists a finite number N_1^* ($1 \leq N_1^* < \infty$). However, it would be difficult to find analytically N_1^* which minimizes $L_1(N)$ in (26). Putting $T = S/N$ in (26) and rewriting it by the function T,
$$L_1(T) = S\left(1 + \frac{C_1}{T}\right)e^{2\lambda T} \qquad (0 < T \leq S). \tag{28}$$

It is evident that $L_1(0) = \lim_{T \to 0} L_1(T) = \infty$ and $L_1(S)$ is given by (27). Thus, there exists an optimal $\widetilde{T}_1$ ($0 < \widetilde{T}_1 \leq S$) which minimizes $L_1(T)$ in (28). Differentiating $L_1(T)$ with respect to T and setting it equal to zero,
$$T^2 + C_1 T - \frac{C_1}{2\lambda} = 0. \tag{29}$$

Solving it with T,
$$\widetilde{T}_1 = \frac{C_1}{2}\left[\sqrt{1 + \frac{2}{\lambda C_1}} - 1\right]. \tag{30}$$

Therefore, we have the following optimal interval number N_1^* [9]:

(i) If $\widetilde{T}_1 < S$, we put $[S/\widetilde{T}_1] = N$, where $[x]$ denotes the greatest integer contained in x, and calculate $L_1(N)$ and $L_1(N+1)$ from (26). If $L_1(N) \leq L_1(N+1)$ then $N_1^* = N$ and $T_1^* = S/N_1^*$, and conversely, if $L_1(N+1) < L_1(N)$ then $N_1^* = N+1$.

(ii) If $\widetilde{T}_1 \geq S$, i.e., we should make no checkpoint until time S then $N_1^* = 1$, and the mean time is given in (27).

Note that $\widetilde{T}_1$ in (30) does not depend on S. Thus, if S is very large, is changed greatly or is unclear, then we may adopt $\widetilde{T}_1$ as an approximate checkpoint time.

Further, the mean time for one checkpoint interval per this interval is
$$\widetilde{L}_1(T) \equiv \frac{L_1(1)}{T} = \left(1 + \frac{C_1}{T}\right)e^{2\lambda T}. \tag{31}$$

Thus, the optimal time which minimizes $\widetilde{L}_1(T)$ also agrees with $\widetilde{T}_1$ in (30).

Next, we consider a redundant system of a majority decision with $(2n+1)$ modules as an error masking system, i.e., $(n+1)$-out-of-$(2n+1)$ system $(n = 1, 2, \cdots)$. If more than n results of $(2n+1)$ modules agree, the system is correct. Then, the probability that the system is correct during $(0, T]$ is

$$\overline{F}_{n+1}(T) = \sum_{k=n+1}^{2n+1} \binom{2n+1}{k} \left(e^{-\lambda T}\right)^k \left(1 - e^{-\lambda T}\right)^{2n+1-k}. \qquad (32)$$

Thus, the mean time to completion of the process is

$$L_{n+1}(N) = \frac{N(T + C_{n+1})}{\overline{F}_{n+1}(T)} \qquad (n = 1, 2, \cdots), \qquad (33)$$

where C_{n+1} is the overhead of a majority decision of $(2n+1)$ modules.

Table 2 Optimal checkpoint number N_1^*, interval λT_1^* and mean time $\lambda L_1(N_1^*)$ for a double modular system when $\lambda S = 10^{-1}$

$\lambda C_1 \times 10^3$	$\lambda \widetilde{T}_1 \times 10^2$	N_1^*	$\lambda L_1(N_1^*) \times 10^2$	$\lambda T_1^* \times 10^2$
0.5	1.556	6	10.650	1.67
1.0	2.187	5	10.929	2.00
1.5	2.665	4	11.143	2.50
2.0	3.064	3	11.331	3.33
3.0	3.726	3	11.715	3.33
4.0	4.277	2	11.936	5.00
5.0	4.756	2	12.157	5.00
10.0	6.589	2	13.435	5.00
20.0	9.050	1	14.657	10.00
30.0	10.839	1	15.878	10.00

Example 3.1. We show numerical examples of optimal checkpoint intervals for a double modular system when $\lambda S = 10^{-1}$. Table 2 presents $\lambda \widetilde{T}_1$ in (30), optimal number $N_1^*, \lambda T_1^*$ and $\lambda L_1 T(N_1^*)$ for $\lambda C_1 = 0.5, 1.5, 2, 3, 4, 5, 10, 20, 30(\times 10^{-3})$. For example, when $\lambda = 10^{-2}$ (1/sec), $C_1 = 10^{-1}$(sec) and $S = 10.0$(sec), $N_1^* = 5$, $T_1^* = S/N_1^* = 2.0$ (sec), and $L_1(5) = 10.929$ (sec), which is longer about 9.3 percent than S.

We consider the problem what a majority system is optimal. When the overhead of comparison of two processes is C_1, it is assumed that the overhead C_{n+1} of an $(n+1)$-out-of-$(2n+1)$ system is given by $C_{n+1} \equiv \binom{2n+1}{2} C_1$ $(n = 1, 2, \cdots)$. This is to select and compare 2 from each of $(2n+1)$ processes.

Table 3 presents the optimal number N_{n+1}^* and the resulting mean time $\lambda L_{n+1}(N_{n+1}^*) \times 10^2$ for $n = 1, 2, 3, 4$ when $\lambda C_1 = 0.1 \times 10^{-3}, 0.5 \times 10^{-3}$. When $\lambda C_1 = 0.5 \times 10^{-3}$, $N_3^* = 2$ and $\lambda L_3(2) = 10.37 \times 10^{-2}$ which is the smallest among these systems, that is, a 2-out-of-3 system is optimal. The mean times for $n = 1, 2$ are smaller than 10.65×10^{-2} for a double modular system. □

Table 3 Optimal checkpoint number N_{n+1}^* and mean time $\lambda L_{n+1}(N_{n+1}^*)$ for $(n+1)$-out-of-$(2n+1)$ system when $\lambda C_1 = 0.5 \times 10^{-3}$, $\lambda C_1 = 0.1 \times 10^{-3}$ and $\lambda S = 10^{-1}$

	$\lambda C_1 = 0.1 \times 10^{-3}$		$\lambda C_1 = 0.5 \times 10^{-3}$	
n	N_{n+1}^*	$\lambda L_{n+1}(N_{n+1}^*) \times 10^2$	N_{n+1}^*	$\lambda L_{n+1}(N_{n+1}^*) \times 10^2$
1	3	10.12 *	2	10.37 *
2	1	10.18	1	10.58
3	1	10.23	1	11.08
4	1	10.36	1	11.81

4 Sequential Checkpoint Intervals for Error Detection

This section considers a general modular system of error detection and error masking on a finite process execution: Suppose that checkpoints are placed at sequential times T_k ($k=1, 2, \cdots, N$), where $T_N \equiv S$ (Figure 4). First, it is assumed that error rates during the interval $(T_{k-1}, T_k]$ ($k=1, 2, \cdots, N$) increase with the number k of checkpoints. The mean time to completion of the process are obtained, and optimal checkpoint intervals which minimize them are derived by solving simultaneous equations.

Further, approximate checkpoint intervals are given by denoting that the probability of the occurrence of errors during $(T_{k-1}, T_k]$ is constant. Second, it is assumed that error rates during $(T_{k-1}, T_k]$ increase with the original execution time, irrespective of the number of recoveries. Optimal checkpoint intervals which minimize the mean time to completion of the process are discussed, and their approximate times are shown. Numerical examples of optimal checkpoint times for a double modular system are presented. It is shown numerically that the approximate method is simple and these intervals give good approximations to optimal ones.

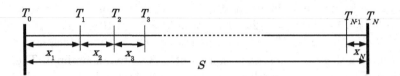

Fig. 4 Sequential checkpoint interval.

4.1 Performance Analysis

It was assumed in Section 3 that the error rate λ is constant and S is divided into an equal part. In general, error rates would be increasing with time, and so that, their intervals should be decreasing with their number. We assume for the simplicity of the model that error rates are increasing with the number of checkpoints.

Suppose that S is a native execution time of the process which does not include the overheads of retries and checkpoint generations. Then, we divide S into N parts and create a checkpoint at sequential times T_k ($k = 1, 2, \cdots, N-1$), where $T_0 \equiv 0$ and $T_N \equiv S$ (Figure 4).

Let us introduce a constant overhead C for the comparison of a modular system. Further, the probability that a modular system has no error during the interval $(T_{k-1}, T_k]$ is $F_k(T_k - T_{k-1})$, irrespective of other intervals and rollback operation. Then, the mean time $L_1(N)$ to completion of the process is the summation of the processing times and the overhead C for the comparison of a modular system.

From the assumption that the system is rolled back to the previous checkpoint when some error has been detected at a checkpoint, the mean execution time of the process for the interval $(T_{k-1}, T_k]$ is

$$L_1(k) = (T_k - T_{k-1} + C)\overline{F}_k(T_k - T_{k-1}) + [T_k - T_{k-1} + C + L_1(k)]F_k(T_k - T_{k-1}), \tag{34}$$

and solving it,

$$L_1(k) = \frac{T_k - T_{k-1} + C}{\overline{F}_k(T_k - T_{k-1})} \qquad (k = 1, 2, \cdots N). \tag{35}$$

Thus, the mean time to completion of the process is

$$L_1(N) \equiv \sum_{k=1}^{N} L_1(k) = \sum_{k=1}^{N} \frac{T_k - T_{k-1} + C}{\overline{F}_k(T_k - T_{k-1})} \qquad (N = 1, 2, \cdots). \tag{36}$$

We find optimal times T_k which minimize $L_1(N)$ for a specified N. Let $f_k(t)$ be a density function of $F_k(t)$ and $r_k(t) \equiv f_k(t)/\overline{F}_k(t)$ that is the failure rate of $F_k(t)$. Then, differentiating $L_1(N)$ with respect to T_k and setting it equal to zero,

$$\frac{1}{\overline{F}_k(T_k - T_{k-1})}[1 + (T_k - T_{k-1} + C)r_k(T_k - T_{k-1})]$$
$$= \frac{1}{\overline{F}_{k+1}(T_{k+1} - T_k)}[1 + (T_{k+1} - T_k + C)r_{k+1}(T_{k+1} - T_k)]. \quad (37)$$

Setting that $x_k \equiv T_k - T_{k-1}$ and rewriting (37) as a function of x_k,

$$\frac{1}{\overline{F}_k(x_k)}[1 + (x_k + C)r_k(x_k)] = \frac{1}{\overline{F}_{k+1}(x_{k+1})}[1 + (x_{k+1} + C)r_{k+1}(x_{k+1})]$$
$$(k = 1, 2, \cdots, N - 1). \quad (38)$$

Next, Suppose that $\overline{F}_k(t) = e^{-\lambda_k t}$, i.e., an error rate during $(T_{k-1}, T_k]$ is constant λ_k which increases with k. Then, (38) is rewritten as

$$\frac{1 + \lambda_{k+1}(x_{k+1} + C)}{1 + \lambda_k(x_k + C)} - e^{(\lambda_k x_k - \lambda_{k+1} x_{k+1})} = 0. \quad (39)$$

It is easily noted that $\lambda_{k+1} x_{k+1} \le \lambda_k x_k$, and hence, $x_{k+1} \le x_k$ since $\lambda_{k+1} \le \lambda_k$.

In particular, when $\lambda_k \equiv \lambda$ for $k = 1, 2, \cdots, N$, (39) becomes

$$\frac{1 + \lambda(x_{k+1} + C)}{1 + \lambda(x_k + C)} - e^{\lambda(x_k - x_{k+1})} = 0. \quad (40)$$

Since $x_{k+1} \le x_k$, we have that $x_{k+1} \ge x_k$ from (40), i.e., it is easily proved that a solution to satisfy (40) is restricted only to $x_{k+1} = x_k \equiv T$, irrespective of the interval number k. Then, the mean time to completion of the process is

$$L_1(N) = (S + NC)e^{\lambda S/N}. \quad (41)$$

If $\lambda_{k+1} > \lambda_k$, then $x_{k+1} < x_k$ from (39). Let $Q(x_{k+1})$ be the left-hand side of (40) for a fixed x_k. Then, $Q(x_{k+1})$ is strictly increasing from

$$Q(0) = \frac{1 + \lambda_{k+1} C}{1 + \lambda_k(x_k + C)} - e^{\lambda_k x_k}$$

to $Q(x_k) > 0$. Thus, if $Q(0) < 0$, then an optimal $x^*_{k+1}(0 < x^*_{k+1} < x_k)$ to satisfy (39) exists uniquely, and if $Q(0) \ge 0$, then $x^*_{k+1} = S - T_k$.

Therefore, noting that $T_0 = 0$ and $T_N = S$, we have the following result:

(i) When $N = 1$ and $T_1 = S$, the mean time is

$$L_1(1) = (S + C)e^{\lambda_1 S}. \quad (42)$$

(ii) When $N = 2$, from (38),

$$[1 + \lambda_1 (x_1 + C)] e^{\lambda_1 x_1} - [1 + \lambda_2 (S - x_1 + C)] e^{\lambda_2 (S - x_1)} = 0. \quad (43)$$

Letting $Q_1^*(x_1)$ be the left-hand side of (43), it is strictly increasing from $Q_1^*(0) < 0$ to

$$Q_1(S) = [1 + \lambda_1 (S + C)] e^{\lambda_1 S} - (1 + \lambda_2 C).$$

Hence, if $Q_1^*(S) > 0$, then $x_1^* = T_1^*$ ($0 < T_1^* < S$) to satisfy (43) exists uniquely, and conversely, if $Q_1^*(S) \leq 0$ then $x_1^* = T_1^* = S$.

(iii) When $N = 3$, we compute x_k^* ($k = 1, 2$) which satisfy the simultaneous equations:

$$[1 + \lambda_1 (x_1 + C)] e^{\lambda_1 x_1} = [1 + \lambda_2 (x_2 + C)] e^{\lambda_2 x_2}, \quad (44)$$

$$[1 + \lambda_2 (x_2 + C)] e^{\lambda_2 x_2} = [1 + \lambda_3 (S - x_1 - x_2)] e^{\lambda_3 (S - x_1 - x_2)}. \quad (45)$$

(iv) When $N = 4, 5, \cdots$, we compute x_k^* and $T_k = \sum_{j=1}^{k} x_j^*$ similarly.

Example 4.1. We compute sequential checkpoint intervals T_k ($k = 1, 2, \cdots, N$) for a double modular system. It is assumed that $\lambda_k = 2[1 + \alpha(k-1)]\lambda$ ($k = 1, 2, \cdots$), i.e., an error rate increases by $100\alpha\%$ of an original rate λ of one module. Table 4 presents optimal sequential intervals λT_k and the resulting mean times $\lambda L_1(N)$ for $N = 1, 2, \cdots, 9$ when $\alpha = 0.1$, $\lambda S = 10^{-1}$ and $\lambda C = 10^{-3}$. In this case, the mean time is the smallest when $N = 5$, i.e., the optimal checkpoint number is $N^* = 5$ and the checkpoint times T_k^* ($k = 1, 2, 3, 4, 5$) should be placed at 2.38, 4.53, 6.50, 8.32, 10.00(sec) for $\lambda = 10^{-2}$(1/sec), and the mean time 11.009 is about 10% longer than an original execution time $S = 10$. Further, all values of $x_k = T_k - T_{k-1}$ decrease with k because error rates increase with the number of checkpoints. □

It is very troublesome to solve simultaneous equations. We consider the following approximate checkpoint times: It is assumed that the probability that a modular system has no error during $(T_{k-1}, T_k]$ is constant, i.e., $\overline{F}_k(T_k - T_{k-1}) \equiv q$ ($k = 1, 2, \cdots, N$). From this assumption, we derive $T_k - T_{k-1} \equiv \overline{F}_k^{-1}(q)$ as a function of q. Substituting this $T_k - T_{k-1}$ into (36), the mean time to completion of the process is

$$L_1(N) = \sum_{k=1}^{N} \frac{\overline{F}_k^{-1}(q) + C}{q}. \quad (46)$$

We discuss an optimal q which minimizes $L_1(N)$.

For example, when $\overline{F}_k(t) = e^{-\lambda_k t}$,
$$e^{-\lambda_k(T_k - T_{k-1})} = q \equiv e^{-\widetilde{q}},$$
and hence,
$$T_k - T_{k-1} = \frac{\widetilde{q}}{\lambda_k}.$$
Since
$$\sum_{k=1}^{N}(T_k - T_{k-1}) = T_N = S = \widetilde{q}\sum_{k=1}^{N}\frac{1}{\lambda_k},$$
we have
$$L_1(N) = e^{\widetilde{q}}\left[\widetilde{q}\sum_{k=1}^{N}\frac{1}{\lambda_k} + NC\right] = e^{\widetilde{q}}(S + NC). \tag{47}$$

Table 4 Checkpoint intervals λT_k and mean time $\lambda L_1(N)$ when $\lambda_k = 2[1+0.1(k-1)]\lambda$, $\lambda S = 10^{-1}$ and $\lambda C = 10^{-3}$

N	1	2	3	4	5
$\lambda T_1 \times 10^2$	10.00	5.24	3.65	2.85	2.38
$\lambda T_2 \times 10^2$		10.00	6.97	5.44	4.53
$\lambda T_3 \times 10^2$			10.00	7.81	6.50
$\lambda T_4 \times 10^2$				10.00	8.32
$\lambda T_5 \times 10^2$					10.00
$\lambda L_1(N) \times 10^2$	12.33617	11.32655	11.07923	11.00950	11.00887

N	6	7	8	9
$\lambda T_1 \times 10^2$	2.05	1.83	1.65	1.52
$\lambda T_2 \times 10^2$	3.91	3.48	3.15	2.89
$\lambda T_3 \times 10^2$	5.62	4.99	4.52	4.15
$\lambda T_4 \times 10^2$	7.19	6.39	5.78	5.31
$\lambda T_5 \times 10^2$	8.65	7.68	6.95	6.38
$\lambda T_6 \times 10^2$	10.00	8.88	8.03	7.37
$\lambda T_7 \times 10^2$		10.00	9.05	8.31
$\lambda T_8 \times 10^2$			10.00	9.18
$\lambda T_9 \times 10^2$				10.00
$\lambda L_1(N) \times 10^2$	11.04228	11.09495	11.15960	11.23220

Therefore, we compute $\widetilde{q}$ and $L_1(N)$ for a specified N. Comparing $L_1(N)$ for $N = 1, 2, \cdots$, we obtain an optimal $\widetilde{N}$ which minimizes $L_1(N)$ and $\widetilde{q} = S/\sum_{k=1}^{N}(1/\lambda_k)$. Lastly, we may compute $\widetilde{T}_k = \widetilde{q}\sum_{j=1}^{k}(1/\lambda_j)$ ($k = 1, 2, \cdots, \widetilde{N} - 1$) for an approximate optimal $\widetilde{N}$ which minimizes $L_1(N)$.

Example 4.2. Table 5 presents $\widetilde{q} = S/\sum_{k=1}^{N}(1/\lambda_k)$ and $\lambda L_1(N)$ in (47) for $N = 1, 2, \cdots, 9$ under the same assumptions as those in Table 4. In this case, $\widetilde{N} = 5 = N^*$ and the mean time $L_1(5)$ is a little longer than that in Table 4. When $\widetilde{N} = 5$, approximate checkpoint times are $\lambda \widetilde{T}_k \times 10^2 = 2.37$, 4.52, 6.49, 8.31, 10.00 that are a little shorter than those in Table 4. Such computations are much easier than to solve simultaneous equations.

It would be sufficient to adopt approximate checkpoint intervals as optimal ones in actual fields. □

Table 5 Mean time $\lambda L_1(N)$ for $\widetilde{q}$ when $\lambda S = 10^{-1}$ and $\lambda C = 10^{-3}$

N	$\widetilde{q}$	$\lambda L_1(N) \times 10^2$
1	0.2000000	12.33617
2	0.1047619	11.32655
3	0.0729282	11.07923
4	0.0569532	11.00951
5	0.0473267	11.00888
6	0.0408780	11.04229
7	0.0362476	11.09496
8	0.0327555	11.15962
9	0.0300237	11.23222

4.2 Modified Model

It has been assumed until now that error rates increase with the number of checkpoints. We assume for the simplicity of the model that the probability that a modular system has no error during the interval $(T_{k-1}, T_k]$ is $\overline{F}(T_k)/\overline{F}(T_{k-1})$, irrespective of rollback operation. Then, the mean execution time of the process for the interval $(T_{k-1}, T_k]$ is given by a renewal equation

$$L_2(k) = (T_k - T_{k-1} + C)\frac{\overline{F}(T_k)}{\overline{F}(T_{k-1})}$$

$$+ \left[T_k - T_{k-1} + C + L_2(k)\right] \frac{F(T_k) - F(T_{k-1})}{\overline{F}(T_{k-1})}, \qquad (48)$$

and solving it,

$$L_2(k) = \frac{(T_k - T_{k-1} + C)\overline{F}(T_{k-1})}{\overline{F}(T_k)} \qquad (k = 1, 2, \cdots, N). \qquad (49)$$

Thus, the mean time to completion of the process is

$$L_2(N) = \sum_{k=1}^{N} \frac{(T_k - T_{k-1} + C)\overline{F}(T_{k-1})}{\overline{F}(T_k)} \qquad (N = 1, 2, \cdots). \qquad (50)$$

We find optimal times T_k which minimize $L_2(N)$ for a specified N. Let $f(t)$ be a density function of $F(t)$ and $r(t) \equiv f(t)/\overline{F}(t)$ be the failure rate of $F(t)$. Then, differentiating $L_2(N)$ with respect to T_k and setting it equal to zero,

$$\frac{\overline{F}(T_{k-1})}{\overline{F}(T_k)}\left[1 + r(T_k)(T_k - T_{k-1} + C)\right]$$
$$= \frac{\overline{F}(T_k)}{\overline{F}(T_{k+1})}\left[1 + r(T_k)(T_{k+1} - T_k + C)\right] \quad (k = 1, 2, \cdots N-1). \qquad (51)$$

Therefore, we have the following result:

(i) When $N = 1$ and $T_1 = S$, the mean time is

$$L_2(1) = \frac{S + C}{\overline{F}(S)}. \qquad (52)$$

(ii) When $N = 2$, from (51)

$$\frac{1}{\overline{F}(T_1)}\left[1 + r(T_1)(T_1 + C)\right] - \frac{\overline{F}(T_1)}{\overline{F}(S)}\left[1 + r(T_1)(S - T_1 + C)\right] = 0. \qquad (53)$$

Letting $Q_2^*(T_1)$ be the left-hand side of (53),

$$Q_2(0) = 1 + r(0)C - \frac{1}{\overline{F}(S)}\left[1 + r(0)(S + C)\right] < 0,$$
$$Q_2(S) = \frac{1}{\overline{F}(S)}\left[1 + r(S)(S + C)\right] - \left[1 + r(S)C\right] > 0.$$

Thus, there exists a T_1 that satisfies (53).

(iii) When $N = 3$, we compute $T_k(k = 1, 2)$ which satisfy the simultaneous equations:

$$\frac{1}{\overline{F}(T_1)}\left[1+r(T_1)(T_1+C)\right] = \frac{\overline{F}(T_1)}{\overline{F}(T_2)}\left[1+r(T_1)(T_2-T_1+C)\right], \tag{54}$$

$$\frac{\overline{F}(T_1)}{\overline{F}(T_2)}\left[1+r(T_2)(T_2-T_1+C)\right] = \frac{\overline{F}(T_2)}{\overline{F}(S)}\left[1+r(T_2)(S-T_2+C)\right]. \tag{55}$$

(iv) When $N = 4, 5, \cdots$, we compute T_k similarly.

Example 4.3. We compute sequential checkpoint intervals $T_k(k = 1, 2, \cdots, N)$ when error rates increase with the original execution time. It is assumed that $\overline{F}(t) = e^{-2(\lambda t)^m}$ ($m > 1$), $\lambda C = 10^{-3}$ and $\lambda S = 10^{-1}$.

Table 6 presents optimal sequential intervals λT_k and the resulting mean times $\lambda L_2(N)$ for $N = 1, 2, \cdots, 9$ when $\overline{F}(t) = \exp[-2(\lambda t)^{1.1}]$, $\lambda S = 10^{-1}$ and $\lambda C = 10^{-3}$. In this case, the mean time is the smallest when $N = 4$, i.e., $N^* = 4$ and the checkpoint times T_k^* ($k = 1, 2, 3, 4$) should be placed at 2.67, 5.17, 7.60, 10.00 (sec) for $\lambda = 10^{-2}$(1/sec), and the mean time 10.8207 is about 8% longer than an original execution time $S = 10$. □

Next, we consider the approximate method similar to that of the previous model. It is assumed that the probability that a modular system has no error during $(T_{k-1}, T_k]$ is constant, i.e., $\overline{F}(T_k)/\overline{F}(T_{k-1}) = q(k = 1, 2, \cdots, N)$. When $\overline{F}(t) = e^{-2(\lambda t)^m}$,

$$\frac{\overline{F}(T_k)}{\overline{F}(T_{k-1})} = e^{-2[(\lambda T_k)^m - (\lambda T_{k-1})^m]} = q \equiv e^{-\widetilde{q}},$$

and hence,

$$2(\lambda T_k)^m - 2(\lambda T_{k-1})^m = \widetilde{q} \qquad (k = 1, 2, \cdots, N).$$

Thus,

$$(\lambda T_k)^m = \frac{k\widetilde{q}}{2},$$

i.e.,

$$\lambda T_k = \left(\frac{k\widetilde{q}}{2}\right)^{1/m} \qquad (k = 1, 2, \cdots, N-1),$$

and

$$\lambda T_N = \lambda S = \left(\frac{N\widetilde{q}}{2}\right)^{1/m}.$$

Table 6 Checkpoint intervals when $\lambda C = 10^{-3}$ and $\lambda S = 10^{-1}$

N	1	2	3	4	5
$\lambda T_1 \times 10^2$	10.00	5.17	3.51	2.67	2.16
$\lambda T_2 \times 10^2$		10.00	6.80	5.17	4.18
$\lambda T_3 \times 10^2$			10.00	7.60	6.15
$\lambda T_4 \times 10^2$				10.00	8.09
$\lambda T_5 \times 10^2$					10.00
$\lambda L_2(N) \times 10^2$	11.83902	11.04236	10.85934	10.82069	10.83840

N	6	7	8	9
$\lambda T_1 \times 10^2$	1.81	1.57	1.38	1.23
$\lambda T_2 \times 10^2$	3.51	3.03	2.67	2.39
$\lambda T_3 \times 10^2$	5.17	4.46	3.93	3.51
$\lambda T_4 \times 10^2$	6.80	5.87	5.17	4.62
$\lambda T_5 \times 10^2$	8.41	7.26	6.39	5.71
$\lambda T_6 \times 10^2$	10.00	8.63	7.60	6.80
$\lambda T_7 \times 10^2$		10.00	8.81	7.87
$\lambda T_8 \times 10^2$			10.00	8.94
$\lambda T_9 \times 10^2$				10.00
$\lambda L_2(N) \times 10^2$	10.88391	10.94517	11.01622	11.09376

Therefore,
$$L_2(N) = e^{\widetilde{q}}(S + NC) = e^{2(\lambda S)^m/N}(S + NC). \tag{56}$$

Forming the inequality $L_2(N+1) - L_2(N) \geq 0$,

$$C \geq (S + NC)\left\{e^{2(\lambda S)^m/[N(N+1)]} - 1\right\}. \tag{57}$$

It is easily proved that the right-hand side of (57) is strictly decreasing to 0. Thus, an optimal $\widetilde{N}$ to minimize $L_2(N)$ in (56) is given by a unique minimum which satisfies (57).

Example 4.4. Table 7 presents $\widetilde{q} = 2(\lambda S)^m/N$ and $\lambda L_2(N)$ in (56) for $N = 1, 2, \cdots, 9$ under the same assumptions in Table 6. In this case, $\widetilde{N} = 4 = N^*$ and approximate checkpoint times are $\lambda \widetilde{T}_k \times 10^2 = 2.84, 5.33, 7.70, 10.00$, that are a little longer than those of Table 6. □

Table 7 Mean time $\lambda L_2(N)$ for $\tilde{q}$ when $\lambda S = 10^{-1}$ and $\lambda C = 10^{-3}$

N	$\tilde{q}$	$\lambda L_2(N) \times 10^2$
1	0.1588656	11.83902
2	0.0794328	11.04326
3	0.0529552	10.86014
4	0.0397164	10.82136
5	0.0317731	10.83897
6	0.0264776	10.88441
7	0.0226951	10.94561
8	0.0198582	11.01661
9	0.0176517	11.09411

5 Random Checkpoint Models

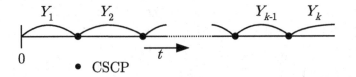

Fig. 5 Task execution for Scheme 1.

Suppose that we have to execute the successive tasks with a processing time $Y_k (k=1, 2, \cdots)$ (Figure 5). A double modular system of error detection for the processing of each task is adopted. Then, introducing two types of checkpoints; compare-and-store checkpoint (CSCP) and compare-checkpoint (CCP) [7], we consider the following three checkpoint schemes:

1) CSCP is placed at each end of tasks.
2) CSCP is placed at the Nth end of tasks.
3) CCP is placed at each end of tasks and CSCP is placed at the Nth end of tasks.

The mean execution times per one task for each scheme are obtained, and optimal numbers N^* that minimize them for Schemes 2 and 3 are derived analytically and are compared numerically. This is one of applied models with random maintenance times [8, 18] to checkpoint models. Such schemes would be useful when it is better to place checkpoints at the end of tasks than those on one's way. Further, we extend a double modular system to a majority decision system in Section 3.

5.1 Performance Analysis

Suppose that task k has a processing time $Y_k (k = 1, 2, \cdots)$ with an identical distribution $G(t) \equiv \Pr\{Y_k \le t\}$ and finite mean $\mu = \int_0^\infty [1 - G(t)]\, dt < \infty$, and is executed successively. To detect errors, we provide two independent modules where they compare two states at checkpoint times. Further, it is assumed that some errors occur at a constant rate $\lambda (\lambda > 0)$, i.e., the probability that two modules have no error during $(0, t]$ is $e^{-2\lambda t}$.

(1) Scheme 1

CSCP is placed at each end of task k: When two states of modules match with each other at the end of task k, the process of task k is correct and its state is stored (Figure 5). In this case, two modules go forward and execute task $k + 1$. However, when two states do not match, it is judged that some errors have occurred. Then, two modules go back and make the retry of task k again.

Let C be the overhead for the comparison of two states and Cs be the overhead for their store. Then, the mean execution time of the process of task k is given by a renewal equation:

$$\widetilde{L}_1(1) = \int_0^\infty \left\{ e^{-2\lambda t}(C + Cs + t) + \left(1 - e^{-2\lambda t}\right)\left[C + t + \widetilde{L}_1(1)\right] \right\} dG(t). \tag{58}$$

Solving (58) for $\widetilde{L}_1(1)$,

$$\widetilde{L}_1(1) = \frac{C + \mu + CsG^*(2\lambda)}{G^*(2\lambda)},$$

where $G^*(s)$ is the Laplace-Stieltjes (LS) transform of $G(t)$, i.e., $G^*(s) \equiv \int_0^\infty e^{-st} dG(t)$ for $s > 0$. Therefore, the mean execution time per one task is

$$L_1(1) \equiv \widetilde{L}_1(1) = \frac{C + \mu}{G^*(2\lambda)} + Cs. \tag{59}$$

(2) Scheme 2

CSCP is placed only at the end of task N (Figure 6): When two states of all task $k(k = 1, 2, \cdots, N)$ match at the end of task N, its state is stored and two modules execute task $N + 1$. When two states do not match, two modules go back in the first task 1 and make their retries. By the method

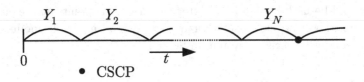

Fig. 6 Task execution for Scheme 2.

similar to obtaining (58), the mean execution time of the process of all task $k(k = 1, 2, \cdots, N)$ is

$$\widetilde{L}_2(N) = \int_0^\infty \left\{ e^{-2\lambda t} \left(NC + Cs + t \right) \right.$$
$$\left. + \left(1 - e^{-2\lambda t}\right) \left[NC + t + \widetilde{L}_2(N) \right] \right\} dG^{(N)}(t), \quad (60)$$

where $G^{(N)}(t)$ is the N-fold Stieltjes convolution of $G(t)$ with itself, i.e., $G^{(N)}(t) \equiv \int_0^t G^{(N-1)}(t-u) \, dG(u)$ ($N = 1, 2, \cdots$), and $G^{(0)}(t) \equiv 1$ for $t \geq 0$ and $G^{(1)}(t) = G(t)$. Solving (60) for $\widetilde{L}_2(N)$,

$$\widetilde{L}_2(N) = \frac{NC + N\mu + Cs \left[G^*(2\lambda)\right]^N}{\left[G^*(2\lambda)\right]^N}.$$

Therefore, the mean execution time per one task is

$$L_2(N) \equiv \frac{\widetilde{L}_2(N)}{N} = \frac{C + \mu}{\left[G^*(2\lambda)\right]^N} + \frac{Cs}{N} \quad (N = 1, 2, \cdots). \quad (61)$$

When $N = 1$, $L_2(1)$ agrees with (59).

We find an optimal number N_2^* that minimizes $L_2(N)$. There exists a finite $N_2^* (1 \leq N_2^* < \infty)$ because $\lim_{N \to \infty} L_2(N) = \infty$. From the inequality $L_2(N+1) - L_2(N) \geq 0$,

$$\frac{N(N+1)\left[1 - G^*(2\lambda)\right]}{\left[G^*(2\lambda)\right]^{N+1}} \geq \frac{Cs}{C + \mu} \quad (N = 1, 2, \cdots). \quad (62)$$

The left-hand side of (62) is strictly increasing to ∞ in N. Thus, there exists a finite and unique minimum $N_2^* (1 \leq N_2^* < \infty)$ which satisfies (62). If

$$\frac{\left[1 - G^*(2\lambda)\right]}{\left[G^*(2\lambda)\right]^2} \geq \frac{Cs}{2(C + \mu)},$$

then $N_2^* = 1$. When $G(t) = 1 - e^{-t/\mu}$, (62) is rewritten as

$$N(N+1)2\lambda\mu(2\lambda\mu+1)^N \geq \frac{Cs}{C+\mu} \quad (N = 1, 2, \cdots). \tag{63}$$

Example 5.1. Table 8 presents the optimal number N_2^* and the resulting execution time $L_2(N_2^*)/\mu$ and $L_2(1)/\mu$ in (59) for $\lambda\mu$ and C/μ when $Cs/\mu = 0.1$. This indicates that N_2^* decrease with $\lambda\mu$ and increase with C/μ. For example, when $\lambda\mu = 0.005$ and $C/\mu = 0.1$, $N_2^* = 10$ and $L_2(N_2^*)/\mu$ is 1.167 that is about 4% shorter than $L_2(1)/\mu = 1.211$ for Scheme 1. □

Table 8 Optimal number N_2^* and the resulting execution time $L_2(N_2^*)/\mu$ for Scheme 2 when $Cs/\mu = 0.1$

$\lambda\mu$	$C/\mu = 0.5$			$C/\mu = 0.1$		
	N_2^*	$L_2(N_2^*)/\mu$	$L_2(1)/\mu$	N_2^*	$L_2(N_2^*)/\mu$	$L_2(1)/\mu$
0.1	1	1.900	1.900	1	1.420	1.420
0.05	1	1.750	1.750	1	1.310	1.310
0.01	2	1.611	1.630	2	1.194	1.222
0.005	3	1.579	1.615	3	1.167	1.211
0.001	6	1.535	1.603	7	1.130	1.202
0.0005	8	1.525	1.602	10	1.121	1.201
0.0001	18	1.511	1.600	21	1.109	1.200

Next, we consider the case of increasing error rate. Let $L_1(k)$ be the mean execution time from task k to the completion of task N. Because the probability that no error of two modules for task k occurs is

$$\int_0^\infty e^{-2\lambda_k t} dG(t) = G^*(2\lambda_k) \quad (k = 1, 2, \cdots, N). \tag{64}$$

Thus, we have a renewal equation

$$\widetilde{L}_3(N) = (NC + N\mu + Cs)\prod_{k=1}^N G^*(2\lambda_k)$$

$$+ \left[NC + N\mu + \widetilde{L}_3(N)\right]\left[1 - \prod_{k=1}^N G^*(2\lambda_k)\right]. \tag{65}$$

Solving (65) for $\widetilde{L}_3(N)$,

$$\widetilde{L}_3(N) = \frac{NC + N\mu}{\prod_{k=1}^N G^*(2\lambda_k)} + Cs. \tag{66}$$

Therefore, the mean execution time per one task is

$$L_3(N) \equiv \frac{\widetilde{L}_3(N)}{N} = \frac{C+\mu}{\prod_{k=1}^{N} G^*(2\lambda_k)} + \frac{Cs}{N} \qquad (N=1,2,\cdots). \qquad (67)$$

When $N=1$, $L_3(1)$ agrees with (59).

We find an optimal number N_3^* that minimizes $L_3(N)$. There exists a finite $N_3^*(1 \leq N_3^* < \infty)$ because $\lim_{N\to\infty} L_3(N) = \infty$. From the inequality $L_3(N+1) - L_3(N) \geq 0$,

$$\frac{N(N+1)\left[1 - G^*(2\lambda_{N+1})\right]}{\prod_{k=1}^{N+1} G^*(2\lambda_k)} \geq \frac{Cs}{C+\mu}. \qquad (68)$$

From the assumption that $\lambda_k \leq \lambda_{k+1}$, $G^*(2\lambda_{k+1}) \leq G^*(2\lambda_k)$, i.e., $1 - G^*(2\lambda_k) \leq 1 - G^*(2\lambda_{k+1})$. Thus, it is clearly noted that the left-hand side of (68) is strictly increasing to ∞ in N. Therefore, there exists a finite and unique minimum $N_3^*(1 \leq N_3^* < \infty)$ that satisfies (68). If

$$\frac{1 - G^*(2\lambda_2)}{G^*(2\lambda_1) G^*(2\lambda_2)} \geq \frac{Cs}{2(C+\mu)},$$

then $N^*=1$.

When $G(t) = 1 - e^{-t/\mu}$, (68) is rewritten as

$$\frac{N(N+1)\left[2\lambda_{N+1}\mu/(2\lambda_{N+1}\mu + 1)\right]}{\prod_{k=1}^{N+1} \left[1/(2\lambda_k\mu + 1)\right]} \geq \frac{Cs}{C+\mu}. \qquad (69)$$

Example 5.2. It is assumed that $\lambda_k = [1 + \alpha(k-1)]\lambda$, i.e., an error rate increases by $100\alpha\%$ of an original rate λ. Then, we compute an optimal number N_3^* which satisfies (69). Table 9 presents optimal N_3^*, the resulting execution time $L_3(N_3^*)/\mu$ and $L_3(1)/\mu$ for $\lambda\mu$ and C/μ when $\alpha=0.1$ and $Cs/\mu = 0.1$. This indicates that N_3^* decrease with both $\lambda\mu$ and C/μ. For example, when $\lambda\mu = 0.005$ and $C/\mu = 0.5$, $N_3^* = 2$ and $L_3(N_3^*)$ is 1.582 that is about 2% shorter than $L_3(1) = 1.615$ for Scheme 1. □

(3) Scheme 3

CSCP is placed at the end of task N and CCP is placed only at the end of task $k(k=1,2,\cdots,N-1)$ between CSCPs (Figure 7): When two states of task $k(k=1,2,\cdots,N-1)$ match at the end of task k, two modules execute task $k+1$. When two states of task $k(k=1,2,\cdots,N)$ do not match, two modules go back in the first task 1. When two states of task N match, the

Table 9 Optimal number N_3^* and the resulting execution time $L_3(N_3^*)/\mu$ for Scheme 2 when $Cs/\mu = 0.1$

	$C/\mu = 0.5$			$C/\mu = 0.1$		
$\lambda\mu$	N_3^*	$L_3(N_3^*)/\mu$	$L_3(1)/\mu$	N_3^*	$L_3(N_3^*)/\mu$	$L_3(1)/\mu$
0.1	1	1.900	1.900	1	1.420	1.420
0.05	1	1.750	1.750	1	1.310	1.310
0.01	2	1.614	1.630	2	1.197	1.222
0.005	2	1.582	1.615	3	1.170	1.211
0.001	5	1.538	1.603	6	1.133	1.202
0.0005	6	1.528	1.602	7	1.124	1.201
0.0001	12	1.514	1.600	14	1.112	1.200

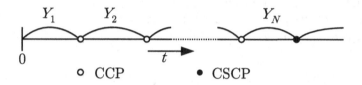

○ CCP   ● CSCP

Fig. 7 Task execution for Scheme 3.

process of all tasks N is completed, and its state is stored. Two modules execute task $N+1$.

Let $\widetilde{L}_4(k)$ be the mean execution time from task k to the completion of task N. Then, by the method similar to obtaining (60),

$$\widetilde{L}_4(k) = \int_0^\infty \left\{ e^{-2\lambda t}\left[C+t+\widetilde{L}_4(k+1)\right] + \left(1-e^{-2\lambda t}\right)\left[C+t+\widetilde{L}_4(k+1)\right] \right\} dG(t)$$
$$(k = 1, 2, \cdots, N-1), \quad (70)$$

$$\widetilde{L}_4(N) = \int_0^\infty \left\{ e^{-2\lambda t}(C+t+Cs) + \left(1-e^{-2\lambda t}\right)\left[C+t+\widetilde{L}_4(1)\right] \right\} dG(t). \quad (71)$$

Solving (70) and (71) for $\widetilde{L}_4(1)$,

$$\widetilde{L}_4(1) = \frac{(C+\mu)\left\{1-[G^*(2\lambda)]^N\right\}}{[1-G^*(2\lambda)][G^*(2\lambda)]^N} + Cs.$$

Therefore, the mean execution time per one task is

$$L_4(N) \equiv \frac{\widetilde{L}_4(1)}{N} = \frac{(C+\mu)\left\{1 - [G^*(2\lambda)]^N\right\}}{N[1 - G^*(2\lambda)][G^*(2\lambda)]^N} + \frac{Cs}{N} \qquad (N = 1, 2, \cdots). \tag{72}$$

When $N = 1$, $L_4(1)$ agrees with (59). By comparing (61) with (72), Scheme 3 is better than Scheme 2.

It can be clearly seen that a finite $N_4^*(1 \leq N_4^* < \infty)$ that minimizes $L_4(N)$ exists. From the inequality $L_4(N+1) - L_4(N) \geq 0$,

$$\frac{1}{[G^*(2\lambda)]^{N+1}} \sum_{j=1}^{N} \left\{1 - [G^*(2\lambda)]^j\right\} \geq \frac{Cs}{C+\mu} \qquad (N = 1, 2, \cdots). \tag{73}$$

The left-hand side of (73) is strictly increasing to ∞ in N. Thus, there exists a finite and unique minimum $N_4^*(1 \leq N_4^* < \infty)$ which satisfies (73). If $(C+\mu)[1 - G^*(2\lambda)] \geq Cs[G^*(2\lambda)]^2$, then $N_4^* = 1$. By comparing (73) with (62), it can be easily seen that $N_4^* \geq N_3^*$. When $G(t) = 1 - e^{-t/\mu}$, (73) is

$$(2\lambda\mu + 1)^{N+1} \sum_{j=1}^{N} \left[1 - \left(\frac{1}{2\lambda\mu + 1}\right)^j\right] \geq \frac{Cs/\mu}{C/\mu + 1} \qquad (N = 1, 2, \cdots). \tag{74}$$

Example 5.3. Table 10 presents the optimal number N_4^* and the resulting execution time $L_4(N_4^*)/\mu$ in (72) for $\lambda\mu$ and C/μ when $Cs/\mu = 0.1$. Clearly, Scheme 3 is better than Scheme 2 and $N_4^* \geq N_3^*$. However, in general, the overhead C for Scheme 2 would be less than that for Scheme 3. In such case, Scheme 2 might be better than Scheme 3. □

Next, we consider the case of increase error rate. Let $\widetilde{L}_5(k)$ be the mean execution time from task k to the completion of task N. Then, by the method similar to obtaining (65),

$$\widetilde{L}_5(k) = \int_0^\infty \left\{e^{-2\lambda_k t}\left[C + t + \widetilde{L}_5(k+1)\right] + \left(1 - e^{-2\lambda_k t}\right)\left[C + t + \widetilde{L}_5(1)\right]\right\} dG(t)$$

$$(k = 1, 2, \cdots, N-1), \tag{75}$$

$$\widetilde{L}_5(N) = \int_0^\infty \left\{e^{-2\lambda_N t}(C + t + Cs) + \left(1 - e^{-2\lambda_N t}\right)\left[C + t + \widetilde{L}_5(1)\right]\right\} dG(t). \tag{76}$$

Table 10 Optimal number N_4^* and the resulting execution time $L_4(N_4^*)/\mu$ for Scheme 3 when $Cs/\mu = 0.1$

$\lambda\mu$	$C/\mu = 0.5$		$C/\mu = 0.1$	
	N_4^*	$L_4(N_4^*)/\mu$	N_4^*	$L_4(N_4^*)/\mu$
0.1	1	1.900	1	1.420
0.05	1	1.750	1	1.310
0.01	3	1.594	3	1.178
0.005	4	1.563	4	1.153
0.001	8	1.526	9	1.122
0.0005	12	1.518	13	1.115
0.0001	26	1.508	30	1.107

Solving (75) and (76) for $\widetilde{L}_5(1)$,

$$\widetilde{L}_5(1) = \frac{(C+\mu)\sum_{j=0}^{N-1}\left[\prod_{k=1}^{j}G^*(2\lambda_k)\right]}{\prod_{k=1}^{N}G^*(2\lambda_k)} + Cs. \tag{77}$$

Therefore, the mean execution time per one task is

$$L_5(N) \equiv \frac{\widetilde{L}_5(1)}{N} = \frac{(C+\mu)\sum_{j=0}^{N-1}\left[\prod_{k=1}^{j}G^*(2\lambda_k)\right]}{N\prod_{k=1}^{N}G^*(2\lambda_k)} + \frac{Cs}{N} \quad (N=1,2,\cdots), \tag{78}$$

where $\prod_{k=1}^{0} \equiv 1$. When $N=1$, $L_5(1)$ agrees with (59).

We find an optimal N_5^* that minimizes $L_5(N)$. From the inequality $L_5(N+1) - L_5(N) \geq 0$,

$$\frac{N\sum_{j=0}^{N}\left[\prod_{k=1}^{j}G^*(2\lambda_k)\right] - (N+1)G^*(2\lambda_{N+1})\sum_{j=0}^{N-1}\left[\prod_{k=1}^{j}G^*(2\lambda_k)\right]}{\prod_{k=1}^{N+1}G^*(2\lambda_k)} \geq \frac{Cs}{C+\mu},$$

i.e.,

$$N+1+\frac{[N-(N+1)G^*(2\lambda_{N+1})]\sum_{j=0}^{N}\left[\prod_{k=1}^{j}G^*(2\lambda_k)\right]}{\prod_{k=1}^{N+1}G^*(2\lambda_k)} \geq \frac{Cs}{C+\mu}$$

$$(N=1,2,\cdots). \qquad (79)$$

First, note that

$$N-(N+1)G^*(2\lambda_{N+1})$$

is increasing because

$$N-(N+1)G^*(2\lambda_{N+1})-(N-1)+NG^*(2\lambda_N)$$
$$=1-G^*(2\lambda_{N+1})+N[G^*(2\lambda_N)-G^*(2\lambda_{N+1})]>0.$$

Furthermore, denoting the left-hand side of (79) by $Q(N+1)$,

$$Q(N+1)-Q(N)$$
$$=\frac{1}{\prod_{k=1}^{N+1}G^*(2\lambda_k)}\left\{\prod_{k=1}^{N+1}G^*(2\lambda_k)+[N-(N+1)G^*(2\lambda_{N+1})]\sum_{j=0}^{N}\left[\prod_{k=1}^{j}G^*(2\lambda_k)\right]\right.$$
$$\left.-[N-1-NG^*(2\lambda_N)]G^*(2\lambda_{N+1})\sum_{j=0}^{N-1}\left[\prod_{k=1}^{j}G^*(2\lambda_k)\right]\right\}$$
$$>\frac{1}{\prod_{k=1}^{N+1}G^*(2\lambda_k)}\left\{\prod_{k=1}^{N+1}G^*(2\lambda_k)+[N-1-NG^*(2\lambda_N)]G^*(2\lambda_{N+1})\prod_{k=1}^{N}G^*(2\lambda_k)\right\}$$

$$=N[1-G^*(2\lambda_N)]>0.$$

Thus, the left-hand side of (79) is strictly increasing in N. Therefore, if a finite N^* to satisfy (79) exists, it is a finite and unique minimum such that (79). If

$$(C+\mu)[1+G^*(2\lambda_1)-2G^*(2\lambda_2)] \geq CsG^*(2\lambda_1)G^*(2\lambda_2)$$

then $N_5^*=1$.

By comparing (67) with (78), Scheme 3 is better than Scheme 2.

Example 5.4. Table 11 presents optimal N_5^* and the resulting execution time $L_5(N_5^*)/\mu$ for $\lambda\mu$ and C/μ when $\alpha=0.1$ and $Cs/\mu=0.1$. This indicates that N_5^* decrease with $\lambda\mu$ and C/μ. Compared with Table 9, Scheme 3 is better than Scheme 2 and $N_5^* \geq N_4^*$. □

Table 11 Optimal number N_5^* and the resulting execution time $L_5(N_5^*)/\mu$ for Scheme 3 when $Cs/\mu = 0.1$

$\lambda\mu$	$C/\mu = 0.5$		$C/\mu = 0.1$	
	N_5^*	$L_5(N_5^*)/\mu$	N_5^*	$L_5(N_5^*)/\mu$
0.1	1	1.900	1	1.420
0.05	1	1.750	1	1.310
0.01	2	1.598	3	1.184
0.005	3	1.568	4	1.158
0.001	6	1.531	7	1.127
0.0005	8	1.522	9	1.120
0.0001	15	1.511	17	1.110

5.2 Majority Decision System

We take up a majority decision system with $(2n+1)$ modules as an error masking system, i.e., $(n+1)$-out-of-$(2n+1)$ system $(n = 1, 2, \cdots)$. If more than n states of $(2n+1)$ modules match, the process of task k is correct and its state is stored. In this case, the probability that the process is correct during $(0, t]$ is, from (32),

$$\overline{F}_{n+1}(t) = \sum_{k=n+1}^{2n+1} \binom{2n+1}{k} \sum_{i=0}^{2n+1-k} \binom{2n+1-k}{i} (-1)^i \left(e^{-\lambda t}\right)^{k+1}$$
$$(n = 1, 2, \cdots). \qquad (80)$$

Thus, the mean execution time of the process of task k is

$$\widetilde{L}_6(1) = \int_0^\infty \left[\overline{F}_{n+1}(t)(C + Cs + t)\right] dG(t) + F_{n+1}(t)\left[C + t + \widetilde{L}_6(1)\right] dG(t). \qquad (81)$$

Therefore, by the method similar to obtaining (59), the mean time execution time per one task is

$$L_6(1) \equiv \widetilde{L}_6(1) = \frac{C + \mu}{\sum_{k=n+1}^{2n+1} \binom{2n+1}{k} \sum_{i=0}^{2n+1-k} \binom{2n+1-k}{i} (-1)^i G^*[(k+i)\lambda]} + Cs. \qquad (82)$$

Next, we consider the case of error rate is increasing. In this case, the probability that the process of task k for Scheme 1 is correct during $(0, t]$ is, from (80),

$$\overline{F}_{n+1}(t) = \sum_{m=n+1}^{2n+1} \binom{2n+1}{m} \sum_{i=0}^{2n+1-m} \binom{2n+1-m}{i} (-1)^i e^{-(m+1)\lambda_1 t}$$
$$(n = 1, 2, \cdots). \qquad (83)$$

Thus, by the method similar to obtaining (58), the mean execution time of the process of task k is

$$L_7 = \int_0^\infty \left[\overline{F}_{n+1}(t)\,(C+C_s+t) + F_{n+1}(t)\,(C+t+L_7)\right] dG(t). \tag{84}$$

Solving (84) for L_7,

$$L_7 = \frac{C+\mu}{\sum_{m=n+1}^{2n+1}\binom{2n+1}{m}\sum_{i=0}^{2n+1-m}\binom{2n+1-m}{i}(-1)^i G^*[(m+i)\lambda_1]} + Cs. \tag{85}$$

For example, when $n = 1$, i.e., the system is composed of a 2-out-of-3 system,

$$L_7 = \frac{C+\mu}{3G^*(2\lambda_1) - 2G^*(3\lambda_1)} + Cs. \tag{86}$$

Similarly, the mean execution time per one task for Scheme 2 is, from (67)

$$L_7(N) = \frac{C+\mu}{\prod_{k=1}^{N}\left[\sum_{m=n+1}^{2n+1}\binom{2n+1}{m}\sum_{i=0}^{2n+1-m}\binom{2n+1-m}{i}(-1)^i G^*[(m+i)\lambda_k]\right]}$$

$$+ \frac{Cs}{N} \qquad (N=1,2,\cdots), \tag{87}$$

and the mean time for Scheme 3 is, from (78),

$$L_7(N) = \frac{(C+\mu)\sum_{j=0}^{N-1}\left\{\prod_{k=1}^{j}\left[\sum_{m=n+1}^{2n+1}\binom{2n+1}{m}\sum_{i=0}^{2n+1-m}\binom{2n+1-m}{i}(-1)^i G^*[(m+i)\lambda_k]\right]\right\}}{N\prod_{k=1}^{N}\left[\sum_{m=n+1}^{2n+1}\binom{2n+1}{m}\sum_{i=0}^{2n+1-m}\binom{2n+1-m}{i}(-1)^i G^*[(m+i)\lambda_k]\right]}$$

$$+ \frac{Cs}{N} \qquad (N=1,2,\cdots). \tag{88}$$

Example 5.5. Suppose that $C \equiv \binom{2n+1}{2}C_1$ as same as in Example 3.2. Table 12 presents the optimal numbers N_7^* which minimize $L_7(N)$ in (87) and its resulting execution times $L_7(N_7^*)/\mu$ when $\lambda\mu = 0.01$ and $Cs/\mu = 10$. This indicates that N_7^* decrease with n and C/μ, and $L_7(N_7^*)$ increase with n and C/μ. Thus, from this table, an optimal decision system is a 2-out-of-3 system. However, the overhead C might not increase generally with the order of n^2. We could determine an optimal decision system by estimating C, Cs and the other parameters from actual data. □

Table 12 Optimal number N_7^* and the resulting execution time $L_7(N_7^*)/\mu$ when $\lambda\mu = 0.01$ and $Cs/\mu = 10$

n	$C/\mu = 0.5$		$C/\mu = 0.1$	
	N_7^*	$L_7\left(N_7^*\right)/\mu$	N_7^*	$L_7\left(N_7^*\right)/\mu$
1	1	15.000	2	10.203
2	1	22.000	2	13.005
3	1	33.000	1	16.200
4	1	48.000	1	19.200

6 Conclusion

It has been assumed in this chapter that the overheads for the generation of checkpoints and the probability of error occurrences are already known. However, it is important in practical applications to identify what a type of distribution fits the collected data and to estimate several kinds of overheads from the observation of actual models. If such distributions and overheads are given, we can determine optimal policies for recovery models and apply to real systems by modifying them.

Recently, most systems consist of distributed systems as computer network technologies have developed rapidly. A general model of distributed systems is a mobile network system [1]. Coordinated and uncoordinated protocols to achieve checkpointing in such distributed processes have been introduced [1]: Uncoordinated protocols allow each process to take its local checkpoint independently and coordinated protocols force each process to coordinate with other processes to take consistent checkpoints. Two protocols have one's own advantages. A typical advantage of coordinated protocols is to avoid the domino effect. From such viewpoints, a number of techniques of checkpoint protocols have been proposed and their performance have been evaluated [1,12,13]. However, there are little research papers to study theoretically optimal policies for checkpoint intervals. Using the methods and techniques used in this thesis, we could analyze optimal intervals of checkpoints for distributed systems.

References

1. Abd-El-Barr, M. (2007). *Reliable and Fault-Tolerant* (Imperial Colledge Press, London).

2. Fukumoto, S., Kaio, N. and Osaki, S. (1992). A study of checkpoint generations for a database recovery mechanism, *Computers & Mathematics with Applications* , 24, pp. 63–70.
3. Kim, H. and Shin, K. G. (1996). Design and analysis of an optimal instruction-retry policy for tmr controller computers, *IEEE Transactions on Computers* , 45, pp. 1217–1225.
4. Lee, P. A. and Anderson, T. (1990). *Fault Tolerance Principles and Practice* (Springer, Wien).
5. Ling, Y., Mi, J. and Lin, X. (2001). A variational calculus approach to optimal checkpoint placement, *IEEE Transactions on Computers* , 50, pp. 699–707.
6. Nakagawa, S., Fukumoto, S. and Ishii, N. (1998). Optimal checkpoint interval for redundant error detection and masking systems, in *Proceeding of the First Euro-Japanese Workshop on Stochastic Risk Modeling for Finance, Insurance, Production and Reliability*, Vol. II.
7. Nakagawa, S., Fukumoto, S. and Ishii, N. (2003). Optimal checkpointing intervals of three error detection schemes by a double modular redundancy, *Mathematical and Computing Modeling* , 38, pp. 1357–1363.
8. Nakagawa, T. (2005). *Maintenance Theory of Reliability* (Springer, London).
9. Nakagawa, T., Yasui, K. and Sandoh, H. (2004). Note on optimal partition problems in reliability models, *Journal of Quality in Maintenance Engineering* , 10, pp. 282–287.
10. Nanya, T. (1991). *Fault Tolerant Computer* (Ohm Co., Tokyo).
11. Naruse, K. (2008). *Studies on Optimal Checkpoint Intervals for Computer Systems*, Ph.D. thesis, aichi institute of technology.
12. Ohara, M., Arai, M., Fukumoto, S. and Iwasaki, K. (2005). On the optimal checkpoint interval for uncoordinated checkpointing with a limited number of checkpoints and bound rollbacks, in *Proceedings of International Workshop on Recent Advances Stochastic Operations Research*, pp. 187–194.
13. Ohara, M., Suzuki, R., Arai, M., Fukumoto, S. and Iwasaki, K. (2006). Analytical Model on Hybrid State Saving with a Limited Number of Checkpoints and Bound Rollbacks, *IEICE TRANSACTIONS on Fundamentals of Electronics, Communications and Computer Sciences* **89**, 9, pp. 2386–2395.
14. Osaki, S. (1992). *Applied Stochastic System Modeling* (Springer, Berlin).
15. Pradhan, D. K. and Vaidya, N. H. (1994a). Roll-forward and rollback recovery: Performance-reliability trade-off, in *Proceeding of the 24nd International Symposium on Fault-Tolerant Computings*, pp. 186–195.
16. Pradhan, D. K. and Vaidya, N. H. (1994b). Roll-forward checkpointing scheme: A novel fault-tolerant architecture, *IEEE Transaction on Computers* **43**, pp. 1163–1174.
17. Ssu, K. F., Yao, B., Fuchs, W. F. and Neves, N. F. (1999). Adaptive checkpointing with storage management for mobile environments, *IEEE Transactions on Reliability* **48**, pp. 315–323.
18. Sugiura, T., Mizutani, S. and Nakagawa, T. (2004). Optimal random replacement polices, in *Tenth ISSAT International Conference on Reliability and Quality in Design*, pp. 99–103.

19. Vaidya, N. H. (1995). A case for two-level distributed recovery schemes, in *Proceedings ACM SIGMETRICS Conference Measurement and Modeling of Computer Systems*, pp. 64–73.
20. Vaidya, N. H. (1998). A case for two-level recovery schemes, *IEEE Transactions on Computers* **47**, pp. 656–666.
21. Yasui, K., Nakagawa, T. and Sandoh, H. (2002). *Reliability models in data communication systems* (Springer, Berlin), pp. 282–301.
22. Ziv, A. and Bruck, J. (1997). Performance optimization of checkpointing schemes with task duplication, *IEEE Transactions on Computers* **46**, pp. 1381–1386.
23. Ziv, A. and Bruck, J. (1998). Analysis of checkpointing schemes with task duplication, *IEEE Transactions on Computers* **47**, pp. 222–227.

PART 3
RELIABILITY APPLICATIONS

Chapter 9

Maintenance Models of Miscellaneous Systems

KODO ITO

Institute of Consumer Sciences and Human Life
Kinjo Gakuin University
1723 Omori 2-chome, Moriyama, Nagoya 463-8521, Japan
E-mail: itokodo@g-mail.com

1 Introduction

Miscellaneous systems such as social infrastructures and security forces, sustain our comfortable daily lives and secure our estates. For the steady operation of these systems without any serious troubles, suitable maintenances have to be undergone. However, maintenance budgets become extremely expensive in most advanced nations because of high personnel costs. Today, conflicts between the demand of budget cut and the demand of sufficient maintenances become serious social problems and the establishing cost-effective maintenance have become an important key technology to solve them.

The maintenance is classified into preventive maintenance (PM) and corrective maintenance (CM): PM is a maintenance policy in which we undergoes some maintenance on a specific schedule before failure, and CM is a maintenance policy after failure [1, 2]. Many researchers have studied optimal PM policies because the CM cost is usually much higher than the PM one and the optimal PM policy differs in every different system. Therefore, detailed investigations of target system characteristics must be performed for the consideration of cost-effective PM policies.

In this chapter, we survey optimal maintenance models for five different systems such as missile, phased array radar, FADEC, co-generation system and aged fossil-fired power plant based on our original works: Missiles and phased array radars are most representative military systems. During the Cold War era, defense budgets had the exceptional priority in most nations. Now in the post-Cold War era, there is no such exceptional priority and maintenance costs of missiles have to be designed to be minimum throughout their lifecycles from a primary design phase [3].

In Section 2, we consider the missile maintenance: A missile spends almost all of its whole lifetime in storage condition and its operational feature is unique compared with other military and ordinary industrial systems [4]. A missile during storage condition degrades gradually and its failure cannot be detected except for a function test. The test during storage is implemented periodically and an optimal test interval which minimizes the maintenance cost and satisfies the required system reliability, must be established.

In Section 3, we take up the phased array radar maintenance: A phased array radar is the latest radar system and its antenna consists of a large number of uniform tiny antennas [5]. The radar is designed to tolerate a certain amount of failed antennas because the increase of failed ones degrades its performance. Failed antennas are detected only by the function test which reduces the system availability. Therefore, an optimal test interval which maximizes the system availability must be discussed.

Section 4 is devoted to the self-diagnosis for FADEC. The FADEC (Full Authority Digital Electronic Control system) is widely utilized as the fuel controller of gas turbine engine because it can realize the complicated and delicate control compared with traditional hydro mechanical controller (HMC) [6]. As the FADEC of industrial gas turbine engine system, PLCs (programmable logic controllers) are utilized because they are tiny, high capacity and reasonably priced. Gas turbine makers which utilize PLCs for FADECs, have to guarantee the high reliability of FADECs and establish high reliable FADEC systems adopting the redundant design. A high-performance self-diagnosis of such redundant FADEC systems must be initiated.

Section 5 takes up the co-generation system maintenance: A co-generation system is the power plant which can generate electricity and steam simultaneously, and is one of applicational examples of a gas turbine engine system [7]. As the power plant resource, a gas turbine engine has superiority compared with other internal-combustion ones because of its tiny size, its emitted gas cleanness and its low vibration. A gas turbine engine is damaged when it is operated, and it has to undergo overhaul forthwith when the total damage is greater than a prespecified level. From the viewpoint of co-generation system users, the overhaul would be implemented at special periods such as some vacations. So that, a system user should establish a managerial level which is lower than overhaul level, and the system undergoes overhaul when its total damage exceeds a managerial level. An optimal managerial level which minimizes the operational cost must be considered.

Finally, Section 6 allocates to the aged fossil-fired power plant maintenance: Aged fossil-fired power plants are on the great increase in Japan. A plant consists of a wide variety of mechanical and electrical components. Various kinds of severe failures inherent in these components and their occurrence probabilities are different. A system fails when the total suffered damage exceeds a peculiar level, *i.e.*, these components have their peculiar damage levels. The PM plan should be established considering such levels of plant components. N kinds of peculiar cumulative damage levels K_i $(i = 1 \cdots N)$ are considered and a system fails when the total damage exceeds these damage levels. The PM is performed when the total damage exceeds the managerial level $k(< K_i)$. The expected cost per unit time between maintenances is secured, and the optimal managerial level k^* which minimizes it must be discussed.

We obtain the expected costs or the availability of each model as an objective function and derive optimal maintenance policies which minimize them, using reliability techniques. Furthermore, we give shortly some comments regarding the limitation and possible extensions of the above models.

2 Missile Maintenance

A system such as missiles is in storage for a long time from the delivery to the actual usage and has to hold a high mission reliability when it is used. Figure 1 shows an example of a service life cycle of missiles [4]: After a system is transported to each firing operation unit via depot, it is installed on launcher and is storaged in warehouse for a great part of its lifetime, and waits for its operation. So that, a missile is often called a dormant system. However, the reliability of a storage system goes down with time because some kinds of electronic and electric parts of a system degrade with time [8]. For example, Menke [8] confirmed by the accelerated test

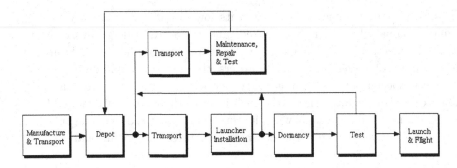

Fig. 1   Service life cycle of missiles [4]

that integrated circuits of a system might deteriorate in storage condition and it might be impossible to operate when it is necessary. Therefore, we should inspect and maintain a storage system at periodic times to hold a high reliability, because it is impossible to inspect whether a storage system can operate normally or not. Optimal inspection policies which minimize the expected cost until detection of failure were summarized. Martinez [9] discussed the periodic test of an electronic equipment in storage for a long period and showed how to compute its reliability after 10 years of storage.

In the above previous studies, it has been assumed that the function test can clarify all of system failures. However, a missile is exposed to very severe flight environment after launch and some kinds of failures are revealed only in such severe conditions. That is, some failures of a missile cannot be detected by the function test on the ground. To solve this problem, we assume that a system is divided into two independent units: Unit 1 becomes new after every inspections because all failures of unit 1 are detected by the function test and are removed completely by maintenance. While, unit 2 degrades steadily with time from delivery to overhaul because all failures of unit 2 cannot be detected by any tests. The reliability of a system deteriorates gradually with time as the reliability of unit 2 deteriorates steadily. A schematic diagram of a missile is given in Figure 2.

This section presents a system in storage which is required to have a higher reliability than a prespecified level q $(0 < q \leq 1)$ [1, 10–12]. To hold the reliability, a system is tested and is maintained at periodic times NT $(N = 1, 2, \cdots)$, and is overhauled if the reliability becomes equal to or lower than q. An inspection number N^* and the time $N^*T + t_0$ until overhaul, are derived when a system reliability is just equal to q. Using them, the expected cost $C(T)$ until overhaul is obtained, and an optimal inspection time T^* which minimizes it is computed.

2.1 Expected Cost

A system consists of unit 1 and 2, where unit i has a hazard rate function $H_i(t)$ ($i = 1, 2$). When a system is inspected at periodic times NT ($N = 1, 2, \cdots$), unit 1 is maintained and is like new after every inspection, and unit 2 is not done, *i.e.*, its hazard rate remains unchanged by any inspections.

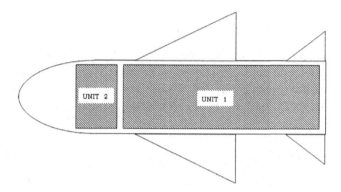

Fig. 2   Schematic diagram of a missile

From the above assumptions, the reliability function $R(t)$ of a system with no inspection is

$$R(t) = e^{-H_1(t)-H_2(t)}. \qquad (1)$$

If a system is inspected and maintained at time t, the reliability just after the inspection is

$$R(t_{+0}) = e^{-H_2(t)}. \qquad (2)$$

Thus, the reliabilities just before and after the Nth inspection are, respectively,

$$R(NT_{-0}) = e^{-H_1(T)-H_2(NT)}, \qquad (3)$$
$$R(NT_{+0}) = e^{-H_2(NT)}. \qquad (4)$$

Next, suppose that the overhaul is performed if the system reliability is equal to or lower than q. Then, if

$$e^{-H_1(T)-H_2(NT)} > q \geq e^{-H_1(T)-H_2[(N+1)T]}, \qquad (5)$$

then the time to overhaul is $NT + t_0$, where t_0 ($0 < t_0 \leq T$) satisfies

$$e^{-H_1(t_0) - H_2(NT+t_0)} = q. \tag{6}$$

This shows that the reliability is greater than q just before the Nth inspection and is equal to q at time $NT + t_0$.

Defining the time interval $[0, NT + t_0]$ as one cycle, the expected cost rate until overhaul is [1]

$$C(T) = \frac{Nc_1 + c_2}{NT + t_0}, \tag{7}$$

where cost c_1 is an inspection cost and c_2 is an overhaul cost.

2.2 Optimal Policies

We consider two particular cases where hazard rate functions $H_i(t)$ are exponential and Weibull ones. An inspection number N^* which satisfies (5) and t_0 which satisfies (6) are computed. Using these quantities, we compute the expected cost $C(T)$ until overhaul and seek an optimal inspection time T^* which minimizes it.

(1) Exponential Case

Suppose that the system obeys an exponential distribution, $i.e.$, $H_i(t) = \lambda_i t$. Then, (5) is rewritten as

$$\frac{1}{Na+1} \ln \frac{1}{q} \leq \lambda T < \frac{1}{(N-1)a+1} \ln \frac{1}{q}, \tag{8}$$

where

$$\lambda \equiv \lambda_1 + \lambda_2, \quad a \equiv \frac{H_2(T)}{H_1(T) + H_2(T)} = \frac{\lambda_2}{\lambda}, \tag{9}$$

and a represents an efficiency of inspection, and is adopted widely in practical reliability calculation of a storage system [4].

When an inspection time T is given, an inspection number N^* which satisfies (8) is determined. Particularly, if $\ln 1/q \leq \lambda T$ then $N^* = 0$, and N^* diverges as λT tends to 0. In this case, (6) is

$$N^* \lambda_2 T + \lambda t_0 = \ln \frac{1}{q}. \tag{10}$$

From (10), we can compute t_0 easily.

Thus, the total time to overhaul is

$$N^* T + t_0 = N^*(1-a)T + \frac{1}{\lambda} \ln \frac{1}{q}, \tag{11}$$

and the expected cost rate is

$$C(T) = \frac{N^* c_1 + c_2}{N^*(1-a)T + \frac{1}{\lambda}\ln\frac{1}{q}}. \tag{12}$$

When an inspection time T is given, we compute N^* from (8) and $N^*T + t_0$ from (11). Substituting these values into (12), we have $C(T)$. Changing T from 0 to $\ln(1/q)/[\lambda(1-a)]$, we can compute an optimal T^* which minimizes $C(T)$. In particular case of $\lambda T \geq \ln(1/q)/(1-a)$, $N^* = 0$ and the expected cost rate becomes constant, i.e.,

$$C(T) = \frac{c_2}{t_0} = -\frac{\lambda c_2}{\ln q}. \tag{13}$$

(2) Weibull Case

Suppose that the system obeys a Weibull distribution, i.e., $H_i(t) = (\lambda_i t)^m$ ($i = 1, 2$). Equations (5) and (6) are rewritten as, respectively,

$$\left\{\frac{1}{a[(N+1)^m - 1] + 1}\ln\frac{1}{q}\right\}^{\frac{1}{m}} \leq \lambda T < \left[\frac{1}{a(N^m - 1) + 1}\ln\frac{1}{q}\right]^{\frac{1}{m}}, \tag{14}$$

$$(1-a)t_0^m + a(NT + t_0)^m = \frac{1}{\lambda^m}\ln\frac{1}{q}, \tag{15}$$

where

$$\lambda^m \equiv \lambda_1^m + \lambda_2^m,$$

$$a \equiv \frac{H_2(T)}{H_1(T) + H_2(T)} = \frac{\lambda_2^m}{\lambda_1^m + \lambda_2^m}. \tag{16}$$

When an inspection time T is given, N^* and t_0 are computed from (14) and (15). Substituting these values into (7), we have $C(T)$, and changing T from 0 to $[\ln(1/q)/(1-a)]^{1/m}/\lambda$, we can compute an optimal T^* which minimizes $C(T)$.

Next, suppose that unit 1 obeys a Weibull distribution with order 2 and unit 2 obeys an exponential distribution, i.e., $H_1(t) = (\lambda_1 t)^2$ and $H_2(t) = \lambda t$. Then, from (5) and (6), respectively,

$$\frac{1}{2(1-a)^2}\left\{-(N+1)a + \sqrt{(N+1)^2 a^2 - 4(1-a)^2 \ln q}\right\}$$

$$\leq \lambda T < \frac{1}{2(1-a)^2}\left\{-Na + \sqrt{N^2 a^2 - 4(1-a)^2 \ln q}\right\}, \tag{17}$$

$$[(1-a)\lambda t_0]^2 + a\lambda(NT + t_0) + \ln q = 0, \tag{18}$$

where a and λ are given in (9).

When an inspection time T is given, an inspection number N^* which satisfies (17) is computed. Then, the total time to overhaul is

$$N^*T + t_0 = N^*T + \frac{1}{2(1-a)^2\lambda}\left\{-a + \sqrt{a^2 - 4(1-a)^2(N^*a\lambda T + \ln q)}\right\}. \quad (19)$$

The expected cost until overhaul is, from (7)

$$C(T) = \frac{N^*c_1 + c_2}{N^*T + \frac{1}{2(1-a)^2\lambda}\left\{-a + \sqrt{a^2 - 4(1-a)^2(N^*a\lambda T + \ln q)}\right\}}. \quad (20)$$

In particular, if

$$\lambda T \geq \frac{-a + \sqrt{a^2 - 4(1-a)^2 \ln q}}{2(1-a)^2}, \quad (21)$$

then $N^* = 0$, and the expected cost is

$$C(T) = \frac{2(1-a)^2\lambda c_2}{-a + \sqrt{a^2 - 4(1-a)^2 \ln q}}. \quad (22)$$

Therefore, when an inspection time T is given, we compute N^* from (17) or (21), and $N^*T + t_0$ from (19). Substituting them into (20) or (22), we compute $C(T)$, and changing T from 0 to

$$\frac{-a + \sqrt{a^2 - 4(1-a)^2 \ln q}}{2(1-a)^2\lambda},$$

we can determine T^* which minimizes $C(T)$.

2.3 Concluding Remarks

This fundamental deterministic model could be applied to various kinds of storage systems which are required to have a high reliability and in which some failures are not be detected by the function test. When all failures can be clarified by the function test, *i.e.*, an efficiency a of inspection is equal to zero, this maintenance model becomes the standard one.

The variation of function tests must be considered for future studies. For example, two types of function test equipments are utilized for the missile maintenance: A function test equipment installed in the warehouse of missile has to perform detailed tests for sizable machines. Another one at the battlefields has to perform simplified tests for small sizes. Inspection efficiencies of such test equipments would differ considerably.

3 Phased Array Radar Maintenance

A phased array radar (PAR) is the radar which steers the electromagnetic wave direction electrically. Comparing with conventional radars which steer their electromagnetic wave direction by moving their antennas mechanically, a PAR has no mechanical portion to steer its wave direction, and hence, it can steer very quickly. Most anti-aircraft missile systems and early warning systems have presently adopted PARs because they can acquire and track multiple targets simultaneously.

A PAR antenna consists of a large number of small and homogeneous element antennas which are arranged flatly and regularly, and steers its electromagnetic wave direction by shifting signal phases of waves which are radiated from these individual elements [13].

The increase in the number of failed elements degrades the radar performance, and at last, this may cause an undesirable situation such as the omission of targets [5]. The detection, diagnosis, localization and replacement of failed elements of a PAR antenna are indispensable to hold a certain required level of radar performance. A digital computer system controls a whole PAR system, and it detects, diagnosis and localizes failed elements. However, such maintenance actions intermit the radar operation and decrease its availability. So that, the maintenance should not be made so frequently. From the above reasons, it would be important to decide an optimal maintenance policy for a PAR antenna, by comparing the downtime loss caused by its maintenance with the degradational loss caused by its performance downgrade.

Recently, a new method of failure detection for PAR antenna elements has been proposed by measuring the electromagnetic wave pattern [14]. This method could detect some failed elements even when a radar system is operating, *i.e.*, it could be applied to the detection of confined failure modes such as power on-off failures. However, it would be generally necessary to stop the PAR operation for the detection of all failed elements.

Keithley [15] showed by Monte Carlo simulation that the maintenance time of PAR with 1024 elements had a strong influence on its availability. Hevesh [16] discussed the following three maintenances of PAR in which all failed elements could be detected immediately, and calculated the average times to failures of its equipments and its availability in immediate maintenance:

1) Immediate maintenance: Failed elements are detected, localized and replaced immediately.

2) Cyclic maintenance: Failed elements are detected, localized and replaced periodically.

3) Delayed maintenance: Failed elements are detected and localized periodically, and replaced when their number has exceeded a predesignated one.

Further, Hesse [17] analyzed the field maintenance data of U.S. Army prototype PAR, and clarified that the repair times have a log-normal distribution. In the actual maintenance, the immediate maintenance is rarely adopted because frequent maintenances degrade a radar system availability. Either cyclic or delayed maintenances is commonly adopted.

We have studied the comparison of cyclic and delayed maintenances of PAR considering the financial optimum [18]: We derived the expected cost rates and discussed the optimal policies which minimize them analytically in these two maintenances, and concluded that the delayed maintenance is better than the cyclic one in suitable conditions by comparing these two costs numerically. Although the financial optimum takes priority for non-military systems and military systems in the non-combat condition, the operational availability should take more priority than economy for military systems in the combat condition. Therefore, maintenance policies which maximize availability should be considered.

We perform the periodic detection of failed elements of a PAR where it is consisted of N_0 elements and failures are detected at scheduled time interval [19]: If the number of failed elements has exceeded a specified number N ($0 < N \leq N_0$), a PAR cannot hold a required level of radar performance, and it causes the operational loss such as the target oversight to a PAR. We assume that failed elements occur at a Poisson process, and consider cyclic and delayed maintenances. Applying the method [20] to such maintenances, the availability is obtained, and optimal policies which maximize them are analytically discussed in cyclic and delayed maintenances.

3.1 Cyclic Maintenance

We consider the following cyclic maintenance of a PAR [19]:

1) A PAR is consisted of N_0 elements which are independent and homogeneous on all plains of PAR, and have an identical constant hazard rate λ_0. The number of failed elements at time t has a binomial distribution with mean $N_0[1 - \exp(-\lambda_0 t)]$. Since N_0 is large and λ_0 is very small, it might be assumed that failures of elements occur approximately at a

Poisson process with mean $\lambda \equiv N_0 \lambda_0$. That is, the probability that j failures occur during $(0, t]$ is

$$P_j(t) \equiv \frac{(\lambda t)^j e^{-\lambda t}}{j!} \quad (j = 0, 1, 2, \cdots).$$

2) When the number of failed elements has exceeded a specified number N, a PAR cannot hold a required level of radar performance such as maximum detection range and resolution.

3) Failed elements cannot be detected during operation and can be ascertained only according to the diagnosis software executed by a PAR system computer. Failed elements are usually detected at periodic diagnosis. The diagnosis is performed at time interval T and a single diagnosis spends time T_0.

4) All failed elements are replaced by new ones at the Mth diagnosis or at the time when the number of failed elements has exceeded N, whichever occurs first. The replacement spends time T_1.

When the number of failed elements is below N at the Mth diagnosis, the expected effective time until replacement is

$$MT \sum_{j=0}^{N-1} p_j(MT). \tag{23}$$

When the number of failed elements exceeds N at the $i\,(i = 1, 2, \cdots M)$th diagnosis, the expected effective time until replacement is

$$\sum_{i=1}^{M} \sum_{j=0}^{N-1} p_j[(i-1)T] \sum_{k=N-j}^{\infty} \int_{(i-1)T}^{iT} t \, dp_k[t - (i-1)T]. \tag{24}$$

Thus, from (23) and (24), the total expected effective time until replacement is

$$T \sum_{i=0}^{M-1} \sum_{j=0}^{N-1} p_j(iT) - \sum_{i=0}^{M-1} \sum_{j=0}^{N-1} p_j(iT) \sum_{k=N-j+1}^{\infty} \frac{k-N+j}{\lambda} p_k(T). \tag{25}$$

Next, when the number of failed elements is below N at the Mth diagnosis, the expected time between two adjacent regeneration points is

$$\sum_{j=0}^{N-1} p_j(MT)[M(T+T_0) + T_1]. \tag{26}$$

When the number of failed elements exceeds N at the $i\,(i = 1, 2, \cdots M)$th diagnosis, the expected time between two adjacent regeneration points is

$$\sum_{i=1}^{M} \sum_{j=0}^{N-1} p_j[(i-1)T] \sum_{k=N-j}^{\infty} [i(T+T_0) + T_1] p_k(T). \tag{27}$$

Thus, from (26) and (27), the total expected time between two adjacent regeneration points is

$$T_1 + (T + T_0) \sum_{i=0}^{M-1} \sum_{j=0}^{N-1} p_j(iT). \tag{28}$$

Therefore, by dividing (25) by (28), the availability of cyclic maintenance $A_1(M)$ is

$$A_1(M) = \frac{\text{Effective time between regeneration points}}{\text{Total time between regeneration points}}$$

$$= \frac{T \sum_{i=0}^{M-1} \sum_{j=0}^{N-1} p_j(iT) - \sum_{i=0}^{M-1} \sum_{j=0}^{N-1} p_j(iT) \sum_{k=N-j+1}^{\infty} (k-N+j) p_k(T)/\lambda}{T_1 + (T+T_0) \sum_{i=0}^{M-1} \sum_{j=0}^{N-1} p_j(iT)}$$

$$= \frac{T}{T+T_0}\left[1 - \frac{T_1/(T+T_0) + \sum_{i=0}^{M-1}\sum_{j=0}^{N-1} p_j(iT) \times \sum_{k=N-j+1}^{\infty}(k-N+j)p_k(T)/(\lambda T)}{T_1/(T+T_0) + \sum_{i=0}^{M-1}\sum_{j=0}^{N-1} p_j(iT)}\right]$$

$$\equiv \frac{T}{T+T_0}[1 - \overline{A_1}(M)] \qquad (M = 1, 2, \cdots). \tag{29}$$

Because maximizing an availability $A_1(M)$ is equal to minimizing an unavailability $\overline{A_1}(M)$ from (29), we derive M^* which minimizes an unavailability $\overline{A_1}(M)$. Forming the inequality $\overline{A_1}(M+1) - \overline{A_1}(M) \geq 0$,

$$L_1(M)\left[\frac{T_1}{T+T_0} + \sum_{i=0}^{M-1}\sum_{j=0}^{N-1} p_j(iT)\right]$$

$$- \sum_{i=0}^{M-1}\sum_{j=0}^{N-1} p_j(iT) \sum_{k=N-j+1}^{\infty} \frac{k-N+j}{\lambda} p_k(T) \geq \frac{T_1 T}{T+T_0}, \tag{30}$$

where

$$L_1(M) \equiv \frac{\sum_{j=0}^{N-1} p_j(MT) \sum_{k=N-j+1}^{\infty}(k-N+j)p_k(T)/\lambda}{\sum_{j=0}^{N-1} p_j(MT)}. \tag{31}$$

Letting $Q_1(M)$ denote the left-hand side of (30),

$$Q_1(M+1) - Q_1(M) = [L_1(M+1) - L_1(M)]\left[\frac{T_1}{T+T_0} + \sum_{i=0}^{M}\sum_{j=0}^{N-1} p_j(iT)\right]. \tag{32}$$

Thus, if $L_1(M)$ is strictly increasing in M, then $Q_1(M)$ is strictly increasing in M.

Therefore, we have the following optimal policy:

(i) If $L_1(M)$ is strictly increasing in M and $Q_1(\infty) > T_1T/(T+T_0)$, then there exists a finite and unique M^* which satisfies (31).
(ii) If $L_1(M)$ is strictly increasing in M and $Q_1(\infty) \le T_1T/(T+T_0)$, then $M^* = \infty$.
(iii) If $L_1(M)$ is decreasing in M, then $M^* = 1$ or $M^* = \infty$.

3.2 Delayed Maintenance

We consider the delayed maintenance of a PAR [19]:

4)' All failed elements are replaced by new ones only when failed elements have exceeded a managerial number N_c ($< N$) at diagnosis. The replacement spends time T_1.

The other assumptions are the same as ones in Section 7.2.1.

When the number of failed elements is between N_c and N, the expected effective time until replacement is

$$\sum_{i=1}^{\infty} \sum_{j=0}^{N_c-1} p_j[(i-1)T] \sum_{k=N_c-j}^{N-j-1} iTp_k(T). \tag{33}$$

When the number of failed elements exceeds N, the expected effective time until replacement is

$$\sum_{i=1}^{\infty} \sum_{j=0}^{N_c-1} p_j[(i-1)T] \sum_{k=N-j}^{\infty} \int_{(i-1)T}^{iT} t \, dp_k[t-(i-1)T]. \tag{34}$$

Thus, the total expected effective time until replacement is, from (33) and (34),

$$T \sum_{i=0}^{\infty} \sum_{j=0}^{N_c-1} p_j(iT) - \sum_{i=0}^{\infty} \sum_{j=0}^{N_c-1} p_j(iT) \sum_{k=N-j+1}^{\infty} \frac{k-N+j}{\lambda} p_k(T). \tag{35}$$

Similarly, when the number of failed elements is between N_c and N, the expected time between two adjacent regeneration points is

$$\sum_{i=1}^{\infty} \sum_{j=0}^{N_c-1} p_j[(i-1)T] \sum_{k=N_c-j}^{N-j-1} [i(T+T_0)+T_1]p_k(T). \tag{36}$$

When the number of failed elements exceeds N, the expected time between two adjacent regeneration points is

$$\sum_{i=1}^{\infty} \sum_{j=0}^{N_c-1} p_j[(i-1)T] \sum_{k=N-j}^{\infty} [i(T+T_0)+T_1]p_k(T). \tag{37}$$

Thus, the total expected time between two adjacent regeneration points is, from (36) and (37),

$$(T+T_0)\sum_{i=0}^{\infty}\sum_{j=0}^{N_c-1} p_j(iT) + T_1. \qquad (38)$$

Therefore, the availability of delayed maintenance $A_2(N_c)$ is, by dividing (35) by (38),

$$A_2(N_c) = \frac{T}{T+T_0}\left[1 - \frac{T_1/(T+T_0) + \sum_{i=0}^{\infty}\sum_{j=0}^{N_c-1} p_j(iT) \times \sum_{k=N-j}^{\infty}[1-(N-j)/(k+1)]p_k(T)}{T_1/(T+T_0) + \sum_{i=0}^{\infty}\sum_{j=0}^{N_c-1} p_j(iT)}\right]$$

$$\equiv \frac{T}{T+T_0}[1 - \overline{A_2}(N_c)]. \qquad (39)$$

Forming the inequality $\overline{A_2}(N_c+1) - \overline{A_2}(N_c) \geq 0$,

$$\frac{E_{N_c}}{D_{N_c} - E_{N_c}}\sum_{j=0}^{N_c-1}(D_j - E_j) - \sum_{j=0}^{N_c-1} E_j \geq \frac{T_1}{T+T_0}. \qquad (40)$$

where $D_j \equiv \sum_{i=0}^{\infty} p_j(iT)$ and $E_j \equiv \sum_{i=0}^{\infty} p_j(iT)\sum_{k=N-j}^{\infty}[1-(N-j)/(k+1)]p_k(T)$. Letting $Q_2(N_c)$ denote the left-hand side of (40) and $L_2(N_c) \equiv E_{N_c}/(D_{N_c} - E_{N_c})$,

$$Q_2(N_c+1) - Q_2(N_c) = [L_2(N_c+1) - L_2(N_c)]\sum_{j=0}^{N_c}(D_j - E_j). \qquad (41)$$

As $D_j - E_j > 0$, the sign of $Q_2(N_c+1) - Q_2(N_c)$ depends on $L_2(N_c+1) - L_2(N_c)$.

Therefore, we have the following optimal policy:

(i) If $L_2(N_c)$ is strictly increasing in N_c and $Q_2(N) > T_1/(T+T_0)$ then there exists a finite and unique N_c^* ($1 \leq N_c^* < N$) which satisfies (41).
(ii) If $L_2(N_c)$ is strictly increasing in N_c and $Q_2(N) \leq T_1/(T+T_0)$ then $N_c^* = N$, i.e., the planned maintenance should not be done.

3.3 Concluding Remarks

It has been assumed that failures of elements occur at a homogeneous Poisson process. The degradation of elements might be caused by moisture and saline of cooling air. In such case, failures of elements would not occur at a unified distribution function and failures would not occur at a non-homogeneous Poisson process.

We have considered cyclic and delayed maintenances. Such maintenances can be extended to the much complicated and more cost-effective ones. For example, particular condition data of degraded elements can be receipt at each maintenance and some condition monitoring maintenances should be designed utilizing these data.

4 Self-diagnosis for FADEC

The original idea of gas turbine engines was represented by Barber in England at 1791, and they were firstly realized in 20-th century. After that, they had advanced greatly during World War II. Today, gas turbine engines have been widely utilized as main engines of airplanes, high performance mechanical pumps, emergency generators and cogeneration systems because they can generate high power comparing with their sizes, their start times are very short and no coolant water is necessary for operation [6].

Gas turbine engines are mainly constituted with three parts, $i.e.$, compressor, combustor and turbine. The engine control is performed by governing the fuel flow to engine. When gas turbine engines are operating, dangerous phenomena, such as surge, stool and over-temperature of exhaust gas, should be paid attention because they may cause serious damage to engine. To prevent them, the turbine speed, inlet temperature and pressure, and exhaust gas temperature of gas turbine engines are monitored, and engine controller should determine appropriate fuel flow by checking these data.

The gas turbine engine has to be operating in serious environment and hydro mechanical controller (HMC) is adopted as engine controller for a long period because of its high reliability, durability and excellent responsibility. However, the performance of gas turbine engines has advanced and customers need to decrease the operation cost. So that, HMC could not meet these advanced demands and the engine controller has been electrified. The first electric engine controller, which was a support unit of HMS, was adopted as J47-17 turbo jet engine of F86D fighter at the late 1940-th. The evolution of devices, from vacuum tube to transistor and transistor to IC, has changed the roll of electric engine controller from the assistant of HMS to the full authority controller because of the reliability growth. In 1960-th, the analogue full authority controller could not meet the accuracy demand of engines, and the full authority digital engine controller (FADEC) was greatly developed [6].

FADEC is an electric engine controller which can perform the complicated signal processes of digitized engine data. Aircraft FADECs must generally build up duplicated and triplicated systems because they are expected high mission reliability and are needed to decrease weight, hardware complication and electric consumption [21]. Industrial gas turbine engines have introduced advanced technologies which were established for aircraft ones. FADECs, which were originally developed for aircrafts, have also been adopted as industrial gas turbine engines now. Comparing between general industrial gas turbine FADECs and aircraft FADECs, the following differences are recognized:

1) Aircraft gas turbine FADECs have to perform high-speed data processing because the rapid response for aircraft body movement is necessary and inlet pressure and temperature change greatly depending on height. On the other hand, industrial gas turbine FADECs are not required such high performance comparing to aircraft ones because they operate at steady speed on ground.
2) Aircraft gas turbine FADECs have to be reliable and fault tolerable, and so that, they adopt duplicated and triplicated systems because their malfunction in operation may cause serious damage to aircrafts and crews. Industrial gas turbine FADECs also have to be reliable and fault tolerable, and still be low cost because they have to be competitive in the market.

Depending on the advance of microelectronics, small, high performance, low cost programmable logic controllers (PLC) have widely distributed in the market. They were originally developed as the substitute for bulky electric relay logic sequencers of industrial automatic systems. Applying the numerical calculation ability of microprocessors, and analogue-digital and digital-analogue converters, these PLCs can perform numerical control. Appropriating such PLCs, very high performance and low cost FADEC systems can be realized. However, these PLCs are developed as general industrial controllers and PLC makers might not permit them for applying high pressurized and hot fluid controllers. Then, gas turbine makers which apply these PLCs to FADECs, have to design some protective mechanism and have to assure high reliability of FADECs.

We consider self-diagnosis policies for dual, triple and N redundant gas turbine engine FADECs, and discuss the diagnosis intervals [22].

4.1 Double Module System

Consider the following self-diagnosis policy for a hot standby double module FADEC system: Figure 5 illustrates an example of the FADEC construction.

1) The FADEC system consists of two independent channels and reliabilities of channel i at time t are $\overline{F}_i(t)$ ($i = 1, 2$).
2) The control calculation of each channel is performed at time interval T_0, and the self-diagnosis and cross-diagnosis are performed synchronously between two channels at every nth calculation. The coverage of these diagnosis is 100%.
3) When the number n decreases, the diagnosis calculation per unit of time increases and degrades the quality of control. It is assumed that the degradation of control is represented as $c_1/(n + T_1)$, where c_1 is constant and T_1 is the percentage of diagnosis time divided by T_0.
4) When n increases, the time interval from occurrence of failure to its detection is prolonged and it causes the damage of gas turbine engine because the extraordinary fuel control signal may incur overspeed or overtemperature of engines. The damage of engine is represented as $c_2(nT_0 - t)$, where t is the time that failure occurs and c_2 is the system loss per unit of time.
5) Initially, channel 1 is in active and channel 2 is in hot standby. When channel 1 fails, it changes to standby and channel 2 changes to active if it does not fail. It is assumed that these elapsed times for changing are negligible. When both channels 1 and 2 have failed, the system makes an emergency stop.

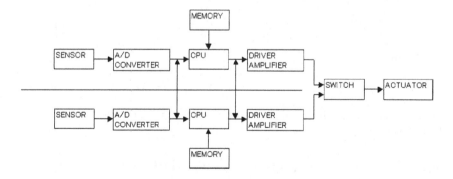

Fig. 3 Double module FADEC construction

When channel i fails at time t_i ($i = 1, 2$), the following two mean times from failure to its detection are considered:

a) When $t_2 \leq t_1 < t_m$ or $t_{m-1} < t_1 < t_2 \leq t_m$, the mean time is

$$\sum_{m=1}^{\infty} F_2(t_m) \int_{t_{m-1}}^{t_m} (t_m - t_1) dF_1(t_1), \qquad (42)$$

where $t_m = mnT_0$ ($m = 1, 2, 3 \cdots$).

b) When $t_1 \leq t_{m-1} < t_2 \leq t_m$, the mean time is

$$\sum_{m=2}^{\infty} \sum_{k=1}^{m-1} \int_{t_{k-1}}^{t_k} (t_k - t_1 + t_m - t_2) dF_1(t_1) \int_{t_{m-1}}^{t_m} dF_2(t_2). \qquad (43)$$

The total mean time from failure to its detection is the summation of (42) and (43), and is

$$\sum_{m=0}^{\infty} \left\{ \int_{t_m}^{t_{m+1}} [F_1(t) - F_1(t_m)] dt + F_1(t_m) \int_{t_m}^{t_{m+1}} [F_2(t) - F_2(t_m)] dt \right\}, \qquad (44)$$

where $F_1(0) \equiv 0$. Thus, the total expected cost of dual redundant FADEC until the system stops is

$$C_2(n) = \frac{c_1}{n + T_1} + c_2 \sum_{m=0}^{\infty} \left\{ \int_{t_m}^{t_{m+1}} [F_1(t) - F_1(t_m)] dt \right.$$

$$\left. + F_1(t_m) \int_{t_m}^{t_{m+1}} [F_2(t) - F_2(t_m)] dt \right\}. \qquad (45)$$

Assuming $F_i(t) = 1 - \exp(-\lambda_i t)$ ($i = 1, 2$), (45) is

$$C_2(n) = \frac{c_1}{n+T_1} + c_2 \left\{ nT_0 \left[\frac{1}{1 - e^{-\lambda_1 nT_0}} + \frac{1}{1 - e^{-\lambda_2 nT_0}} - \frac{1}{1 - e^{-(\lambda_1+\lambda_2)nT_0}} \right] \right.$$

$$\left. + \frac{1 - e^{-\lambda_2 nT_0}}{\lambda_2(1 - e^{-(\lambda_1+\lambda_2)nT_0})} - \frac{1}{\lambda_1} - \frac{1}{\lambda_2} \right\}. \qquad (46)$$

We easily find that

$$C_2(0) = \frac{c_1}{T_1}, \quad C_2(\infty) \equiv \lim_{n \to \infty} C_2(n) = \infty. \qquad (47)$$

Therefore, there exist a finite n_2^* ($< \infty$) which minimizes $C_2(n)$.

When $\lambda_1 = \lambda_2 \equiv \lambda$, (46) is

$$C_2(n) = \frac{c_1}{n + T_1} + c_2 \left(\frac{2}{1 - e^{-n\lambda T_0}} - \frac{1}{1 - e^{-2n\lambda T_0}} \right) \left(nT_0 - \frac{1 - e^{-n\lambda T_0}}{\lambda} \right). \qquad (48)$$

Supposing $x = nT_0$ and $C_2(x)$ is a continuous function of x, (48) is

$$C_2(x) = \frac{c_1 T_0}{x + T_1 T_0} + c_2 \left(\frac{2}{1 - e^{-\lambda x}} - \frac{1}{1 - e^{-2\lambda x}} \right) \left(x - \frac{1 - e^{-\lambda x}}{\lambda} \right). \quad (49)$$

Differentiating $C_2(x)$ with respect to x, and putting it equal to zero,

$$\left(\frac{2}{1 - e^{-\lambda x}} - \frac{1}{1 - e^{-2\lambda x}} \right)(1 - e^{-\lambda x})$$

$$- 2 \left[\frac{e^{-\lambda x}}{(1 - e^{-\lambda x})^2} - \frac{e^{-2\lambda x}}{(1 - e^{-2\lambda x})^2} \right] (\lambda x - 1 + e^{-\lambda x}) = \frac{c_1 T_0}{c_2 (x + T_1 T_0)^2}. \quad (50)$$

We compute optimal x^* which minimizes $C_2(x)$, and using these values, we can obtain optimal n^* which minimizes $C_2(n)$ in (48).

4.2 Triple Module System

Consider the self-diagnosis policy for a hot standby triple module FADEC system: We make following assumptions 1') and 5') instead of 1) and 5) in Section 4.1, respectively.

1') The FADEC system consists of three independent channels and the reliabilities of channel i at time t are $\overline{F}_i(t)$ ($i = 1, 2, 3$).

5') Initially, channel 1 is in active and channel 2 and 3 are in hot standby. When channel 1 fails, channel 1 changes to standby and channel 2 changes to active if it does not fail. Furthermore, when both channels 1 and 2 have failed, they change to standby and channel 3 changes to active if it does not fail. It is assumed that these elapsed times for changing are negligible. When all channels 1, 2 and 3 have failed, the system makes an emergency stop.

When channel i fails at time t_i ($i = 1, 2, 3$), the following four mean times from failure to its detection are considered:

a) When $t_2, t_3 \leq t_1$ or $t_{m-1} < t_1 < t_2 < t_3 \leq t_m$, the mean time is

$$\sum_{m=1}^{\infty} F_2(t_m) F_3(t_m) \int_{t_{m-1}}^{t_m} (t_m - t_1) dF_1(t_1). \quad (51)$$

b) When $t_{m-1} < t_1 < t_2 \leq t_m < t_3$ or $t_2 \leq t_1 \leq t_m < t_3$, the mean time is

$$\sum_{m=2}^{\infty} \sum_{l=1}^{m-1} F_2(t_l) \int_{t_{l-1}}^{t_l} (t_l - t_1 + t_m - t_3) dF_1(t_1) \int_{t_{m-1}}^{t_m} dF_3(t_3). \quad (52)$$

c) When $t_{m-1} < t_l < t_3 \leq t_m < t_2$, $t_1 < t_m < t_2 < t_3 t_{m+1}$, or $t_1 < t_m < t_3 < t_2$, the mean time is

$$\sum_{m=2}^{\infty} \sum_{l=1}^{m-1} F_3(t_m) \int_{t_{l-1}}^{t_l} (t_l - t_1 + t_m - t_2) \mathrm{d}F_1(t_1) \int_{t_{m-1}}^{t_m} \mathrm{d}F_2(t_2). \quad (53)$$

d) When $t_1 < t_m < t_2 < t_{m+1} < t_3$, the mean time is

$$\sum_{m=3}^{\infty} \sum_{l=2}^{m-1} \sum_{k=1}^{l-1} \int_{t_{k-1}}^{t_k} (t_k - t_1 + t_l - t_2 + t_m - t_3) \mathrm{d}F_1(t_1)$$
$$\times \int_{t_{l-1}}^{t_l} \mathrm{d}F_2(t_2) \int_{t_{m-1}}^{t_m} \mathrm{d}F_3(t_3). \quad (54)$$

The total mean time is the summation of (51)–(54) and is

$$\sum_{m=0}^{\infty} \left\{ \int_{t_m}^{t_{m+1}} [F_1(t) - F_1(t_m)] \mathrm{d}t + F_1(t_m) \int_{t_m}^{t_{m+1}} [F_2(t) - F_2(t_m)] \mathrm{d}t \right.$$
$$\left. + F_1(t_m) F_2(t_m) \int_{t_m}^{t_{m+1}} [F_3(t) - F_3(t_m)] \mathrm{d}t \right\}. \quad (55)$$

Thus, the total expected cost of triple redundant FADEC until the system stops is

$$C_3(n) = \frac{c_1}{n+T_1} + c_2 \sum_{m=0}^{\infty} \left\{ \int_{t_m}^{t_{m+1}} [F_1(t) - F_1(t_m)] \mathrm{d}t \right.$$
$$+ F_1(t_m) \int_{t_m}^{t_{m+1}} [F_2(t) - F_2(t_m)] \mathrm{d}t$$
$$\left. + F_1(t_m) F_2(t_m) \int_{t_m}^{t_{m+1}} [F_3(t) - F_3(t_m)] \mathrm{d}t \right\}. \quad (56)$$

Assuming $F_i(t) = 1 - \exp(-\lambda_i t)$ ($i = 1, 2, 3$) and $\lambda_1 = \lambda_2 = \lambda_3 \equiv \lambda$, (56) is

$$C_3(n) = \frac{c_1}{n+T_1} + c_2 \left(\frac{3}{1 - e^{-n\lambda T_0}} - \frac{3}{1 - e^{-2n\lambda T_0}} + \frac{1}{1 - e^{-3n\lambda T_0}} \right)$$
$$\times \left(nT_0 - \frac{1 - e^{-n\lambda T_0}}{\lambda} \right) \quad (n = 1, 2, \dots). \quad (57)$$

We easily find that

$$C_3(0) = \frac{c_1}{T_1}, \quad C_3(\infty) \equiv \lim_{n \to \infty} C_3(n) = \infty. \quad (58)$$

Therefore, there exist a finite n_3^* ($< \infty$) which minimizes $C_3(n)$.

Supposing $x = nT_0$ and $C_3(x)$ is a continuous function of x, (57) is

$$C_3(x) = \frac{c_1 T_0}{x + T_1 T_0} + c_2 \left(\frac{3}{1-e^{-\lambda x}} - \frac{3}{1-e^{-2\lambda x}} + \frac{1}{1-e^{-3\lambda x}} \right) \left(x - \frac{1-e^{-\lambda x}}{\lambda} \right). \quad (59)$$

Differentiating $C_3(x)$ with respect to x and putting it to zero,

$$\left(\frac{3}{1-e^{-\lambda x}} - \frac{3}{1-e^{-2\lambda x}} + \frac{1}{1-e^{-3\lambda x}} \right) (1 - e^{-\lambda x})$$

$$-3 \left[\frac{e^{-\lambda x}}{(1-e^{-\lambda x})^2} - \frac{2e^{-2\lambda x}}{(1-e^{-2\lambda x})^2} + \frac{e^{-3\lambda x}}{(1-e^{-3\lambda x})^2} \right] (\lambda x - 1 + e^{-\lambda x})$$

$$= \frac{c_1 T_0}{c_2 (x + T_0 T_1)^2}. \quad (60)$$

We can compute an optimal n_3^* which minimizes $C_3(n)$, using x^* which satisfies (60).

4.3 N Module System

Consider a hot standby N module FADEC system, i.e., a FADEC system which consists of N independent channels and channels i have the reliabilities $\overline{F_i}(t)$ ($i = 1, 2, \cdots, N$). By the similar method of Sections 4.1 and 4.2, the total expected cost is

$$C_N(n) = \frac{c_1}{n + T_1} + c_2 \sum_{m=0}^{\infty} \left\{ \int_{t_m}^{t_{m+1}} [F_1(t) - F_1(t_m)] dt \right.$$

$$+ F_1(t_m) \int_{t_m}^{t_{m+1}} [F_2(t) - F_2(t_m)] dt + \cdots$$

$$\left. \cdots + F_1(t_m) F_2(t_m) \cdots F_{N-1}(t_m) \int_{t_m}^{t_{m+1}} [F_N(t) - F_N(t_m)] dt \right\}$$

$$= \frac{c_1}{n + T_1}$$

$$+ c_2 \sum_{m=0}^{\infty} \sum_{k=1}^{N} F_1(t_m) \cdots F_{k-1}(t_m) \int_{t_m}^{t_{m+1}} [F_k(t) - F_k(t_m)] dt, \quad (61)$$

where $F_0(t) \equiv 1$. The total expected costs $C_N(n)$ agree with (45) and (56) for $N = 2, 3$, respectively.

4.4 Concluding Remarks

It has been assumed that the degradation of control is denoted as $c_1/(n+T_1)$. In case of hot standby multiple module systems, it becomes difficult to synchronize their operation because the clock frequency of CPU has raised so rapidly. When multiple systems operate asynchronously, the assumption cannot be approved and this FADEC maintenance model might not be applicable.

Furthermore, we assume that the degradation of gas turbine engine is relative to the operation time. Strictly, the surplus fuel flow raises the combustor temperature and a high temperature of combustor conducts to compressor and turbine blades. Because these blades are under a high tensile stress, the creep fatigue of blades may occur and stretched blades may harm the inside of casing structure. When such degradation mechanisms are simulated precisely, $c_1/(n+T_1)$ might be exchanged to more realistic form.

5 Co-generation System Maintenance

A co-generation system produces coincidentally both electric power and process heat in a single integrated system, and today, is exploited as the distributed power plant [7]. Various kinds of generators, such as steam turbine, gas turbine engine, gas engine, and diesel engine, are adopted as the power sources of co-generation systems. A gas turbine engine has some attractive advantages as compared with other power sources, because its size is the smallest, its exhaust gas emission is the cleanest, and both its noise and its vibration level are the lowest in all power sources of the same

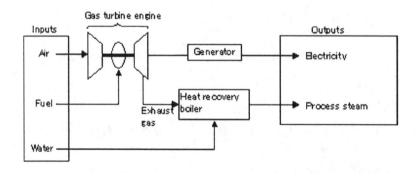

Fig. 4 Schematic diagram of gas turbine co-generation system

power output. So, gas turbine co-generation systems are now widely utilized in factories, hospitals, and intelligent buildings to reduce costs of fuel and electricity. A schematic diagram of gas turbine engine co-generation system is shown in Figure 4.

Maintenance is essential to uphold system availability, however, its cost may oppress customers financially. System suppliers should propose the effective maintenance plan to minimize the financial load on customers. Because the maintenance cost of gas turbine engine dominates mostly the maintenance costs of a whole system, an efficient maintenance policy should be established.

Cumulative damage models have been proposed by many authors [23]. We discuss the maintenance plan of gas turbine engine utilizing cumulative damage models [23]. The engine is overhauled when its cumulative damage exceeds a managerial damage level. The expected cost rate is obtained and an optimal damage level which minimizes it is derived [24].

5.1 Model and Assumptions

Customers have to operate their co-generation system based on their respective operation plans. A gas turbine engine suffers the mechanical damage when it is turned on and operated, and it is assured to hold its required performance in a prespecified number of cumulative turning on and a certain cumulative operating period. So, the engine has to be overhauled before it exceeds the number of cumulative turning on or the cumulative operating period, whichever occurs first. When a co-generation system is continuously operated throughout the year, the occasion to perform overhaul is restricted strictly, such as Christmas vacation period, because the overhaul needs a definite period and customers want to avoid the loss of unoperation.

We consider the following policies:

1) The jth turning on and operation time of the system arises an amount W_j of damage, where random variables W_j have an identical probability distribution $G(x)$ with finite mean, independent of the number of operation, where $\overline{G}(x) \equiv 1 - G(x)$. These damages are assumed to be accumulated to the current damage level, and the cumulative damage $Z_j \equiv \sum_{i=1}^{j} W_i$ up to the jth turning on and operation time has

$$\Pr\{Z_j \leq x\} = G^{(j)}(x) \quad (j = 0, 1, 2, \cdots), \tag{62}$$

where $Z_0 \equiv 0$, $G^{(0)}(x) \equiv 0$ for $x < 0$ and 1 for $x \geq 0$, and in general, $\Phi^{(j)}(x)$ is the j-fold Stieltjes convolution of $\Phi(x)$ with itself.

2) When the cumulative damage exceeds a prespecified level K at which the engine vendor prescribes, the customer of co-generation system performs the engine overhaul immediately, because the assurance of engine performance expires. A cost c_K is needed for the sum of the engine overhaul cost and the intermittent loss of operation.

3) The customer performs the massive system maintenance annually, and checks all major items of the system precisely in several weeks. When the cumulative damage at such maintenance exceeds a managerial level k ($0 \leq k < K$) at which the customer prescribes, the customer performs the engine overhaul. A cost $c(z)$ is needed for the engine overhaul cost at the cumulative damage z ($k \leq z < K$). It is assumed that $c(0) > 0$ and $c(K) < c_K$, because it is not required to consider the loss of operational interruption.

5.2 Expected Cost

The probability that the cumulative damage is less than k at the jth turning on and operation, and between k and K at the $j+1$th is

$$\int_0^k \left[\int_{k-u}^{K-u} dG(x) \right] dG^{(j)}(u). \qquad (63)$$

The probability that the cumulative damage is less than k at the jth turning on and operation, and more than K at the $j+1$th is

$$\int_0^k \overline{G}(K-u) dG^{(j)}(u). \qquad (64)$$

It is evident that $(63) + (64) = G^{(j)}(k) - G^{(j+1)}(k)$.

When the cumulative damage is between k and K, the expected maintenance cost is, from (63),

$$\sum_{j=0}^{\infty} \int_0^k \left[\int_{k-u}^{K-u} c(x+u) dG(x) \right] dG^{(j)}(u) = \int_0^k \left[\int_{k-u}^{K-u} c(x+u) dG(x) \right] dM(u), \qquad (65)$$

where $M(x) \equiv \sum_{j=0}^{\infty} G^{(j)}(x)$. Similarly, when the cumulative damage is more than K, the expected maintenance cost is, from (64),

$$c_K \int_0^k \overline{G}(K-u) dM(u). \qquad (66)$$

Next, we define a random variable X_j as the time interval from the $(j-1)$th to the jth turning on and operation, and its distribution as $\Pr\{X_j \leq t\} \equiv F(t)$ $(j = 1, 2, \cdots)$ with finite mean $1/\lambda$. Then, the probability that the jth turning on and operation occurs until time t is

$$\Pr\left\{\sum_{i=1}^{j} X_i \leq t\right\} = F^{(j)}(t). \quad (67)$$

From (67), the mean time that the cumulative damage exceeds k at the jth turning on and operation, is

$$\sum_{j=1}^{\infty} \int_0^{\infty} t\,[G^{(j-1)}(k) - G^{(j)}(k)]\mathrm{d}F^{(j)}(t) = \frac{M(k)}{\lambda}. \quad (68)$$

Therefore, the expected cost rate $C(k)$ is [1]

$$\frac{C(k)}{\lambda} = \frac{\int_0^k \left[\int_{k-u}^{K-u} c(x+u)\mathrm{d}G(x)\right]\mathrm{d}M(u) + c_K \int_0^k \overline{G}(K-u)\mathrm{d}M(u)}{M(k)}. \quad (69)$$

Especially, the expected costs at $k = 0$ and $k = K$ are, respectively,

$$\frac{C(0)}{\lambda} = \int_0^K c(x)\mathrm{d}G(x) + c_K \overline{G}(K), \quad (70)$$

$$\frac{C(K)}{\lambda} = \frac{c_K}{M(K)}. \quad (71)$$

5.3 Optimal Policy

We find an optimal damage level k^* which minimizes the expected cost rate $C(k)$ in (69). Differentiating $C(k)$ with respect to k and setting it equal to zero,

$$[c_K - c(K)]\int_{K-k}^{K} M(K-x)g(x)\mathrm{d}x$$
$$+ \int_0^k \left[\int_k^K g(x-u)\mathrm{d}c(x)\right] M(u)\mathrm{d}u - c(k) = 0, \quad (72)$$

where $g(x) \equiv \mathrm{d}G(x)/\mathrm{d}x$ which is a density function of $G(x)$. Denoting the left-hand side of (72) as $Q(k)$,

$$Q(0) = -c(0) < 0, \quad Q(K) = [c_K - c(K)]M(K) - c_K. \quad (73)$$

Thus, if $Q(K) > 0$, i.e., $M(K) > c_K/[c_K - c(K)]$, then there exists a finite k^* $(0 < k^* < K)$ which minimizes $C(k)$, and the resulting cost is

$$\frac{C(k^*)}{\lambda} = \int_0^{K-k^*} [c(k^* + x) - c(k^*)]\,\mathrm{d}G(x) + [c_K - c(k^*)]\overline{G}(K - k^*). \quad (74)$$

When $c(z) = c_1 z + c_0$ $(k \leq z < K)$, where $c_1 K + c_0 < c_K$, (69) and (70) are rearranged as, respectively,

$$\frac{C(k)}{\lambda} = \frac{\int_0^k \left[\int_{k-u}^{K-u}(c_1(u+x)+c_0)\,\mathrm{d}G(x)\right]\mathrm{d}M(u) + c_K \int_0^k \overline{G}(K-u)\,\mathrm{d}M(u)}{M(k)} \quad (75)$$

$$\frac{C(0)}{\lambda} = \int_0^K (c_1 x + c_0)\mathrm{d}G(x) + c_K \overline{G}(K), \quad (76)$$

and $C(K)/\lambda$ is equal to (71).

Differentiating $C(k)$ with respect to k and putting it equal to zero,

$$(c_K - c_1 K - c_0)\int_{K-k}^{K} M(K-u)\mathrm{d}G(u) - c_1 \int_{K-k}^{K} M(K-u)\overline{G}(u)\mathrm{d}u = c_0. \quad (77)$$

Letting denote the left-hand side of (77) by $T(k)$,

$$T(0) = 0, \quad T(K) = (c_K - c_1 K - c_0)[M(K) - 1] - c_1 K. \quad (78)$$

Thus, if $T(K) > c_0$, i.e., $M(K) > c_K/(c_K - c_1 K - c_0)$, then there exists a finite k^* $(0 < k^* < K)$ which minimizes $C(k)$.

Next, suppose that $G(x) = 1 - \exp(-\mu x)$, i.e., $M(x) = \mu x + 1$. Then, if $\mu K + 1 > c_K/(c_K - c_1 K - c_0)$, i.e., $\mu > (c_1 + c_0/K)/(c_K - c_1 K - c_0)$, then there exists a finite k^* $(0 < k^* < K)$. Further, differentiating $T(k)$ with respect to k,

$$T'(k) = (\mu k + 1)e^{-\mu(K-k)}(c_K - c_1 K - c_0)\left(\mu - \frac{c_1}{c_K - c_1 K - c_0}\right) > 0, \quad (79)$$

because $(c_1 + c_0/K)/(c_K - c_1 K - c_0) > c_1/(c_K - c_1 K - c_0)$.

Therefore, we have the following optimal policy:

(i) If $\mu K > (c_1 K + c_0)/(c_K - c_1 K - c_0)$, then there exists a finite and unique k^* $(0 < k^* < K)$ which satisfies

$$ke^{-\mu(K-k)} = \frac{c_0}{\mu(c_K - c_1 K - c_0) - c_1}, \quad (80)$$

and the resulting cost rate is

$$\frac{C(k^*)}{\lambda} = \frac{c_1}{\mu}(1 - e^{-\mu(K-k^*)}) + (c_K - c_1 K - c_0)e^{-\mu(K-k^*)}. \quad (81)$$

(ii) If $\mu K \leq (c_1 K + c_0)/(c_K - c_1 K - c_0)$, then $k^* = K$ and $C(K)/\lambda = c_K/(\mu K + 1)$.

5.4 Concluding Remarks

We have discussed the optimal policy in only the case of $c(z) = c_1 z + c_0$ for $k \leq z < K$. We could consider easily several cost structures according to those of actual systems. For example, when the maintenance cost increases discretely with every step of the amount of cumulative damage, i.e. $c(z) = c_j$ for $k_{j-1} \leq z < k_j$ $(j = 1, 2, \cdots, n)$ and c_K for $z \geq K$ where $k_0 \equiv k$, $k_n \equiv K$, $c_{n+1} \equiv c_K$ and $c_j < c_{j+1}$ $(j = 1, 2, \cdots, n)$, the expected cost rate is, from (69),

$$\frac{C(k)}{\lambda} = \frac{c_1 + \sum_{j=1}^{n}(c_{j+1} - c_j) \int_0^k \overline{G}(k_j - u) \, \mathrm{d}M(u)}{M(k)}. \tag{82}$$

In particular, when $n = 1$,

$$\frac{C(k)}{\lambda} = \frac{c_1 + (c_K - c_1) \int_0^k \overline{G}(K - u) \, \mathrm{d}M(u)}{M(k)}. \tag{83}$$

It has been assumed that cumulative damage models can apply to the maintenance of a gas turbine engine. Cumulative damage models are based on Palmgren-Miner rule which is extended to some derivative methods such as Corten-Dolan method, Freudenthal-Heller method, Haibach method and others [25]. When cumulative damage models cannot be approved and these derivative methods are adopted, the co-generation system maintenance model might not be applicable.

Furthermore, we assume that the amount of damage has varied with distribution $G(x)$. In future studies, both amount and occurrence probability of damage have distribution functions. Let $Z(t)$ be the total amount of damage at time t. Then, the distribution of $Z(t)$ is

$$\Pr\{Z(t) \leq x\} = \sum_{j=0}^{\infty} \Pr\{Z_j \leq x\} H_j(t), \tag{84}$$

where $H_j(t)$ denotes the occurrence probability of j damages during $(0, t]$.

6 Aged Fossil-fired Power Plant Maintenance

A number of aged fossil-fired power plants are increasing in Japan. For example, 33% of these plants are currently operated from 150,000 to 199,999 hours (from 17 to 23 years), and 26% of them are above 200,000 hours (23 years) [26]. Although Japanese government eliminates regulations of electric power industry, most industries restrain from the investment for new plants and are prefer to operate current plants efficiently because of the long-term recession in Japan.

The deliberative maintenance plans are indispensable to operate these aged plants without serious trouble such as the emergency stop of operation. The importance of maintenance for aged plants is much higher than that for new ones, because occurrence probabilities of severe troubles increase and new failure phenomena might unexpectedly appear according to the degradation of plants. Furthermore, actual life spans of plant components are mostly different from predicted ones because they are affected by various kinds of factors such as material qualities and operational circumstances [27]. So that, maintenance plans should be established considering occurrence probabilities of miscellaneous components.

The maintenance is classified into the preventive maintenance (PM) and the corrective maintenance (CM). Many authors have studied PM policies for systems because the CM cost at failure is much higher than the PM one and the consideration of effective PM is significant [2], [28–36]. The occurrence of failure is discussed by utilizing the cumulative damage model, where a system suffers damage due to shocks and fails when the total damage exceeds a failure level. Such cumulative damage models generate a cumulative process theoretically [37]. Some aspects of damage models from the reliability viewpoint were discussed by [38]. The PM policies where a system is replaced before failure at time T [39], at shock N [40,41], and at damage K [42,43] were considered. Nakagawa and Kijima [44] applied the periodic replacement with minimal repair at failure to a cumulative damage model and obtained optimal values T^*, N^* and K^* which minimize the expected cost.

The PM plan should be established considering various kinds of severe criticalities (risks) of failure phenomena which are caused by individual components of a plant. These criticalities are quantitated by FMECA (Failure Mode, Effects and Criticality Analysis) [45]. Applying FMECA, the criticality is denoted as the product of the severity of failure and the occurrence probability of failure. So that, even if criticalities are the same, occurrence probabilities of failures may be different. The cumulative damage model assumes that a failure occurs when the total damage exceeds a peculiar level. A plant consists of a wide variety of components such as mechanical ones (piping, power boiler, steam and gas turbine) and electrical ones (wiring, relay, electromagnetic valve, electric pump and fan). Various kinds of severe failures immanent in these components and their occurrence probabilities are different, *i.e.*, they appear at their peculiar damage levels.

In past PM studies and cumulative damage models, a unique failure phenomenon of component is considered, *i.e.*, a peculiar cumulative damage level is regarded. Actually, there exist diversified cumulative damage levels of miscellaneous components and they have to be considered when the PM policy of a plant is established.

In this section, we consider a system with N kinds of multiechelon cumulative damage levels and the failure occurs when the cumulative damage of a plant surpasses these levels. Shocks occur at a renewal process and each shock causes a random amount of damage to the plant. It is assumed that these damages are accumulated to the current damage level. Suppose that the above plant undergoes PM at time T $(0 < T \leq \infty)$ or when the total damage exceeds a managerial level k, whichever occurs first. It is assumed that the time returns to zero after each PM. Then, the expected cost per unit time between maintenances is obtained, and an optimal managerial damage level k^* which minimizes it is analytically discussed [46].

6.1 Model 1

Consider the system which operates for an infinite time span and assume:

1) Successive shocks occur at time intervals X_j $(j = 1, 2, \cdots)$ which have an identical distribution $F(t) \equiv \Pr\{X_j \leq t\}$ with finite mean $1/\lambda$. That is, shocks occur at a renewal process with distribution $F(t)$, and hence, the probability that shocks occur exactly j times during $(0, t]$ is $H_j(t) \equiv F^{(j)}(t) - F^{(j+1)}(t)$, where, in general, $\Phi^{(j)}(t)$ $(j = 1, 2, \cdots)$ is the j-fold convolution of $\Phi(t)$ and $\Phi^{(0)} \equiv 1$ for $t \geq 0$.

2) An amount Y_j of damage due to the jth shock has an identical distribution $G(x) \equiv \Pr\{Y_j \leq x\}$ $(j = 1, 2, \cdots)$ with finite mean, and each damage is additive. Then, the total damage $Z_j \equiv \sum_{i=1}^{j} Y_i$ has an distribution $\Pr\{Z_j \leq x\} = G^{(j)}(x)$, where $Z_0 \equiv 0$.

3) Assume N kinds of failures. When the total damage has exceeded a failure level $K_i(< K_{i+1})$ $(i = 0, 1, 2, \cdots, N)$, the CM is performed, and its cost is $c_i(< c_{i+1})$, where $K_{N+1} = \infty$.

4) The PM is performed when the total damage level exceeds a managerial level $k(\equiv K_0)$ $(0 \leq k \leq K_1)$, and its cost is $c_k(\equiv c_0 < c_1)$.

Let $Z(t)$ denote the total damage at time t. Then, the probability is [37]

$$\Pr\{Z(t) \leq x\} = \sum_{j=0}^{\infty} G^{(j)}(x) H_j(t). \tag{85}$$

The probability that the system undergoes PM when the total damage is between k and K_1 at some shock is

$$P_k = \sum_{j=0}^{\infty} \int_0^k [G(K_1 - x) - G(k - x)] \, \mathrm{d}G^{(j)}(x), \qquad (86)$$

and the probability P_i $(i = 1, 2, \cdots, N)$ that it undergoes CM when the total damage is between K_i and K_{i+1} at some shock is

$$P_i = \sum_{j=0}^{\infty} \int_0^k [G(K_{i+1} - x) - G(K_i - x)] \, \mathrm{d}G^{(j)}(x) \quad (i = 1, 2, \cdots, N). \quad (87)$$

It is noted that $P_k + \sum_{i=1}^{N} P_i = 1$. Thus, the mean time $E\{U\}$ to PM or CM is

$$E\{U\} = \sum_{j=1}^{\infty} [G^{(j-1)}(k) - G^{(j)}(k)] \int_0^{\infty} t \, \mathrm{d}F^{(j)}(t) = \frac{1}{\lambda} \sum_{j=0}^{\infty} G^{(j)}(k). \quad (88)$$

The total expected cost $E\{C\}$ until PM or CM is, from (86) and (87),

$$E\{C\} = c_k P_k + \sum_{i=1}^{N} c_i P_i = c_k + \sum_{i=1}^{N} (c_i - c_k) P_i. \qquad (89)$$

Therefore, by dividing (89) by (88), the expected cost rate is

$$\frac{C_1(k)}{\lambda} = \frac{c_k + \sum_{i=1}^{N}(c_i - c_k) \sum_{j=0}^{\infty} \int_0^k [G(K_{i+1} - x) - G(K_i - x)] \, \mathrm{d}G^{(j)}(x)}{\sum_{j=0}^{\infty} G^{(j)}(k)}$$

$$= \frac{c_k + \sum_{i=1}^{N}(c_i - c_{i-1})\left[\overline{G}(K_i) + \int_0^k \overline{G}(K_i - x) \, \mathrm{d}M(x)\right]}{1 + M(k)}, \qquad (90)$$

where $G^{(j)}(K_{N+1} - x) = 1$ from $K_{N+1} \equiv \infty$, $M(x) \equiv \sum_{j=1}^{\infty} G^{(j)}(x)$ and $\overline{G}^{(j)}(x) \equiv 1 - G^{(j)}(x)$.

We find an optimal k^* which minimizes $C_1(k)$. Differentiating $C_1(k)$ with respect to k and setting it equal to zero,

$$\sum_{i=1}^{N}(c_i - c_{i-1}) \int_{K_i - k}^{K_i} [1 + M(K_i - x)] \, \mathrm{d}G(x) = c_k. \qquad (91)$$

Letting $L_1(k)$ be the left-hand side of (91),

$$L_1(0) = 0, \qquad (92)$$

$$L_1(K_1) = \sum_{i=1}^{N}(c_i - c_{i-1}) \int_{K_i - K_1}^{K_i} [1 + M(K_i - x)] \, \mathrm{d}G(x), \qquad (93)$$

$$L_1'(k) = \sum_{i=1}^{N}(c_i - c_{i-1}) \, g(K_i - k)[1 + M(k)] > 0, \qquad (94)$$

where $g(x) \equiv \mathrm{d}G(x)/\mathrm{d}x$. Thus, $L_1(k)$ is strictly increasing in k from 0 to $L_1(K_1)$. Therefore, we have the following optimal policy:

(i) If $L_1(K_1) > c_k$, then there exists a finite and unique k^* ($0 < k^* < K_1$) which satisfies (91), and the resulting cost rate is

$$\frac{C_1(k^*)}{\lambda} = \sum_{i=1}^{N}(c_i - c_{i-1})\overline{G}(K_i - k^*). \tag{95}$$

(ii) If $L_1(K_1) \le c_k$ then $k^* = K_1$, i.e., a system undergoes only CM, and the expected cost rate is

$$\frac{C_1(K_1)}{\lambda} = \frac{c_1 + \sum_{i=2}^{N}(c_i - c_{i-1})\left[\overline{G}(K_i) + \int_0^{K_1} \overline{G}(K_i - x)\,\mathrm{d}M(x)\right]}{1 + M(K_1)}. \tag{96}$$

Since $L_1(K_1) \le \sum_{i=1}^{N}(c_i - c_{i-1})M(K_i)$, if $\sum_{i=1}^{N}(c_i - c_{i-1})M(K_i) \le c_k$, then $k^* = K_1$.

When $N = 1$, (90) and (91) are rewritten as, respectively,

$$\frac{C_2(k)}{\lambda} = \frac{c_k + (c_1 - c_k)\left[\overline{G}(K_1) + \int_0^{k} \overline{G}(K_1 - x)\,\mathrm{d}M(x)\right]}{1 + M(k)}, \tag{97}$$

$$\int_{K_1-k}^{K_1}[1 + M(K_1 - x)]\,\mathrm{d}G(x) = \frac{c_k}{c_1 - c_k}. \tag{98}$$

Therefore, the optimal policy is simplified as follows [43]:

(i) If $M(K_1) > c_k/(c_1 - c_k)$ then there exists a finite and unique k^* ($0 < k^* < K_1$) which minimizes $C(k)$, and its expected cost is

$$\frac{C_2(k^*)}{\lambda} = (c_1 - c_k)\overline{G}(K_1 - k^*). \tag{99}$$

(ii) If $M(K_1) \le c_k/(c_1 - c_k)$, then $k^* = K_1$ and its expected cost rate is

$$\frac{C_2(K_1)}{\lambda} = \frac{c_1}{1 + M(K_1)}. \tag{100}$$

When $G(x) \equiv 1 - e^{-\mu x}$, $M(x) = \mu x$, and (90) and (91) are rewritten as, respectively,

$$\frac{C_3(k)}{\lambda} = \frac{c_k + \sum_{i=1}^{N}(c_i - c_{i-1})e^{-\mu(K_i - k)}}{1 + \mu k}, \tag{101}$$

$$\mu k \sum_{i=1}^{N}(c_i - c_{i-1})e^{-\mu(K_i - k)} = c_k. \tag{102}$$

Therefore, the optimal policy is rewritten as:

(i) If $\sum_{i=1}^{N} \mu K_i(c_i - c_{i-1}) > c_k$ then there exists a finite and unique k^* $(0 < k^* < K_1)$ which satisfies (102), and its expected cost rate is

$$\frac{C_3(k^*)}{\lambda} = \sum_{i=1}^{N}(c_i - c_{i-1})e^{-\mu(K_i - k^*)}. \tag{103}$$

(ii) If $\sum_{i=1}^{N} \mu K_i(c_i - c_{i-1}) \leq c_k$ then $k^* = K_1$ and its expected cost rate is

$$\frac{C_3(K_1)}{\lambda} = \frac{c_1 + \sum_{i=2}^{N}(c_i - c_{i-1})e^{-\mu(K_i - K_1)}}{1 + \mu K_1}. \tag{104}$$

6.2 Model 2

Assumptions from 1) to 4) are the same as ones in Section 6.1 and the following assumption is attached:

5) The system is operated continuously and the PM is performed when the total operation time exceeds T $(0 < T < \infty)$, and its cost is c_k

The probability P_T that the system undergoes PM when the total damage is below k and total time exceeds T is

$$P_T = \sum_{j=0}^{\infty} G^{(j)}(k) H_j(T). \tag{105}$$

Equations (86) and (87) are rewritten as, respectively,

$$P_k = \sum_{j=0}^{\infty} \int_0^k [G(K_1 - x) - G(k - x)] \, dG^{(j)}(x) F^{(j+1)}(T), \tag{106}$$

$$P_i = \sum_{j=0}^{\infty} \int_0^k [G(K_{i+1} - x) - G(K_i - x)] \, dG^{(j)}(x) F^{(j+1)}(T) \quad (i = 1, 2, \cdots, N), \tag{107}$$

where $P_T + P_k + \sum_{i=1}^{N} P_i = 1$. From (105)–(107), $E\{U\}$ and $E\{C\}$ are, respectively,

$$E\{U\} = \sum_{j=1}^{\infty} [G^{(j-1)}(k) - G^{(j)}(k)] \int_0^T t \, dF^{(j)}(t) + T P_T$$

$$= \sum_{j=0}^{\infty} G^{(j)}(k) \int_0^T H_j(t) \, dt, \tag{108}$$

$$E\{C\} = c_k(P_T + P_k) + \sum_{i=1}^{N} c_i P_i = c_k + \sum_{i=1}^{N}(c_i - c_k)P_i. \tag{109}$$

The expected cost rate for the system in which the PM is performed at time T and at damage level k is, by dividing (109) by (108)

$$C_4(k,T) = \frac{c_k + \sum_{i=1}^{N}(c_i - c_{i-1})\sum_{j=0}^{\infty} F^{(j+1)}(T)\int_0^k \overline{G}(K_i - x)\,dG^{(j)}(x)}{\sum_{j=0}^{\infty} G^{(j)}(k)\int_0^T H_j(t)\,dt},$$
(110)

which agrees with (90) as T is infinity, i.e., Model 1 is a special case of Model 2. Partial differentiating $C_4(k,T)$ with respect to k and setting it equal to zero,

$$\sum_{i=1}^{N}(c_i - c_{i-1})\sum_{j=0}^{\infty} F^{(j+1)}(T)\int_{K_i-k}^{K_i} G^{(j)}(K_i - x)\,dG(x) = c_k. \quad (111)$$

Letting $L_2(k)$ be the left-hand side of (111),

$$L_2(0) = 0, \quad (112)$$

$$L_2(K_1) = \sum_{i=1}^{N}(c_i - c_{i-1})\sum_{j=0}^{\infty} F^{(j+1)}(T)\int_{K_i-K_1}^{K_i} G^{(j)}(K_i - x)\,dG(x), \quad (113)$$

$$L_2'(k) = \sum_{i=1}^{N}(c_i - c_{i-1})\sum_{j=0}^{\infty} F^{(j+1)}(T)G^{(j)}(k)g(K_i - k) > 0. \quad (114)$$

Therefore, we have the following optimal policy:

(i) If $L_2(K_1) > c_k$, then there exists a finite and unique k^* $(0 < k^* < K_1)$ which satisfies (111), and the resulting cost rate is

$$C_4(k^*,T) = \frac{\sum_{i=1}^{N}(c_i - c_{i-1})\sum_{j=0}^{\infty} F^{(j+1)}(T)\overline{G}(K_i - k^*)G^{(j)}(k^*)}{\sum_{j=0}^{\infty} G^{(j)}(k^*)\int_0^T H_j(t)dt}.$$
(115)

(ii) If $L_2(K_1) \leq c_k$, then $k^* = K_1$.

When $G(x) = 1 - e^{-\mu x}$ and $F(t) = 1 - e^{-\lambda t}$, $G^{(j)}(x) = \sum_{i=j}^{\infty}[(\mu x)^i/i!]e^{-\mu x}$, $H_j(t) = [(\lambda t)^j/j!]e^{-\lambda t}$ and $F^{(j)}(t) = \sum_{i=j}^{\infty}[(\lambda t)^i/i!]e^{-\lambda t}$. Equations (110) and (111) are rewritten as, respectively,

$$\frac{C_4(k,T)}{\lambda} = \frac{c_k + \sum_{i=1}^{N}(c_i - c_{i-1})e^{-\mu K_i}\sum_{j=0}^{\infty}\frac{(\mu k)^j}{j!}\sum_{m=j+1}^{\infty}\frac{(\lambda T)^m}{m!}e^{-\lambda T}}{\sum_{j=0}^{\infty}\left[\sum_{i=j+1}^{\infty}\frac{(\mu k)^i}{i!}e^{-\mu k}\right]\left[\sum_{m=j+1}^{\infty}\frac{(\lambda T)^m}{m!}e^{-\lambda T}\right]},$$
(116)

$$\sum_{i=1}^{N}(c_i - c_{i-1})e^{-\mu(K_i-k)}\sum_{j=0}^{\infty}\left[\sum_{i=j+1}^{\infty}\frac{(\mu k)^i}{i!}e^{-\mu k}\right]\left[\sum_{m=j+1}^{\infty}\frac{(\lambda T)^m}{m!}e^{-\lambda T}\right] = c_k.$$
(117)

When T goes to infinity, (116) and (117) are equal to (101) and (102), respectively.

6.3 Concluding Remarks

We assume that the shock occurrence time interval has the identical distribution with a certain finite mean. The occurrence probability of shocks depends on the user operation and it cannot be specified precisely. When the time interval cannot observe the identical distribution with determinated mean, the aged fossil-fired power plant maintenance model might not be applicable.

References

1. Nakagawa, T. (2005). *Maintenance Theory of Reliability*, Springer-Verlag, London.
2. Barlow, R. E. and Proschan, F. (1965). *Mathematical Theory of Reliability*, New York, John Wiley & Sons.
3. Congress, U. S., and Office of Technology Assessment (1992). *After the Cold War: Living With Lower Defense Spending*, OTA-ITE-524.
4. Bauer, J., Cottrell, D. F., Gagnier, T. R. and Kimball, E. W. et al. (1973). *Dormancy and Power ON-OFF Cycling Effects on Electronic Equipment and Part Reliability*, RADAC-TR-73-248 (AD-768 619).
5. Brookner, E. (1985). *Phased-array radars, Scientific American*, Vol.252, pp. 94–102.
6. Robinson, K. (1987). *Digital Controls for Gas Turbine Engines, Presented at the Gas Turbine Conference and Exhibition*, Anaheim, California - May31-June4.
7. Witte, L.C., Schmidt, P.S. and Brown, D.R. (1988). *Industrial Energy Management and Utilization, Hemisphere Publishing Corporation*, New York.
8. Menke, J.T. (1983). *Deterioration of electronics in storage, Proceedings of National SAMPE Symposium*, pp. 966–972.
9. Martinez, E. C. (1984). *Storage Reliability with Periodic Test, Proceedings of Annual Reliability and Maintainability Symposium*, pp. 181–185.
10. Ito, K. and Nakagawa, T. (1992). *Optical Inspection Policies for a System in Storage, Computers Math. Applic.*, Vol.24, No.1/2, pp. 87–90.
11. Ito, K. and Nakagawa, T. (1994). *Optimal Inspection Policies for a System in Storage, The Journal of Reliability Engineering Association of Japan*, Vol.15, No.5, pp. 3–7.
12. Ito, K. and Nakagawa, T. (1995). *An Optimal Inspection Policy for a Storage System with High Reliability, Microelectronics and Reliability*, Vol.35, No.6, June, pp. 875–886

13. Brookner, E. (1991). *Practical Phased-Array Antenna Systems*, Artech House, Boston.
14. Bucci, O.M., Capozzoli, A. and D'elia, G. (2000) *Diagnosis of Array Faults from Far-Field Amplitude-Only Data*, IEEE Transaction on Antennas and Propagation, Vol.48, pp. 647–652.
15. Keithley, H. M. (1966). *Maintainability impact on system design of a phased array radar*, Annual New York Conference on Electronic Reliability, 7th, Vol.9, pp. 1–10.
16. Hevesh, A.H. (1967). *Maintainability of phased array radar systems*, IEEE Transactions on Reliability, Vol.R-16, pp. 61–66.
17. Hesse, J. L. (1975). *Maintainability analysis and prototype operations*, Proceedings 1975 Annual Reliability and Maintainability Symposium, pp. 194–199.
18. Ito, K. and Nakagawa, T. (2004). *Comparison of cyclic and delayed maintenances for a phased array radar*, Journal of the Operations Research Society of Japan, Vol.47, No.1, pp. 51–61.
19. Nakagawa, T. and Ito, K. (2007). *Optimal Availability Models of a Phased Array Radar, Recent Advances in Stochastic Operations Research*, World Scientific Publishing Co., pp. 115–130.
20. Nakagawa, T. (1986). *Modified discrete preventive maintenance policies*, Naval Research Logistics Quarterly, Vol.33, pp. 703–715.
21. Cahill, M.J. and Underwood, F.N. (1987). *Development of Digital Engine Control System for the Harrier* II, Presented at AIAA/SAE/ASME/ASEE 23rd Joint Propulsion Conference, San Diego, California - June 29 - July 2.
22. Ito, K. and Nakagawa, T. (2003). *Optimal Self-Diagnosis Policy for FADEC of Gas Turbine Engines*, Mathematical and Computer Modeling, Vol.38, No.11-13, December, pp. 1243–1248.
23. Nakagawa, T. (2007). *Shock and Damage Models in Reliability Theory*, Springer-Verlag, London.
24. Ito, K. and Nakagawa, T. (2006). *Maintenance of a Cumulative Damage Model and Its Application to Gas Turbine Engine of Co-generation System, Reliability Modeling, Analysis and Optimization*, Series on Quality, Reliability and Engineering Statistics, Vol.9 (Editor H. Pham), chapter 21, World Scientific Publishing Co. Pte. Ltd., pp. 429–438.
25. Li, Z.X., Ko, J.M. and Chen, H.T. (2001). *Modelling of Load Interaction and Overload Effect on Fatigue Damage of Steel Bridges*, Fatigue & Fracture of Engineering Materials and Structures, Vol.24, Issue 6, June, pp. 379–390.
26. Hisano, K. (2000). *Preventive Maintenance and Residual Life Evaluation Technique for Power Plant* (I.Preventive Maintenance) (in Japanese), The Thermal and Nuclear Power, Vol.51, pp. 491–517.
27. Hisano, K. (2001). *Preventive Maintenance and Residual Life Evaluation Technique for Power Plant* (V.Review of Future Advances in Preventive Maintenance Technology)" (in Japanese), The Thermal and Nuclear Power, Vol.52, pp. 363–370.
28. Nakagawa, T. (2005). *Maintenance Theory of Reliability*, Springer-Verlag, London.

29. Nakagawa, T. (1977). *Optimal preventive maintenance policies for repairable system*, IEEE Transactions on Reliability, **R-26**, pp.168-173.
30. Muth, E.J. (1977). *An optimal decision rule for repair vs replacement*, IEEE Transactions on Reliability, **R-26**, pp.179-181.
31. Nakagawa, T. (1979). *Optimum replacement policies for a used unit*, Journal of the Operations Research Society of Japan, **22**, pp. 338–347.
32. Murthy, D. N. P. and Nguyen, D.G. (1981). *Optimal age-policy with imperfect preventive maintenance*, IEEE Transactions on Reliability, **R-30**, pp. 80–81.
33. Nakagawa, T. (1979). *Optimal policies when preventive maintenance is imperfect*, IEEE Transactions on Reliability, **R-28**, pp. 331–332.
34. Lie, C.H. and Chun, Y.H. (1986). *An algorithm for preventive maintenance policy*, IEEE Transactions on Reliability, **R-35**, pp. 71-75.
35. Nakagawa, T. (1980). *A summary of imperfect preventive maintenance policies with minimal repair*, RAIRO Operations Research, **14**, pp. 249–255.
36. Kijima, M. and Nakagawa, T. (1992). *Replacement policies for a shock model with imperfect preventive maintenance*, European Journal of Operational Research, **57**, pp. 100–110.
37. Cox, D. R. (1962). *Renewal Theory*. Methuen, London.
38. Esary, J. D., Marshall, A. W. and Proschan, F. (1973). *Shock models and wear processes*, Annals of Probability, **1**, pp. 627–649.
39. Taylor, H. M. (1975). *Optimal replacement under additive damage and other failure models*, Naval Res. Logist. Quart, **22**, pp. 1–18.
40. Nakagawa, T. (1984). *A summary of discrete replacement policies*, European J. of Operational Research, **17**, pp. 382–392.
41. Qian, C.H., Nakamura, S. and Nakagawa, T. (2003). *Replacement and minimal repair policies for a cumulative damage model with maintenance*, Computers and Mathematics with Applications, **46**, pp. 1111–1118.
42. Feldman, R. M. (1976). *Optimal replacement with semi-Markov shock models*, Journal of Applied Probability, **13**, pp. 108–117.
43. Nakagawa, T. (1976). *On a replacement problem of a cumulative damage model*, Operational Research Quarterly, **27**, pp. 895–900.
44. Nakagawa, T. and Kijima, M. (1989). *Replacement policies for a cumulative damage model with minimal repair at failure*, IEEE Transactions on Reliability, **38**, pp. 581–584.
45. MIL-STD-1629A (1984). *Procedure for Performing a Failure Mode, Effects, and Criticality Analysis*.
46. Nakagawa, T. and Ito, K. (2008). *Optimal Maintenance Policies for a System with Multiechelon Risks* IEEE Transactions on Systems, Man and Cybernetics – Part A: Systems and Humans, Vol.38, No.2, pp. 461–469.

Chapter 10

Management Policies for Stochastic Models with Monetary Facilities

SYOUJI NAKAMURA

Department of Human Life and Information,
Kinjo Gakuin University,
1723 Omori 2-chome, Moriyama-ku, Nagoya 463-8521, Japan
E-mail: snakam@kinjo-u.ac.jp

1 Introduction

In a modern society, the enterprise is faced with its globalization and the fast economical change and has to always provide better risk management. We measure and evaluate the risk in advance for doing the risk management successfully. Such things would be very useful for doing the good and quick decision making of risk management. Especially, it is important to consider the risk in the monetary facility scientifically and to analyze it mathematically.

This chapter considers three problems to need an important decision making in the monetary facility:

1) Maximizing an expected liquidation profit of holdings.
2) Prepayment risk for a monetary facility.
3) Determination of loan interest rate.

In 1), we consider the problem of maximizing an expected liquidation profit of holdings when the market impact of stock price is caused by the holdings sell-off [1,2]. The cumulative damage model in reliability theory [7] is applied to the fluctuations of stock price [30]. We discuss analytically an optimal sell-off interval of holdings to maximize the expected liquidation profit of holdings.

In 2), we form a stochastic model of prepayment risk for a monetary facility, using cumulative damage models [7]: Prepayments occur at random times. A monetary facility takes some counter-measure as its management policy [8,9], considering risks in which the total amount of prepayment capital exceeds a threshold value. We consider two models where the total prepayment capital is estimated at periodic times and linearly with time. A monetary facility takes some counter-measure to avoid risks when the total prepayment capital exceeds a threshold level or at time NT, whichever occurs first [31]. The expected costs of two models are obtained and optimal intervals which minimize them are derived.

In 3), the risk management of financed enterprises has become very important to banks after the collapse of the bubble economy in Japan. To obtain suitable revenues, banks have to apply high interest rates to financed enterprises with high risk. When enterprises, which have made a secure loan from banks, go bankrupt, banks foreclose such enterprises from its mortgage to recover the loss incurred. However, in actual circumstances, it may need much cost and effort to do so. In this section, we proposes a stochastic model to determine an adequate interest rate, taking account of the probabilities of bankruptcy and mortgage collection [24], and their costs. Numerical examples are given to illustrate this model and a loan interest rate is determined [32].

2 Liquidation Profit Policy

When we have to sell off security holdings in a short term on the market, we need to consider a liquidation policy which maximizes the total amount of security holdings in consideration of their influence for the market price.

We formulate the following two stochastic models of liquidation policies for the security holdings: The security holdings S are sold off by one time or by dividing them into n blocks, S/n is sold off. Then, the market price decreases along with the amount of disposition according to an impact function. In addition, the market price also decreases from the supply-demand relation in the market, as the accumulation of selling orders increases gradually. But, the influence degree of the price becomes lowered if the security holdings are broken down into small blocks, however, the dealing cost gradually increases. Conversely, if the security holding are sold off by dividing roughly, the market price greatly falls, however, the dealing cost decreases. That is, the disposition lot and market price of the security holdings have a trade-off relation. Another assumption is that the stock prices rise by selling off the stocks. In addition, the market impact to which stock prices drop sharply is assumed when it reaches the threshold price.

In general, it is not easy to formulate a stochastic model of market impact because it depends on various factors. The consensus of liquidation policies for security holdings has not been obtained yet, although various approaches in academic and business fields have been made. In this section, we consider the market impact when the security holdings are sold off on the market, and derive analytically an optimal liquidation policy which maximizes its total amount.

The following notations are used:

S_0: Nominal value of security holdings at time 0.
$E(n)$: Real value of security holdings at time n.
c_0: Constant dealing cost per transaction.
λ: Parameter of market impact function.
μ: Parameter of price restoration function.

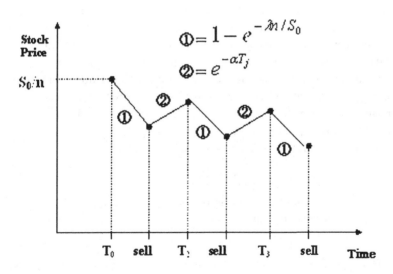

Fig. 1 Market impact for Model 1

Table 1 Relation between amount of contract and transaction fee

Amount of contract(yen)	Transaction fee(yen)
1,000,000	12,495
2,000,000	21,420
5,000,000	47,145
10,000,000	82,845
20,000,000	140,595
30,000,000	198,345

2.1 Model 1

It is assumed that the security holdings S_0 at time 0 have to be sold off in a certain limit time. As liquidation methods, the security holdings S are sold off by one time or S/n sold off by dividing into n every time. Then, the market price decreases along with the amount of disposition according to an impact function (Figure 1). After a certain time has passed, the security price recovers to the price before the previous time. In addition, the market price also decreases from the supply-demand relation in the market, as the accumulation of selling orders increases gradually. But, the influence degree of the price becomes lower if the security holdings are broken down into small blocks, however, the dealing cost gradually increases. Conversely, if the security holding are sold off by dividing roughly, the market price greatly falls, however, the dealing cost decreases (Table. 1).

When the security holdings S_0/n are sold off on the market, the amount of liquidation decreases exponentially and is given by $(S_0/n)(1 - e^{-\lambda n/S_0})$. Further, letting c_0 be a dealing cost of transaction, we evaluate the present values of all costs by using a discount rate α $(0 < \alpha < \infty)$. In addition, we consider the restoration function for an increasing market price rate and the decreasing market price rate by the market impact.

The amount of liquidation of transaction at time 0 is

$$\frac{S_0}{n}(1 - e^{-\lambda n/S_0}) - c_0. \qquad (1)$$

This liquidation is restored exponentially and is given by $Q_1 \equiv (S_0/n)(1 - e^{-\lambda n/S_0})e^{\mu T}$. Then, the amount of liquidation of transaction at time T is

$$[Q_1(1 - e^{-\lambda/Q_1}) - c_0]e^{-\alpha T}. \qquad (2)$$

Further, letting denote $Q_2 \equiv (S_0/n)(1 - e^{-\lambda n/S_0})e^{\mu T}$, the amount of liquidation at time $2T$ is

$$\left[Q_2(1 - e^{-\lambda/Q_2}) - c_0\right]e^{-\alpha 2T}. \qquad (3)$$

Table 2 Total amount of liquidation of security holdings $S_0 = 10^8$, $\lambda = 1.5 \times 10^8$ and $\alpha = 0.01$

μ	n^*	$E_1(n^*)$
0.0002	33	92,943,840
0.0004	34	92,647,601
0.0006	35	93,289,357
0.0008	36	93,949,124
0.0010	38	94,638,556
0.0012	39	95,367,121

Repeating the above procedures, we have generally

$$Q_{j+1} = Q_j(1 - e^{-\lambda/Q_j})e^{\mu T} \quad (j = 0, \ldots, n-2), \qquad (4)$$

where $Q_0 \equiv S_0/n$. Hence, the amount of liquidation of n transaction times is

$$E_1(n) = \sum_{j=0}^{n-1} \left[Q_j(1 - e^{-\lambda/Q_j}) - c_0 \right] e^{-\alpha j T}. \qquad (5)$$

Thus, substituting (4) into (5), the total expected amount of liquidation per unit of security holdings is

$$Z_1(n) = \frac{E_1(n)}{S_0} \quad (n = 1, 2, \ldots). \qquad (6)$$

Example 2.1. In Table 2, we compute the optimal number n^* numerically when $S_0 = 10^8$ yen, $\lambda = 1.5 \times 10^8$, $\alpha = 0.01$ and a transaction cost is $c_0 = S_0 \times 0.002625 + 119595.0 yen$. For example, when $\mu = 0.006$, $n^* = 35$ blocks and $E_2(n^*) = 93,289.357$ yen. □

2.2 Model 2

The stock is sold off for T period, and its prices is assumed to rise gradually. However, as the market impact, the price drops sharply when it reaches the threshold price. The cumulative damage model [7] is applied to this model by replacing the change of stock prices with damage.

When the clearance of the stock is continuously exercised, the stock price $Z(0) = z_0$ at time 0 is assumed to rise to $Z(t) = at + z_0 (a > 0)$. When the stock price reaches a threshold K, it drops sharply and becomes 0 (Figure 2). In this case, a threshold K is a random variable with a distribution function $K(x)$.

The probability that the stock price drops sharply at time jT is denoted by $\Pr\{jaT + z_0 \geq K\} = K(jaT + z_0)$, and the probability that the stock price drops sharply at time nT is

$$\sum_{j=1}^{n} K(jaT + z_0) \prod_{i=1}^{j-1} \overline{K}(iaT + z_0), \tag{7}$$

where $\overline{K} \equiv 1 - K$. Conversely, the probability that the stock prices does not drop sharply until time nT is

$$\prod_{j=1}^{n} \overline{K}(jaT + z_0). \tag{8}$$

It is clearly shown that $(7) + (8) = 1$. Similarly, the mean time to the clearance of the stock is

$$\sum_{j=1}^{n}(jT)K(jaT + z_0) \prod_{i=0}^{j-1} \overline{K}(iaT + z_0) + (nT) \prod_{j=1}^{n} \overline{K}(jaT + z_0)$$

$$= T \sum_{j=0}^{n-1} \left[\prod_{i=0}^{j} \overline{K}(iaT + z_0) \right]. \tag{9}$$

Then, the mean time until the stock price drop is

$$E\{Y\} = T \sum_{j=0}^{\infty} \left[\prod_{i=0}^{j} \overline{K}(iaT + z_0) \right]. \tag{10}$$

The amount of stock is divided equally in n and is sold off by time $nT(n = 1, 2, \ldots)$. That is, the stock S_0/n at each $jT(j = 1, 2, \ldots)$ are sold off and stock price at jT is $jaT + z_0$. In addition, when the stock price drops sharply, the amount of the clearance of the stocks is assumed to be 0. The expected liquidation of S_0 at time nT is

$$\widetilde{C}(n) = \frac{S_0}{n} \left\{ \sum_{j=1}^{n} K(jaT + z_0) \left[\prod_{i=1}^{j-1} \overline{K}(iaT + z_0) \sum_{i=1}^{j-1}(iaT + z_0) \right] \right.$$

$$\left. + \prod_{j=1}^{n} \overline{K}(jaT + z_0) \sum_{i=1}^{n}(iaT + z_0) \right\}$$

$$= \frac{S_0}{n} \sum_{j=1}^{n}(jaT + z_0) \prod_{i=1}^{j} \overline{K}(iaT + z_0) \quad (n = 1, 2, \ldots). \tag{11}$$

Thus, the expected liquidation per unit of time for an infinite interval is, from (9) and (11),

$$C(n) = \frac{S_0}{n} \frac{\sum_{j=1}^{n}(jaT+z_0)\prod_{i=1}^{j}\overline{K}(iaT+z_0)}{T\sum_{j=0}^{n-1}\left[\prod_{i=0}^{j}\overline{K}(iaT+z_0)\right]} \qquad (n=1,2,\dots). \qquad (12)$$

We seek an optimal n^* that maximizes $C(n)$. From $C(n)-C(n+1) \geq 0$,

$$\frac{1}{n}\sum_{j=1}^{n}(jaT+z_0)K_j \geq \frac{1}{n+1}\sum_{j=1}^{n+1}(jaT+z_0)K_j,$$

i.e.,

$$\sum_{j=1}^{n}\{(jaT+z_0)K_j - [(n+1)aT+z_0]K_{n+1}\} \geq 0, \qquad (13)$$

where $K_j \equiv \prod_{i=0}^{j}\overline{K}_j(iaT+z_0)$ $(j=1,2,\dots)$. Letting $L(n)$ denote the left-side of (13),

$$L(n+1) - L(n) = (n+1)\{[(n+1)aT+z_0]K_{n+1} - [(n+2)aT+z_0]K_{n+2}\}$$
$$= (n+1)K_{n+1}\{[(n+1)aT+z_0]-[(n+2)aT+z_0]\overline{K}[(n+2)aT+z_0]\}$$
$$= (n+1)K_{n+1}\{K[(n+2)aT+z_0]-aT\}. \qquad (14)$$

It can be easily seen that $K[(n+2)aT+z_0]$ is strictly increasing to 1 in n. Thus, $L(n)$ is also strictly increasing from $L(1)$ or it increases after decreasing once for $aT < 1$. Therefore, there exists a finite and unique n^* which satisfies (13).

Example 2.2. In particular, when $K(x) = 1 - e^{-\theta x}$, (13) is

$$\sum_{j=1}^{n}\left((jaT+z_0)\exp\left\{-\theta\left[\frac{j(j+1)}{2}aT+z_0\right]\right\}\right.$$
$$\left.-[(n+1)aT+z_0]\exp\left\{-\theta\left[\frac{(n+1)(n+2)}{2}aT+(n+1)z_0\right]\right\}\right) \geq 0. \qquad (15)$$

For example, when $S = 1$, $a = 5$, $z_0 = 450$, $T = 1$ and $\theta = 4.0 \times 10^{-7}$ in Table 3, the optimal sell-off interval is $n^* = 42$, and the maximum profit per unit time for an infinite is $C(n^*) = 13.27$. Further, when θ is large, the optimal sell-off interval is small, and the expected total amount of liquidation $C(n^*)$ is low. This reason is that if the probability of stock price drops sharply is large, the risk grows, and hence, the optimal sell-off interval is small. □

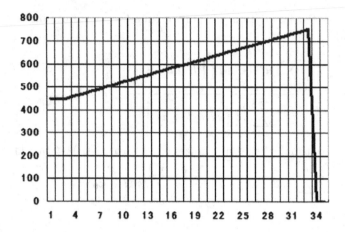

Fig. 2 Market impact for Model 2

Table 3 Total amount of liquidation of security holdings $S_0 = 1, a = 5.0$ and $T = 1$

θ	n^*	$C(n^*)$
4×10^{-7}	42	13.27
8×10^{-7}	27	19.26
12×10^{-7}	18	27.64
16×10^{-7}	14	34.82
20×10^{-7}	11	43.64
24×10^{-7}	10	47.75

3 Prepayment Risk

We consider the repayment before the deadline time that relates to the source of another income of the monetary facility. Debtors can repay a part or all of their housing loan debt at any time before the deadline time. In such cases, the profit of a monetary facility might decrease. On the other hand, a monetary facility should invest the repayment capital of loan at low interest, and so that, cannot obtain earnings at first. Moreover, if the interest rate of the housing loan would be high, debtors who have borrowed the loan at low interest would be not executed the prepayment. Thus, there is a possibility of changing greatly in the cash flow in the near future because debtors can freely select the repayment. A monetary facility should determine an appropriate amount of prepayment capital by the estimating cash flow of the housing loan. It is profitable to estimate the amount of prepayment capital when a monetary facility manages the housing loan risk. It is assumed that the prepayment risk follows probability distributions with some parameters.

We apply the cumulative damage model [7] to the prepayment risk of the housing loan. The cumulative damage model, in which shocks occur in random times and damage incurrs, such as fatigue, wear, crack growth, creep, and dielectric breakdown is additive is considered. It is assumed that the unit fails when the total damage has exceeded a prespecified value $K(0 < K < \infty)$ for the first time. Usually, a failure value K is statistically estimated and already known.

In this section, we apply the cumulative damage model to the housing loan prepayment risk in the monetary facility, by putting *shock* by *prepayment*, *failure* by *a large amount of prepayment* and *damage* by *uneconomic housin loan*. We consider two prepayment models where the total amount of prepayment capital is estimated at periodic intervals nT or linearly with time. A monetary facility takes some counter-measure such as a special low interest rate of a certain period as an appropriate management policy when the total prepayment exceeds a threshold level K or at time NT, whichever occurs first. Introducing two costs in cases where the total prepayment exceeds K or not, we obtain the expected cost rates $C_i(N)$. Further, we discuss analytically optimal N^* which minimize the expected cost rates $C_i(N)$. Finally, two examples are presented when a prepayment capital is exponential during each interval and a linear parameter of the total prepayment is distributed normally.

In a monetary facility, it is a concept of Asset-Liability Management to analyze the expiration of the property and the debt, and to match assets and liabilities according to the period [8,9]. However, the housing loan can be repaid before the repayment completion date. Prepayment is generated by various factors (refinance, seasonality, burnout, economic growth, business cycle). Then, a monetary facility should correspond to the risk of the mismatch of assets and liabilities by the change and take some counter-measure for such risks [8].

The following notations are used:

$Z(t)$: Total amount of prepayment capital at time t.
K: Threshold value of the total prepayment capital.
c_K, c_N: Loss cost when the total prepayment capital exceeds (does not exceed) K, respectively, where $c_K > c_N$.

3.1 Model 1: Interval Estimation

We can estimate for the amount of prepayment capital at each time interval $[(n-1)T, nT]$: The prepayment of a monetary facility occurs in the amount W_n during $[(n-1)T, nT]$. The total prepayment capital is additive. It is assumed that each W_n has an identical probability distribution $G(x) \equiv \Pr\{W_n \leq x\}$ $(n = 1, 2, \ldots)$ with mean μ_T. Then, the total prepayment capital $Z_n \equiv \sum_{i=1}^{n} W_i$ up to the nth interval nT where $Z_0 \equiv 0$ has a distribution

$$\Pr\{Z_n \leq x\} \equiv G^{(n)}(x) \quad (n = 0, 1, 2, \ldots), \tag{16}$$

and $G^{(0)}(x) \equiv 1$ for $x \geq 0$, where $G^{(n)}(x)$ is the n-fold convolution of $G(x)$ with itself. Then, the probability that the total prepayment capital exceeds exactly a threshold value K at the nth prepayment is $G^{(n-1)}(K) - G^{(n)}(K)$.

When the total prepayment capital exceeds a threshold value K at time nT $(n = 1, 2, \ldots)$, this probability is

$$1 - G^{(n)}(K), \tag{17}$$

and its mean time that the total prepayment exceeds K is

$$E\{Y\} = T[1 + M_G(K)], \tag{18}$$

where $M_G(K) \equiv \sum_{i=1}^{\infty} G^{(i)}(K)$. The expected total prepayment at time nT is

$$E\{Z(nT)\} = n\mu_T. \tag{19}$$

Suppose that a momentary facility takes some counter-measure when the total prepayment capital excess a threshold value K or at time NT ($N = 1, 2, \ldots$), whichever occurs first. Then, the expected cost rate is [7, p. 84]

$$C_1(N) = \frac{c_K - (c_K - c_N)G^{(N)}(K)}{T \sum_{n=0}^{N-1} G^{(n)}(K)} \quad (N = 1, 2, \ldots). \tag{20}$$

Clearly,

$$C_1(\infty) = \frac{c_K}{T[1 + M_G(K)]}.$$

We obtain an optimal time N_1^* which minimizes the expected cost rate $C_1(N)$. From $C_1(N+1) - C_1(N) \geq 0$,

$$Q_1(N+1) \sum_{n=0}^{N-1} G^{(n)}(K) - [1 - G^{(N)}(K)] \geq \frac{c_N}{c_K - c_N} \quad (N = 1, 2, \ldots), \tag{21}$$

where

$$Q_1(N) \equiv \frac{G^{(N-1)}(K) - G^{(N)}(K)}{G^{(N-1)}(K)}.$$

3.2 Model 2: Linear Estimation

Suppose that the prepayment capital occurs continuously and its total prepayment capital $Z(t)$ increases linearly with time t, i.e., $Z(t) = At$, where A is a random variable with distribution $L(x) \equiv \Pr\{A \leq x\}$. The probability that the total prepayment capital exceeds K at time nT is

$$\Pr\{Z(nT) > K\} = \Pr\{nAT > K\} = \overline{L}\left(\frac{K}{nT}\right), \tag{22}$$

where $\overline{L}(x) \equiv 1 - L(x)$ and $L(K/0) \equiv 1$. The mean time to some countermeasure is

$$NTL\left(\frac{K}{NT}\right) + \sum_{n=1}^{N} nT\left[L\left(\frac{K}{(n-1)T}\right) - L\left(\frac{K}{nT}\right)\right]$$

$$= T\sum_{n=0}^{N-1} L\left(\frac{K}{nT}\right). \tag{23}$$

The expected cost rate is

$$C_2(N) = \frac{c_K - (c_K - c_N)L(K/(NT))}{T\sum_{n=0}^{N-1} L(K/(nT))} \quad (N = 1, 2, \ldots). \tag{24}$$

From $C_2(N+1) - C_2(N) \geq 0$,

$$Q_2(N+1)\sum_{n=0}^{N-1} L\left(\frac{K}{nT}\right) - \overline{L}\left(\frac{K}{NT}\right) \geq \frac{c_N}{c_K - c_N} \quad (N = 1, 2, \ldots), \tag{25}$$

where $Q_2(N) \equiv \dfrac{L(K/((N-1)T)) - L(K/(NT))}{L(K/((N-1)T))}$.

3.3 Optimal Policies

In case of Model 1, when $G(x) = 1 - e^{-x/\mu_T}$,

$$G^{(N)}(K) = \sum_{n=N}^{\infty} \frac{(K/\mu_T)^n}{n!} e^{-K/\mu_T}. \tag{26}$$

An optimal N_1^* that minimizes $C_1(N)$ is, from (21),

$$\frac{[(K/\mu_T)^N/N!]e^{-K/\mu_T}\sum_{n=0}^{N-1} G^{(n)}(K)}{G^{(N)}(K)} - [1 - G^{(N)}(K)] \geq \frac{c_N}{c_K - c_N}$$

$$(N = 1, 2, \ldots). \tag{27}$$

The left-hand side of (27) is strictly increasing function to K/μ_T. Hence, if $K/\mu_T > c_N/(c_K - c_N)$, then there exists a unique minimum N_1^* ($1 \leq N_1^* < \infty$) which satisfies (27).

In case of Model 2, when A is distributed normally, i.e., A has a normal distribution $N(\mu_A, \sigma^2/t)$,

$$E\{Z(nT)\} = nT\mu_A, \qquad V\{Z(nT)\} = V\{AnT\} = nT\sigma^2.$$

In Model 1, from $E\{Z(nT)\} = n\mu_T$ and $V\{Z(nT)\} = n\mu_T^2$, we set that two models have the same mean and variance;

$$nT\mu_A = n\mu_T, \qquad \mu_A = \frac{\mu_T}{T},$$

$$nT\sigma^2 = n\mu_T^2, \qquad \sigma^2 = \frac{\mu_T^2}{T}.$$

Thus, when A has a normal distribution $N(\mu_T/T, \mu_T^2/T)$,

$$L(K/nT) = \Pr\left\{A \leq \frac{K}{nT}\right\} = \Phi\left(\frac{K/n - \mu_T}{\mu_T\sqrt{T}}\right), \tag{28}$$

where $\Phi(x)$ is a standard normal distribution with mean 0 and variance 1, i.e., $\Phi(x) \equiv (1/\sqrt{2\pi})\int_{-\infty}^{\infty} e^{-u^2/2} du$. An optimal N_2^* that minimizes $C_2(N)$ is, from (25),

$$\frac{\Phi\left(\frac{K/N-\mu_T}{\mu_T\sqrt{T}}\right) - \Phi\left(\frac{K/(N+1)-\mu_T}{\mu_T\sqrt{T}}\right)}{\Phi\left(\frac{K/N-\mu_T}{\mu_T\sqrt{T}}\right)} \sum_{n=0}^{N-1} \Phi\left(\frac{K/n-\mu_T}{\mu_T\sqrt{T}}\right) - \left[1 - \Phi\left(\frac{K/N-\mu_T}{\mu_T\sqrt{T}}\right)\right]$$

$$\geq \frac{c_N}{c_K - c_N} \qquad (N = 1, 2, \ldots). \tag{29}$$

3.4 Numerical Example

We assume $T = 1, \mu_T = 1$, i.e., $\mu_A = 1, \sigma^2 = 1$. Then, from (27), we obtain N_1^* as a unique minimum which satisfies

$$\frac{K^N e^{-K}/N!}{1 - \sum_{n=0}^{N-1}[K^n e^{-K}/n!]} \sum_{n=0}^{N-1}\left[1 - \sum_{j=0}^{n-1}\frac{K^j}{j!}e^{-K}\right] - \sum_{n=0}^{N-1}\frac{K^n}{n!}e^{-K} \geq \frac{c_N}{c_K - c_N}$$

$$(N = 1, 2, \ldots). \tag{30}$$

Similarly, from (29), we obtain N_2^* as a unique minimum which satisfies

$$\frac{\Phi\left(\frac{K}{N}-1\right) - \Phi\left(\frac{K}{N+1}-1\right)}{\Phi\left(\frac{K}{N}-1\right)} \sum_{n=0}^{N-1}\Phi\left(\frac{K}{n}-1\right) - \left[1 - \Phi\left(\frac{K}{N}-1\right)\right] \geq \frac{c_N}{c_K - c_N}$$

$$(N = 1, 2, \ldots). \tag{31}$$

Table 4 Optimal intervals N_1^* and N_2^* when $T = 1$ and $\mu_T = 1$

c_K/c_N	K							
	5		10		30		50	
	N_1^*	N_2^*	N_1^*	N_2^*	N_1^*	N_2^*	N_1^*	N_2^*
2	6	3	10	5	25	11	42	29
5	4	3	7	5	21	11	37	19
10	3	3	6	4	20	10	35	16
20	3	2	5	4	18	9	34	15

Table 5 Optimal N_2^* when $\sigma^2 = b^2$ and $c_K/c_N = 10$

b	K			
	5	10	30	50
0.5	7	10	27	49
0.6	6	8	21	47
0.7	5	6	16	31
0.8	4	5	15	25
0.9	4	4	12	20
1.0	3	4	10	16

For example, when $K = 30$ and $c_K/c_N = 10$ in Table 4, the optimal intervals are $N_1^* = 20$ and $N_2^* = 10$. It would be natural to assume that we set up the campaign of obstructing the prepayment housing loan at one unit of month. Table 4 indicates that we should campaign to obstruct the prepayment at every 20 months in Model 1 and every 10 months in Model 2. Note that $(N_1^*\mu_T/K) \times 100 = 66.7\%$ in Model 1 and $(N_2^*\mu_T)/K = 33.3\%$.

Further, optimal intervals increase with K and decrease with c_k/c_N, and also, $N_i^*\mu_T/K$ decrease with K for a specified c_K/c_N. This shows that if the threshold value K is large, the necessity of the campaign decreases with K, and so that, N_i^*/K also decrease.

In Table 4, N_2^* are less than N_1^* for large c_K/c_N and K. Next, assume that $\sigma^2 = b^2\mu_T^2/T = b^2$ for $0 < b \le 1$, $T = 1$ and $\mu_T = 1$. Table 5 presents optimal N_2^* for b and K when $c_K/c_N = 10$. When $b = 1$, N_2^* are equal to optimal values in Table 4. Compared to Tables 4 and 5, N_2^* are almost the same as N_1^* when $b = 1.0, 0.7, 0.62, 0.61$ for $K = 5, 10, 30, 50$, respectively, which decrease with K. This shows that the variance in Model 2 should be estimated smaller as K becomes larger. In general, Model 2 is simpler than Model 1. We should determine which model would be better by collecting actual data from the prepayment housing loan.

4   Loan Interest Rate

It is important to consider the risk management of financed enterprises, following a well-known law of *high risk and high return* of the market mechanism: Banks have to loan at a high interest rate for financed enterprises with high risk. When financed enterprises go bankrupt, banks have to collect as much of their loans as possible and gain the earnings which correspond to the risk. However, the mortgage collection cost might be sometimes higher than the administrative cost for the commencement of the loans. Moreover, the mortgage might be collected at once or at many times over a period of time.

This section forms the following stochastic model of mortgage collection: The total amount of loans and mortgages can be collected at one time into a batch. A financed enterprise goes bankrupt according to a bankruptcy probability, and its mortgage is collected according to a mortgage collection probability. It is assumed that such two probabilities are estimated and already known.

There have been many papers which treat the determination of a loan interest rate considering default-risk and asset portfolios [13-17,19-31]. However, there have been few papers which study the period of mortgage collection. For example, Mitchner and Peterson [24] determined the optimal pursuit duration of the collection of defaulted loans which maximizes the expected net profit. In this section, we are concerned with both mortgage collection time and loan interest rate when financed enterprises go bankrupt.

Banks have to decide on a margin interest rate to gain earnings, taking into considerations the loss due to bankruptcy and the administrative cost of mortgage collection. In case of no bankruptcy, we could easily decide on a fair rate by charging a margin interest rate for the amount of loans. However, this decision would become difficult and complex by taking into account risk and uncertain factors such as the probabilities of mortgage collection and bankruptcy. In such situations, we formulate a stochastic model, and discuss theoretically and numerically how to decide an adequate interest rate.

Banks finance loans at a constant period for a financed enterprise which may go bankrupt with a certain probability. It is assumed that when a financed enterprise goes bankrupt, banks can make its batch collection of loans according to a probability distribution which is already known from the past data. Then, we need to provide a fair loan interest rate at which banks gain earnings.

We list the following notations,

M_1: All amount of the deposit.

M_2: All amount of loans which can be procured of the deposit, where the deposit is caught as the original capital of financing.

$r_1(t)$: Deposit interest rate which is defined as the spot interest rate of the continuous time. The deposit interest rate during $(0, T]$ is
$$\int_0^T r_1(t)\,dt.$$

$r_2(t)$: Loan interest rate in no consideration of bankruptcy, which is the sum of the deposit interest rate and constant margin rate, i.e., $r_2(t) \equiv r_1(t) + \alpha_N$ ($\alpha_N > 0$), where α_N is a margin interest rate in no consideration of bankruptcy.

$r_F(t)$: Loan interest rate without bankruptcy which is the sum of the deposit interest rate and constant margin rate, i.e., $r_F(t) \equiv r_1(t) + \alpha_F$, where α_F is a margin interest without bankruptcy and $\alpha_F \geq \alpha_N$.

β: Ratio of amount of bankruptcy to the amount of total financing when a financed enterprise has gone bankrupt ($0 \leq \beta \leq 1$), i.e., βM_2 is the amount of claim collection.

$c(t)$: Administrative cost for claim collection during $(0, t]$.

$F(t), f(t)$: Bankruptcy probability distribution and its density function, i.e., $F(t) \equiv \int_0^t f(u)\,du.$

$Z(t), z(t)$: Mortgage collection probability distribution and its density, i.e.,
$$Z(t) \equiv \int_0^t z(u)\,du.$$

$\overline{\Phi}(t) \equiv 1 - \Phi(t)$ for any function $\Phi(t)$.

4.1 Expected Earning without Bankruptcy

It is assumed that all amount of principal in mortgage can be collected at into a batch one time. Further, the distribution $Z(t)$ represents the probability that the principal in mortgage can be collected until time t after a financed enterprise has gone bankrupt. Suppose that any financed enterprise never goes bankrupt and T is a finance period. When the amount M_1 of deposit was at time 0, the total amount $S_N(T)$ of principal and interest is, from the above notations,

$$S_N(T) = M_1 \exp\left[\int_0^T r_1(t)\,dt\right]. \tag{32}$$

Similarly, the total amount $S_L(T)$ of loan M_2 at time T, which is made in banks at momentarily interest rate $r_2(t) \equiv r_1(t) + \alpha_N$, is

$$S_L(T) = M_2 \exp\left[\int_0^T r_2(t)\,dt\right]. \tag{33}$$

Therefore, the bank earning at time T is

$$P_N(T) \equiv S_L(T) - S_N(T) = M_1 \exp\left[\int_0^T r_1(t)\,dt\right][\omega \exp(\alpha_N T) - 1], \tag{34}$$

where $M_2 = \omega M_1$ ($0 < \omega < 1$). The interest rate α_N which satisfies $P_N(T) \geq 0$ is given by

$$\alpha_N \geq -\frac{\ln \omega}{T}. \tag{35}$$

4.2 Expected Earning with Bankruptcy

Suppose that a financed enterprise goes bankrupt according to the probability $F(T)$ in a financed period $(0,T]$, and banks can make only βM_2 of all amount of loans M_2. When a financed enterprise has gone bankrupt at time t_0 ($0 < t_0 < T$), the total amount of mortgage collection $S_F(T)$ of M_2 at time 0 is given by

$$\begin{aligned}
S_F(T) &= \overline{F}(T) M_2 \exp\left\{\int_0^T [r_1(t) + \alpha_F]\,dt\right\} + F(T)\beta M_2 \exp\left\{\int_0^{t_0} [r_1(t) + \alpha_F]\,dt\right\} \\
&= M_2 \exp\left\{\int_0^{t_0} [r_1(t) + \alpha_F]\,dt\right\}\left[\overline{F}(T) \exp\left\{\int_{t_0}^T [r_1(t) + \alpha_F]\,dt\right\} + \beta F(T)\right].
\end{aligned} \tag{36}$$

Therefore, the bank earning is, from (32) and (36),

$$\begin{aligned}
P_F(T) &\equiv S_F(T) - S_N(T) \\
&= M_1 \exp\left\{\int_0^{t_0} r_1(t)\,dt\right\}\left\{[\omega\overline{F}(T)\exp(\alpha_F T) - 1]\exp\left[\int_{t_0}^T r_1(t)\,dt\right]\right. \\
&\quad \left. + \beta\omega F(T)\exp(\alpha_F t_0)\right\}.
\end{aligned} \tag{37}$$

Then, we can obtain a margin interest rate α_F which satisfies $P_F(T) \geq 0$ at time T.

Next, suppose that banks can collect the amount of $\beta M_2 \exp\left[\int_0^{t_0} r_F(v)\,dv\right]$ according to the mortgage collection probability $z(u - t_0)\,d(u - t_0)$ ($t_0 < u \leq T$), when a financed enterprise has gone bankrupt at time t_0 ($0 < t_0 < T$), i.e, its amount is

$$\int_{t_0}^{T} \beta M_2 \exp\left\{\int_0^{t_0} [r_1(v) + \alpha_F]\,dv\right\} z(u - t_0)\,d(u - t_0). \tag{38}$$

In general, it might be unprofitable to continue the mortgage collection until its completion, because we need to consider some cost for its administration. It would be reasonable to assume that the administrative cost is constant regardless of the amount of mortgage and is proportional to the working time. Thus, let $c(t)$ be the constant per unit of time and be proportional to both amount of loans and time of mortgage collection i.e., $c(t) \equiv c_1 t$. Then, the expected earning of mortgage collection, when a financed enterprise has gone bankrupt at time t_0 and its mortgage collection is stopped at time t ($t \geq t_0$), is

$$Q(t \mid t_0) = \int_{t_0}^{t} \left[\beta M_2 \exp\left\{\int_0^{t_0} [r_1(v) + \alpha_F]\,dv\right\} - c(u - t_0)\right] z(u - t_0)\,d(u - t_0)$$

$$- c(t - t_0)\left[1 - \int_{t_0}^{t} z(u - t_0)\,d(u - t_0)\right]$$

$$= \beta M_2 \exp\left\{\int_0^{t_0} [r_1(v) + \alpha_F]\,dv\right\} Z(t - t_0) - c_1 \int_0^{t - t_0} \overline{Z}(u)\,du. \tag{39}$$

We find an optimal time t^* ($t^* \geq t_0$) of mortgage collection which maximizes $Q(t \mid t_0)$ for given t_0. Differentiating $Q(t \mid t_0)$ in (39) with respect to t and setting it equal to zero, we have

$$\frac{z(t - t_0)}{\overline{Z}(t - t_0)} = K(t_0) \quad (t \geq t_0), \tag{40}$$

where

$$K(t_0) \equiv \frac{c_1}{\beta M_2 \exp\left\{\int_0^{t_0} [r_1(v) + \alpha_F]\,dv\right\}}.$$

Let us denote a mortgage collection rate by $r(t) \equiv z(t)/\overline{Z}(t)$, which corresponds to the failure rate in reliability theory[18]. It is assumed that $r(t)$ is continuous and strictly decreasing because it would be difficult with time to make the collection. Then, we have the following optimal policy:

(i) If $r(0) > K(t_0) > r(\infty)$ then there exists a finite and unique t^* ($t_0 < t^* < \infty$) which satisfies (40), and the expected earning is

$$Q(t^* \mid t_0) = c_1 \left[\frac{Z(t^* - t_0)}{r(t^* - t_0)} - \int_0^{t^* - t_0} \overline{Z}(u)\, du \right]. \qquad (41)$$

(ii) If $r(0) \leq K(t_0)$ then $t^* = t_0$, i.e., the mortgage collection should not be made.

(iii) If $r(\infty) \geq K(t_0)$ then $t^* = \infty$, i.e., the mortgage collection should be continued until its completion.

Suppose, for convenience, that $Z(t)$ is a Weibull distribution with shape parameter m, i.e., $Z(t) = 1 - e^{-\lambda t^m}$ and $r(t) = \lambda m t^{m-1}$ ($0 < m < 1$). Then, since $r(t)$ strictly decreases from infinity to zero, from (40),

$$t^* - t_0 = \left[\frac{K(t_0)}{\lambda m} \right]^{1/(m-1)}. \qquad (42)$$

In this case, if $t^* < T$ then the collected capital will be worked again, and oppositely, if $t^* > T$ then the capital will be newly raised. In both cases, the total amount $Q(T \mid t_0)$ at time T, when a financed enterprise has gone bankrupt at time t_0, is

$$Q(T \mid t_0) = \begin{cases} Q(t^* \mid t_0) \exp\left\{ \int_0^{T-t^*} [r_1(v) + \alpha_F]\, dv \right\} & (T \geq t^*), \\ Q(t^* \mid t_0) \exp\left\{ -\int_0^{t^*-T} [r_1(v) + \alpha_F]\, dv \right\} & (T < t^*). \end{cases} \qquad (43)$$

From the above discussions, the expected bank earning at time T, when a financed enterprise has gone bankrupt at time t_0, is, from (32) and (43),

$$P_F(T \mid t_0) \equiv Q(T \mid t_0) - S_N(T)$$
$$= \left\{ \beta M_2 \exp\left[\int_0^{t_0} r_F(v)\, dv \right] Z(t^* - t_0) - c_1 \int_0^{t^*-t_0} \overline{Z}(u)\, du \right\}$$
$$\times \exp\left[\mathrm{sign}(T - t^*) \int_0^{|T-t^*|} r_F(v)\, dv \right] - M_1 \exp\left[\int_0^T r_1(v)\, dv \right], \qquad (44)$$

where $\mathrm{sign}(x)$ denotes the negative or positive sign of x. Therefore, the expected bank earning at time T is, from (34) and (44),

$$P(T) \equiv P_N(T)\overline{F}(T) + \int_0^T P_F(T \mid t_0)\, dF(t_0). \qquad (45)$$

We can obtain a margin interest α_F which satisfies $P(T) \geq 0$ when both distributions of $Z(t)$ and $F(t)$ are given.

4.3 Numerical Examples

Suppose that the bankruptcy probability distribution $F(t)$ is a discrete one: It is assumed that T is equally divided into n, i.e., $nT_1 \equiv T$, and if a financed enterprise goes bankrupt during $((k-1)T_1, kT_1]$ then it goes bankrupt at time kT_1. Then, the distribution is rewritten as

$$F(t) \equiv p_1 + p_2 + \cdots + p_k \quad (k-1)T_1 < t \leq kT_1 \quad (k = 1, 2, \ldots, n). \quad (46)$$

Thus, the expected earning $P(T)$ in (14) is

$$P(T) = P_N(T)[1 - (p_1 + p_2 + \cdots + p_n)] + \sum_{k=1}^{n} P_F(T \mid kT_1) p_k, \quad (47)$$

where $T \equiv nT_1$.

In general, a bankruptcy probability would be greatly affected by business fluctuations: The number of bankruptcies is small when the business is good, is large when it is bad, and becomes constant when the business is stable. This probability is also constant for a short time, except for the influence of economic prospects. Thus, when T is 12 months, we suppose that $p_1 = p_2 = \cdots = p_{12} \equiv p$. From the average bankruptcy probability of one year from 1991 to 1992 of financed enterprises with ranking points $40 \sim 60$ marked by *TEIKOKU DATABANK* in Japan, we can consider that the financed enterprises of such points are normal, and the average bankruptcy probability is about 0.01945. Hence, we put that $p \equiv 0.0016 = 0.01945/12$.

Next, we show the mortgage collection probability: We have not yet made a statistical investigation of mortgage collection data in Japan, and so, consider that they are similar to America. We draw the mortgage collection probability in Figure. 3 by applying a Weibull distribution which we can estimate the parameters as $\lambda = 0.3$ and $m = 0.24$. Further, suppose that a mortgage collection cost c_1 is given by the ratio of administrative cost to a general amount of loans, i.e., $c_1 = M/250$.

In the above conditions, we assume the following three cases of deposit interest spot rates including funding cost:

1) Constant rate 3% per year shown in (1) of Figure 4.
2) Constant rate 3% plus increasing rate 0.12% per year shown in (2) of Figure 4.
3) Constant rate 3% plus decreasing rate 0.12% per year shown in (3) of Figure 4.

In three cases, we compute and draw in Figure 4 a margin interest rate α_F which satisfies $P(T) = 0$ in (47) when the bankruptcy probability changes from 0.0006 to 0.0022 per month.

Figure 4 indicates that the differences in α_F between cases (2), (3) and case (1) are also large as the bankruptcy probability becomes large. For instance, these values of two differences are 0.24% at $p = 10^{-3}$ and 0.36% at $p = 2 \times 10^{-3}$.

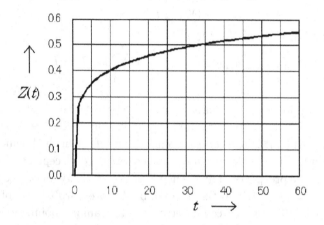

Fig. 3 Mortgage collection probability for a Weibull distribution $Z(t) = 1 - \exp[-\lambda t^m]$ when $\lambda = 0.3$ and $m = 0.24$.

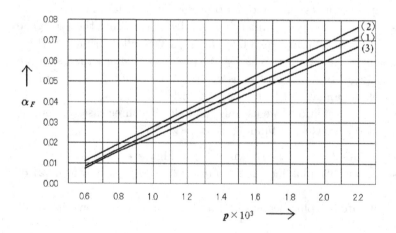

Fig. 4 Loan margin interest rate α_F for bankruptcy probability p.

References

1. Hisata, Y. and Yamai, Y. (2000). *Research toward the practical application of liquidity risk evaluation methods*, Monetary and Economic Studies, Vol.18, pp. 83-127.
2. Dubil, R. (2002). *Optimal Liquidation of Large security Holdings in Thin Markets*, working paper, University of Connecticut, Storrs, USA.
3. Jarrow, R. and Turnbul, S. (1973). *Derivative Securities*, Thomson Learning Company.
4. Nakagawa, T. (2005). *Maintenance Theory of Reliability*, Springer-Verlag, London.
5. Cox, R. D. (1962). *Renewal Theory*, John Wiley & Sons Inc.
6. Duffie, D. and Singleton, J. K. (2003). *Credit Risk*. Princeton University Press, USA.
7. Nakagawa, T. (2005) *Shock and Damage Models in Reliability Theory*, Springer-Verlag, London.
8. Salomon Brothers, Co. Ltd.,(1985). *Anatomy of Prepayments: The Salomon Brothers Prepayment Model*, Salomon Brothers Inc, September 4.
9. Wandey, N. F. (2006). *A Confidence Interval for Default and Prepayment Predictions of Manufactured Housing Seasoned Loans*, Working paper, University of Minnesota.
10. Thomas, O. J. (1998). *Mortgage-Equity Analysis in Contaminated Property Valuation*, The Appraisal J., pp. 46–55.
11. Lee, J. N. and Ong, E. S. (2003) *Prepayment risk of Public housing Mortgages*, J. Real Estate Portfolio Management, Vol.9, pp. 251-264.
12. Allen, L. (1988). *The determination of bank interest margins: A note*, Journal of Financial and Quantitative Analysis, **23**, pp. 1231–235.
13. Altman and Edward, I. (2001). *Bankruptcy, Credit Risk, and High Yield Junk Bonds*, Blackwell Publishers, USA.
14. Ammann, M. (2001). *Credit Risk Valuation: Methods, Models, and Applications*, Springer-Verlag, Berlin Heidelberg.
15. Angbazo, L. (1997). *Commercial bank net interest margins, default risk, interest risk, and off-balance sheet banking*, Journal of Banking & Finance, **21**, pp. 55–87.
16. Athavale, M. and Edmister, O. R. (1999). *Borrowing relationships, monitoring, and the influence on loan rates*, Journal of Financial Research, **22**, pp. 341–352.
17. Barlow, R. E. and Proschan, F. (1965). *Mathematical Theory of Reliability*, John Wiley & Sons, New York.
18. Brigo and Mercurio, D. (2001). *Interest Rate Models - Theory and Practice*, Springer-Verlag, Berlin Heidelberg.
19. Chalasani, P. and Jha, S. (1996). *Stochastic Calculus and Finance*, http://www.stat.berkeley.edu/users/evans/shreve.pdf
20. Ebrahim, S. M. and Mathur, I. (2000). *Optimal entrepreneurial financial contracting*, Journal of Business Financial & Accounting, **27**, pp. 1349–1374.

21. Ho, Y. S. T. and Saunders, A. (1981). *The determinants of bank interest margins: Theory and empirical evidence*, Journal of Financial and Quantitative Analysis, **16**, pp. 581–600.
22. Hull, C. J. (1989). *Options, Futures, and Other Derivatives*, Prentice-Hall, Inc., NJ.
23. James, Jessica, and Webber, N. (2000). *Interest Rate Modeling*, John Wiley & Sons, England.
24. Mitchner, M. and Peterson, P. R. (1957). *An Operations-Research study of the collection of defaulted loans*, Operations Research, **5**, pp. 522–1546.
25. Sealey, W. C. Jr. (1980). *Deposit rate-setting, risk aversion, and the theory of depository financial intermediaries*, The Journal of Finance, **35**, pp. 1139–1154.
26. Slovin, B. M. and Sushka, M. (1983). *A model of commercial loan rate*, The Journal of Finance, **38**, pp. 1583–1596.
27. Steele, M. J. (2001). *Invitation to Stochastic Calculus and Financial Applications*, Springer-Verlag, New York.
28. Wong, P. K. (1997). *On the determinants of bank interest margins under credit and interest rate risks*, Journal of Banking & Finance, **21**, pp. 251–271.
29. Zarruk, R. E. and Madura, J. (1992). *Optimal bank interest margin under capital regulation and deposit insurance*, Journal of Financial and Quantitative Analysis, **27**, pp. 143–149.
30. Nakamura, S, Arafuka, M, Nakagawa, T, and Kondo, H. (2006). *Application for Market Impact of Stock Price Using A Cumulative Damage Model*, Advanced Reliability Modeling II, World Scientific, Singapore, pp. 660–667. io
31. Nakamura, S, Maeji, S, Nakagawa, T, and Nakayama, K. (2008). *Management policy for prepayment risk of a monetary facility*, 14th ISSAT International Conference, pp. 303–306.
32. Nakamura, S, Sandoh, H, and Nakagawa, T. (2002). *determination of loan interest rate considering bankruptcy and mortgage collection costs*, International Transactions in Operational Research, **9**, pp. 695–701.